T0207265

Lecture Notes in Computer Science

Lecture Notes in Bioinformatics 14137

The series Lecture Notes in Bioinformatics (LNBI) was established in 2003 as a topical subseries of LNCS devoted to bioinformatics and computational biology.

The series publishes state-of-the-art research results at a high level. As with the LNCS mother series, the mission of the series is to serve the international R & D community by providing an invaluable service, mainly focused on the publication of conference and workshop proceedings and postproceedings.

Jun Pang · Joachim Niehren
Editors

Computational Methods in Systems Biology

21st International Conference, CMSB 2023
Luxembourg City, Luxembourg, September 13–15, 2023
Proceedings

Springer

Editors
Jun Pang
University of Luxembourg
Esch-sur-Alzette, Luxembourg

Joachim Niehren
Inria Lille
Villeneuve d'Ascq, France

ISSN 0302-9743 ISSN 1611-3349 (electronic)
Lecture Notes in Bioinformatics
ISBN 978-3-031-42696-4 ISBN 978-3-031-42697-1 (eBook)
https://doi.org/10.1007/978-3-031-42697-1

LNCS Sublibrary: SL8 – Bioinformatics

This Springer imprint is published by the registered company Springer Nature Switzerland AG
The registered company address is: Gewerbestrasse 11, 6330 Cham, Switzerland

Preface

This volume contains the papers presented at the 21st International Conference on Computational Methods in Systems Biology (CMSB 2023), held on September 13–15, 2023 in Luxembourg. The conference website can be found at https://cmsb2023.uni.lu/.

The CMSB annual conference series was initiated in 2003. The previous editions of the conference were held in Rovereto, Italy (2003), Paris, France (2004, 2011), Edinburgh, UK (2005, 2007), Trento, Italy (2006, 2010), Rostock, Germany (2008), Bologna, Italy (2009), London, UK (2012), Klosterneuburg, Austria (2013), Manchester, UK (2014), Nantes, France (2015), Cambridge, UK (2016), Darmstadt, Germany, (2017), Brno, Czech Republic (2018), Trieste, Italy (2019), Konstanz, Germany (2020, online), Bordeaux, France (2021, hybrid), and Bucharest, Romania (2022, hybrid).

The conference brings together researchers from across the biological, mathematical, computational, and physical sciences who are interested in the modeling, simulation, analysis, inference, design, and control of biological systems. It covers the broad field of computational methods and tools in systems and synthetic biology and their applications. Topics of interest to the conference include, but are not limited to, methods and tools for biological system analysis, modeling, and simulation; high-performance methods for computational systems biology; identification of biological systems; applications of machine learning; network modeling, analysis, and inference; automated parameter and model synthesis; model integration and biological databases; multi-scale modeling and analysis methods; design, analysis, and verification methods for synthetic biology; methods for biomolecular computing and engineered molecular devices; data-based approaches for systems and synthetic biology; optimality and control of biological systems; and modeling, analysis, and control of microbial communities. The conference welcomes new theoretical results with potential applications to systems and synthetic biology, as well as novel applications and case studies of existing methods, tools, or frameworks.

This year, there were 25 submissions of regular papers to CMSB 2023. Each submission was reviewed by 3 or 4 program committee members. The committee decided to accept 14 regular papers, which are published in these conference proceedings. Furthermore, CMSB 2023 received 4 submissions of tool desciption papers of which 3 were accepted for presentation and short publication in these conference proceedings. Finally, we received 13 extended abstracts for poster presentations (12 accepted, but not included in this volume) and 6 extended abstracts for highlight talks (2 accepted, 2 accepted as posters, but not included in this volume). The conference followed a single-blind review process.

We had four invited talks at CMSB 2023, given by Jorge Goncalves (University of Luxembourg, Luxembourg), Philippe Jacques (University of Liège, Belgium), Mirco Tribastone (IMT School for Advanced Studies Lucca, Italy), and Verena Wolf (Saarland University, Germany). The abstracts of their talks are included in the front matter of this volume.

We would like to thank all the members of the Program Committee, the members of the Tool Evaluation Committee, and our reviewers for their thoughtful and diligent work. We want to thank the organising committee of CMSB 2023, especially Rita Giannini and Tsz Pan Tong, for their suppport and effort in organizing this edition of the conference.

July 2023

Jun Pang
Joachim Niehren

Organization

Program Committee

Tatsuya Akutsu	University of Kyoto, Japan
Claudio Altafini	University of Linköping, Sweden
Daniela Besozzi	University of Milano-Bicocca, Italy
Luca Bortolussi	University of Trieste, Italy
Luca Cardelli	University of Oxford, UK
Milan Ceska	Brno University of Technology, Czech Republic
Eugenio Cinquemani	Inria Grenoble-Rhône-Alpes, France
François Fages	Inria Saclay, France
Maxime Folschette	École Centrale de Lille, France
Anna Gambin	University of Warsaw, Poland
Ashutosh Gupta	Indian Institute of Technology Bombay, India
Monika Heiner	Brandenburg Technical University Cottbus-Senftenberg, Germany
Hillel Kugler	Bar-Ilan University, Israel
Luca Laurenti	TU Delft, The Netherlands
Pedro T. Monteiro	University of Lisbon, Portugal
Joachim Niehren (PC Co-chair)	Inria Lille, France
Jun Pang (PC co-chair)	University of Luxembourg, Luxembourg
Loïc Paulevé	CNRS Bordeaux, France
Andrei Paun	University of Bucharest, Romania
Ion Petre	University of Turku, Finland
Tatjana Petrov	University of Konstanz, Germany
Maria Rodriguez Martinez	IBM Zurich Research Laboratory, Switzerland
Jakob Ruess	Inria Paris, France
Thomas Sauter	University of Luxembourg, Luxembourg
Abhyudai Singh	University of Delaware, USA
Chris Thachuk	University of Washington, USA
Mirco Tribastone	IMT School for Advanced Studies Lucca, Italy
Andrea Vandin	Sant'Anna School of Advanced Studies, Italy
Christoph Zechner	Max Planck Institute of Molecular Cell Biology and Genetics, Germany
David Šafránek	Masaryk University, Czech Republic

Tool Evaluation Committee

Georgios Argyris	Technical University of Denmark, Denmark
Candan Çelik	Institute for Basic Science, South Korea
Andrzej Mizera (Chair)	University of Warsaw, Poland
Samuel Pastva	Masaryk University, Czech Republic
Jaros law Paszek	University of Warsaw, Poland
Misbah Razzaq	INRAE, France
Matej Troják	Masaryk University, Czech Republic

Additional Reviewers

Baj, Giovanni
Ballarin, Emanuele
Ballif, Guillaume
Chung, Neo Christopher
Domżał, Barbara
Gandin, Ilaria
Hemery, Mathieu
Inverso, Omar
Kouye, Henri Mermoz
Martinelli, Julien
Niemyska, Wanda
Pastva, Samuel
Riva, Sara
Tschaikowski, Max
Wathieu, Lancelot
Šmijáková, Eva

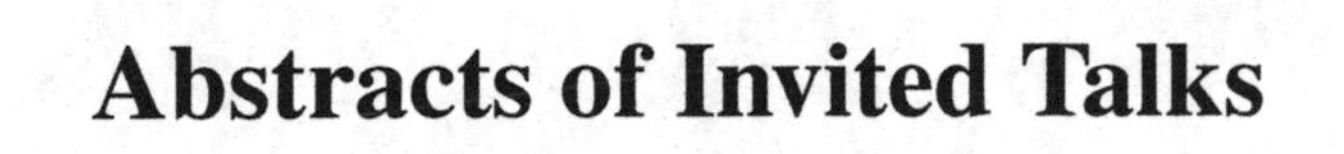

Abstracts of Invited Talks

Causal Dynamical Network Inference: From Theory to Applications in Biology

Jorge Gonçalves[1,2]

[1] Luxembourg Centre for Systems Biomedicine, University of Luxembourg, Belvaux, L-4367, Luxembourg
jorge.goncalves@uni.lu
https://www.en.uni.lu/lcsb/research/systems_control
[2] Department of Plant Sciences, University of Cambridge, Cambridge, CB2 3EA, UK

Abstract. One of the major objectives in systems biology is to extract information from data that can be used to make reliable predictions. In most cases, however, this is very difficult due to the limited amount of data. One major obstacle is the lack of mathematical tools tailored to the specific needs of systems biology. This talk presents both theoretical and practical advancements in building dynamical network models from limited time-series data. The theoretical aspect discusses experimental requirements that guarantee identifiability of dynamical networks. The application part illustrates recent tools developed in our group applied to a wide range of experimental data, including bulk and single cell data [1–5]

Keywords: Systems biology · Dynamical network modelling · System identification and Multiomics

References

1. Aalto, A., Lamoline, F., Gonçalves, J.: Linear system identifiability from single-cell data. Syst. Control Lett. **165**, 105287 (2022). https://doi.org/10.1016/j.sysconle.2022.105287
2. Yeung, E., Kim, J., Yuan, Y., Gonçalves, J., Murray, R.: Data-Ddriven network models for genetic circuits from time-series Data with incomplete measurements. J. Royal Soc. Interface **18**(182), 20210413 (2021). https://doi.org/10.1098/rsif.2021.0413
3. Aalto, A., Viitasaari, L., Ilmonen, P., Mombaerts, L., Gonçalves, J.: Gene regulatory network inference from sparsely sampled noisy data. Nature Commun. **11**(3493), 3493 (2020). https://doi.org/10.1038/s41467-020-17217-1

4. Mauroy. A., Gonçalves, J.: Koopman-based lifting techniques for nonlinear systems identification. IEEE Trans. Autom. Control **65**(6), 2550–2565 (2020). https://doi.org/https://doi.org/10.1109/TAC.2019.2941433
5. Mombaerts, L., et al.: Dynamical differential expression (DyDE) reveals the period control mechanisms of the Arabidopsis circadian oscillator. PLoS Comput. Bio. **15**(1), e1006674 (2019). https://doi.org/10.1371/journal.pcbi.1006674

Optimizing NonRibosomal Peptide Synthesis by Biocomputing Approaches

Philippe Jacques

Microbial Processes and Interactions (MiPI), TERRA Teaching and Research Centre, INRAE Joint Research Unit BioEcoAgro (UMRt 1158), University of Lille, Gembloux Agro-Bio-Tech University of Liege, B-5030 Gembloux
Philippe.jacques@uliege.be

The biosynthesis mechanism of NonRibosomal Peptides (NRPs) was elucidated for the first time in 1965 with the description of the NonRibosomal Peptide Synthetase (NRPS) of the gramicidin from *B. subtilis*. As suggested by their name, these peptides are not assembled on ribosomes but by large multienzymatic proteins that are encoded by gene clusters that could span over tens of kilobase in the genome. More than 500 different monomers can be incorporated in such peptides leading to a high structural biodiversity of NRPs with various interesting biological activities (antimicrobials, immunomodulators, biosurfactants,...) and a high set of applications in different sectors (pharmaceutical, agronomical, chemical, environmental...) [1, 2].

Improving the biosynthesis of NRPs is thus frequently an important objective that can be managed by rational metabolic engineering approaches such as: (i) blocking competitive pathways for building blocks, as well as, those pathways that consume products; (ii) pulling flux through biosynthetic pathways by removing regulatory signals; and (iii) by overexpressing rate-limiting enzymes [3].

One way to develop this metabolic engineering approach is to use knockout of genes which negatively influence the intracellular pool of monomer precursors. To do that, the precursor metabolic pathways have to be modeled as a reaction network taking into account the regulation processes. The choice of those knockouts can be directed by methods from computational biology, to narrow them down and reduce the laboratory time needed. Some prediction methods are related to formal reasoning techniques based on abstract-interpretation of reaction networks with partial kinetic information [4]. While being a general framework for the static analysis of programming languages, abstract interpretation is applied here to steady state equation of the reaction network, in order to enable qualitative reasoning with logical methods. On example of this approach will be developed for the overproduction of a NonRibosomal lipopeptide produced by *Bacillus* strains, the surfactin. It shows strong biosurfactant activities and is able to stimulate plant defense mechanisms [5, 6].

Keywords: NonRibosomal Peptide Synthesis · lipopeptide · metabolic engineering · knock-out · prediction methods · abstract interpretation

References

1. Vassaux et al.: NonRibosomal Peptides in fungal cell factories : from genome mining to optimized heterologous production. Biotechnol. Adv. **37**, 107449 (2019). https://doi.org/10.1016/j.biotechadv.2019.107449
2. Flissi et al.: Norine: update of the nonribosomal peptide resource; Nucl. Acids Res. **48**(D1), D465–D469 (2020). https://doi.org/10.1093/nar/gkz1000
3. Theatre et al.: *Bacillus sp.*: A remarkable source of bioactive lipopeptides. Adv. Biochem. Eng. Biotechnol. **181**, 123–179 (2022). https://doi.org/10.1007/10_2021_182
4. Niehren et al. 2016. Predicting changes of reaction networks with partial kinetic information. *BioSystems* **149**, 113–124. https://doi.org/10.1016/j.biosystems.2016.09.003
5. Coutte et al.: Modeling leucine's metabolic pathway and knockout prediction improving the production of surfactin, a biosurfactant from *Bacillus subtilis*. *Biotechnol. J.* **10**(8), 1216–1234 (2015). https://doi.org/10.1002/biot.201400541
6. Dhali et al.: Genetic engineering of the branched fatty acid metabolic pathway of *Bacillus subtilis* for the overproduction of surfactin C14 isoform. Biotechnol. J. **12**(7), 1–23 (2017). https://doi.org/10.1002/biot.201600574

Integrating Machine Learning and Mechanistic Modeling: Unlocking the Potential of Hybrid Approaches in the Life Sciences

Verena Wolf[1,2]

[1] Saarland Informatics Campus, Saarland University, Germany
verena.wolf@dfki.de
https://www.dfki.de/web/forschung/forschungsbereiche/neuro-mechanistische-modellierun
[2] German Research Center for Artificial Intelligence (DFKI), Saarbruecken, Germany

Abstract. In the life sciences, both machine learning and mechanistic modeling are widely employed methodologies. Machine learning offers data-driven solutions for regression and classification problems, while mechanistic modeling focuses on describing real-world systems and validating hypotheses about them. In this presentation, I will explore the strengths and limitations of machine learning and mechanistic modeling, highlighting instances from the life sciences where a hybrid approach combining these two paradigms has successfully enhanced our understanding of the underlying systems. Furthermore, I will discuss various methods for incorporating mechanistic relationship information into machine learning models, particularly neural networks, and demonstrate the immense potential of these hybrid models.

Contents

A Formal Approach for Tuning Stochastic Oscillators

Paolo Ballarini[(✉)], Mahmoud Bentriou, and Paul-Henry Cournède

MICS, CentraleSupélec, Université Paris-Saclay, Paris, France
{paolo.ballarini,mahmoud.bentriou,paul-henry.cournede}@centralesupelec.fr

Abstract. Periodic recurrence is a prominent behavioural of many biological phenomena, including cell cycle and circadian rhythms. Although deterministic models are commonly used to represent the dynamics of periodic phenomena, it is known that they are little appropriate in the case of systems in which *stochastic noise* induced by small population numbers is actually responsible for periodicity. Within the stochastic modelling settings automata-based model checking approaches have proven an effective means for the analysis of oscillatory dynamics, the main idea being that of coupling a *period detector* automaton with a continuous-time Markov chain model of an alleged oscillator. In this paper we address a complementary aspect, i.e. that of assessing the dependency of oscillation related measure (period and amplitude) against the parameters of a stochastic oscillator. To this aim we introduce a framework which, by combining an Approximate Bayesian Computation scheme with a hybrid automata capable of quantifying how *distant* an instance of a stochastic oscillator is from matching a desired (average) period, leads us to identify regions of the parameter space in which oscillation with given period are highly likely. The method is demonstrated through a couple of case studies, including a model of the popular Repressilator circuit.

Keywords: Stochastic oscillators · Approximate Bayesian Computation · Parameter estimation · Hybrid Automata Stochastic Logic · Statistical Model-Checking

1 Introduction

Oscillations are prominent dynamics at the core of many fundamental biological processes. They occur at different level and concern different time scales, ranging from ion channels regulated transmission of intercellular electrical signal driving the heartbeat (with period in the order of one second), to intracellular calcium oscillations triggering glycogen-to-glucose release in liver cells (with periods ranging from few seconds to few minutes [30]), to gene-expression regulated circadian cycle (with typical period of roughly 24 h).

Mathematical modelling and computational methods have proved fundamental to gain a better understanding of the complexity of the mechanisms that

J. Pang and J. Niehren (Eds.): CMSB 2023, LNBI 14137, pp. 1–17, 2023.
https://doi.org/10.1007/978-3-031-42697-1_1

regulate oscillations [22]. Although continuous-deterministic models (i.e. ODEs) are more often considered in the literature of biological oscillators, with systems characterised by low population numbers (e.g., genetic circuits responsible for *circadian rhythms*), discrete-stochastic models are more appropriate [24]: when few molecules are involved the stochasticity of the system becomes important resulting in noisy periodic behavior. Understanding the effect that an oscillator's parameters have on the quality of the oscillations is a relevant research problem.

Contribution. We introduce a methodology for calibrating stochastic models that exhibit a noisy periodic dynamics. Given a parametric stochastic oscillator the methodology allows to identify regions of the parameter space with a positive probability of matching a given target oscillation period. This entails integrating formal means for noisy periodicity analysis within a parameter inference approach, more specifically, integrating a *period distance meter* (formally encoded as a Hybrid Automaton) within an Approximate Bayesian Computation scheme. Given some period-related requirements (e.g. mean value and variance of the oscillation period) the trajectories sampled from the oscillator are classed through the distance period meter leading us to the approximation of the corresponding posterior distribution, that is, to the identification of the region of the parameter space that are more likely to comply with the considered period-related requirements.

Paper Organisation. The paper is organised as follows: in Sect. 2 we overview the background material our approach relies upon, that includes an overview of HASL model checking and of ABC algorithms. In Sect. 3 we introduce our approach for calibrating stochastic oscillators w.r.t. to the oscillation period. The method is then demonstrated through experiments presented in Sect. 4. Conclusive remarks and future perspective are discussed in Sect. 5.

1.1 Related Work

Temporal Logic Based Analysis of Stochastic Oscillators. The analysis of oscillatory behaviour entails two complementary aspects: determining whether a model exhibits recurrent patterns (detection) and assessing relevant periodicity indicators (e.g. period and amplitude of oscillations). In the continuous-deterministic settings these can be achieved through a combination of mathematical approaches including structural analysis of the corresponding ODE system and stability analysis of its *steady-state* solutions [22]. As those approaches clearly do not apply in the discrete-stochastic settings researchers progressively started looking at alternatives such as the adaptation of model checking to the analysis of periodicity [1]. Detection of sustained oscillations through model checking requires identifying temporal logical formulae that single out infinite cyclic (non-constant) behaviour. Seminal ideas introduced for (non-probabilistic) *transition systems* models [18] and further developed in [8] included Computational Tree Logic (CTL) [19] *qualitative* specifications such as

$$EG(((X_i = k) \Rightarrow EF(X_i \neq k)) \wedge ((X_i \neq k) \Rightarrow EF(X_i = k))) \quad (1)$$

$$AG(((X_i = k) \Rightarrow EF(X_i \neq k)) \wedge ((X_i \neq k) \Rightarrow EF(X_i = k))) \quad (2)$$

which encode *infinite alternation* by demanding that for "at least one path" (1) or for "all paths" (2) of a given model a state condition $X_i = k$[1] can be met and then left (and, inversely, whenever it is not met $X_i \neq k$ it is possible to eventually meet it) infinitely often. In the stochastic, continuous time Markov chains (CTMCs), settings the Continuous Stochastic Logic (CSL) [2] counterpart of (2) leads to the following qualitative probabilistic formulae (introduced in [8]):

$$P_{=1}(((X_i = k) \Rightarrow P_{>0}(X_i \neq k)) \wedge ((X_i \neq k) \Rightarrow P_{>0}(X_i = k))) \quad (3)$$

which identifies those states of a CTMC for which the probability of outgoing paths that infinitely alternate between $X_i = k$ and $X_i \neq k$ states adds up to 1. If formulae such as (3) can be used to rule out CTMCs that do not oscillate sustainably (as they contain at least an absorbing state), they fall short w.r.t. effectively detecting sustained oscillators as they are satisfied by any *ergodic* CTMC regardless whether it actually exhibits sustained oscillations. The problem is that, in the stochastic settings, a logical characterisation such as (3) is too weak to allow for distinguishing between models that gather probability mass on actual oscillatory paths (of given amplitude and period) from those whose probability mass is concentrated on infinite noisy fluctuations which do not correspond with actual oscillations. Further developments included employing the probabilistic version of the linear time logic (LTL) [28] to characterise oscillations based on a notion of *noisy monotonicity* [5]: although an improvement w.r.t. to the limitation of the CSL based characterisation of periodicity this approach still fails to satisfactorily treat the oscillation detection problem.

Automata-Based Analysis of Stochastic Oscillators. To work around the limited suitability of temporal logic approaches Spieler [31] proposed to employ a single-clock deterministic timed automata (DTA) as *noisy period detector*. The idea is that, through synchronisation with a CTMC model, such DTA is used to accept *noisy periodic* paths described as those that infinitely alternate between crossing a lower threshold L, and a higher threshold H (hence corresponding to fluctuations of minimal amplitude $H-L$) and by imposing that such fluctuations should happen with a period falling in a chosen interval $[t_p^{min}, t_p^{max}]$ (with H, L, t_p^{min} and t_p^{max} being parameters of the DTA). The issues that CSL detector formulae such as (3) suffer from are overcome, as by properly settings thresholds L and H the DTA rules out non-oscillating ergodic CTMC models. The detection procedure boils down to computing the probability measure of all CTMC paths that are accepted by the DTA which is achieved through numerical procedures (requiring the construction of the CTMC x DTA product process). Spieler's original idea has then further evolved by resorting to the more expressive hybrid automata as detectors of periodicity [6,7]. This allowed, on one hand, to account for more sophisticated oscillation related indicators such as the variance (other than the mean value) of the period of a stochastic oscillator and also to develop an alternative, *peaks detector*, automaton which, differently from Spieler's *period detector*, does not depend on the chosen L and H thresholds.

[1] The population of species i is $k \in \mathbb{N}$.

Parametric Verification of Stochastic Models. Parametric verification of a probabilistic model is concerned with combining parameter estimation techniques with stochastic model checking, i.e. with studying how the stochastic model checking problem for a property Φ is affected by the parameters θ upon which a probabilistic model $\mathcal{M}_\theta$ depends. A number of approaches have been proposed in the literature such as [15], in which a bounded approximation of parameter space fulfilling a CSL [2] threshold formula is efficiently determined through an adaptation of *uniformisation*, or also [17,23] and, more recently, a novel ABC-based method [27] that is based on observations even solves parameter inference and statistical parameter synthesis in one go.

The framework we introduce in this paper, on the other hand, tackles the *estimation problem* in stochastic model checking and is in line with the works of Bortolussi *et al.* [14], where the so-called smoothed model checking (Smoothed MC) method to estimate the satisfaction probability function of parametric Markov population models is detailed. The goal here is to obtain a good estimate of the so-called satisfaction probability function of the considered property Φ w.r.t. to model's parameter space Θ. In our case we rely on hybrid automata-based adaptations of Approximate Bayesian Computation (ABC) schemes [11,12] to estimate the satisfaction probability function, similarly to [26] where ABC algorithms are combined with statistical model checking to approximate the satisfaction function of non-nested CSL reachability formulae.

2 Background

We briefly overview the notion of chemical reaction network (CRN) models, of Hybrid Automata Stochastic Logic (HASL) model checking and the Approximate Bayesian Computation (ABC) scheme for parameter estimation that Sect. 3 relies on.

Chemical Reaction Network and Discrete-Stochastic Interpretation. We consider models that describe the time evolution of biochemical species $X_1, X_2, \ldots$ which interact through a number of reactions $R_1, R_2, \ldots$ forming a so-called chemical reaction network (CRN) expressed as a system of chemical equations with the following form:

$$R_j : \sum_{i=1}^{n} a_{ij}^{-} X_i \xrightarrow{k_j} \sum_{i=1}^{n} a_{ij}^{+} X_i$$

where a_{ij}^{-} (a_{ij}^{+}), are the stoichiometric coefficients of the *reactant* (*product*), species and k_j is the kinetic constant of the j-th reaction R_j of the CRN. We assume the discrete stochastic interpretation of CRNs models, i.e. species X_i represent molecules counting (rather than concentrations) hence models consist of countable states $x = (x_1, \ldots x_n) \in \mathbb{N}^n$ representing the combined populations count while the state transitions consist of time-delayed jumps governed by a probability distribution function. In case of Exponentially distributed jumps the underlying class of models is that of continuous-time Markov chains (CTMCs), conversely in case of generically distributed jumps we refer to underlying class of models as of discrete event stochastic process (DESP). A path of a DESP model is a (possibly infinite) sequence of jumps denoted $\sigma = x^0 \xrightarrow[R^0]{t_0} x^1 \xrightarrow[R^1]{t_1} x^2 \xrightarrow[R^2]{t_2} \dots$ with x^i being the i-th state, $t_i \in \mathbb{R}_{>0}$ being the sojourn-time in the i-th state and R^i being the reaction whose occurrence issued the $(i+1)$-th jump. Notice that paths of a DESP are *càdlàg* (i.e. step) functions of time (e.g. Fig. 2). A DESP depend on a p-dimensional vector of parameters $\theta = [\theta_1, \dots, \theta_p] \in \Theta \subseteq \mathbb{R}^p$. which, in the context of this paper, affect the kinetic rate of the reaction channels. For $\theta \in \Theta \subseteq \mathbb{R}^p$ we denote $\mathcal{M}_\theta$ the corresponding DESP model. Given $\mathcal{M}_\theta$ we let $Path(\mathcal{M}_\theta)$ denote the set of paths of $\mathcal{M}_\theta$. It is well known that a DESP model $\mathcal{M}_\theta$ induces a probability space over the set of events $2^{Path(\mathcal{M}_\theta)}$, where the probability of a set of paths $E \in 2^{Path(\mathcal{M}_\theta)}$ is given by the probability of their common finite prefix [3]. For $\mathcal{M}_\theta$ an n-dimensional DESP population model, we denote $dom(\mathcal{M}_\theta)$ its state space, and $dom_i(\mathcal{M}_\theta)$ the projection of $dom(\mathcal{M}_\theta)$ along the ith dimension $1 \leq i \leq n$ of $\mathcal{M}_\theta$. For σ a path of an n-dimensional $\mathcal{M}_\theta$ we denote σ_i its projection along the i-th dimension.

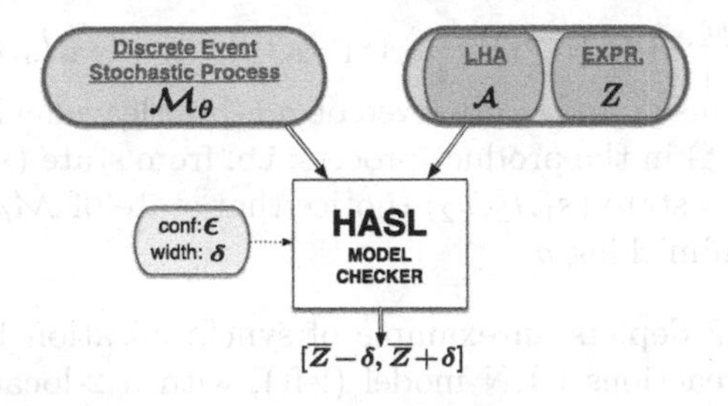

Fig. 1. HASL-SMC schema: sampled paths are filtered by a LHA and the accepted ones used for a confidence interval estimate of the target measure.

2.1 HASL Model Checking

The procedure for tuning stochastic oscillators we introduce in the remainder relies on the HASL statistical model checking (SMC) framework. We quickly overview the basic principles of HASL-SMC referring the reader to the literature [4] for more details. HASL-SMC is a framework for assessing properties of stochastic model expressed by means of a linear hybrid automaton (LHA). It consists of an iterative procedure which takes 3 inputs (Fig. 1): a parametric DESP model $\mathcal{M}_\theta$, a LHA $\mathcal{A}$, and a target expression denoted Z and outputs a confidence interval estimation of the mean value $\hat{Z}$ of the target measure. At each iteration the procedure samples a path $\sigma \in Path(\mathcal{M}_\theta \times \mathcal{A})$ of the product process $\mathcal{M}_\theta \times \mathcal{A}$ (whose formal semantics is given in [4]), that means that a path $\sigma \in Path(\mathcal{M}_\theta)$ is sampled and synchronised *on-the-fly* with $\mathcal{A}$ leading to either acceptance or rejection of σ. In the synchronisation process, relevant statistics of the simulated path are computed and are used to define the target measure Z of interest (see [10]). This let HASL-SMC be a very expressive formalism that allows for sophisticated signal-processing-like analysis of a model.

Synchronisation of a Model with a LHA. The semantics of the synchronised process $\mathcal{M} \times \mathcal{A}$ naturally yields a stochastic simulation procedure which is implemented by the HASL model checker. For the sake of space here we only provide an intuitive (informal) description of such synchronisation based on the example given in Fig. 2. An LHA consists of a set of locations L (with at least one initial and one final location) a set of real-valued variables V whose value may evolve, according to a flow function which map each location to a (real-valued) rate of change to the variables, during the sojourn in a location. Locations change through transitions denoted $l \xrightarrow{\gamma, E', U} l'$ where γ is an enabling guard (an inequality built on top of variables in V), E' is either set of events names (i.e. the transition is *synchronously* traversed on occurrence of any reaction in E' occurring in the path being sampled) or $\sharp$ (i.e. the transition is *autonomously* traversed without synchronisation) and U are the variables' update. A state of $\mathcal{M}_\theta \times \mathcal{A}$ is a 3-tuple (s, l, ν), with s the current state of $\mathcal{M}_\theta$, l the current location of $\mathcal{A}$ and $\nu \in \mathbb{R}^{|V|}$ the current values of $\mathcal{A}$'s variables. Therefore if $\sigma : s \xrightarrow[e_1]{t_1} s_1 \xrightarrow[e_2]{t_2} s_2 \ldots$ is a path of $\mathcal{M}_\theta$ the corresponding path $\sigma \times \mathcal{A} \in Path(\mathcal{M}_\theta \times \mathcal{A})$ may be $\sigma \times \mathcal{A} : (s, l, \nu) \xrightarrow[e_1]{t_1} (s_1, l_1, \nu_1) \xrightarrow[\sharp]{t_1^*} (s_1, l_2, \nu_2) \xrightarrow[e_2]{t_2} (s_2, l_3, \nu_3) \ldots$ where, the sequence of transitions e_1 and e_2 observed on σ is interleaved with an autonomous transition (denoted $\sharp$) in the product process: i.e. from state (s_1, l_1, ν_1) the product process jumps to state (s_1, l_2, ν_2) (notice that state of $\mathcal{M}_\theta$ does not change) before continuing mimicking σ.

Example 1. Figure 2 depicts an example of synchronisation between a path of a toy 3-species, 3-reactions CRN model (left), with a 2-locations LHA (right) designed for assessing properties of it. The LHA locations are l_0 (initial) and l_1 (final) while its variables are $V = \{t, x_1, n_2\}$ with t a clock variable, x_1 a real valued variable (for measuring the average population of A) and n_2 an integer variable (for counting the number of occurrences of the R_1 reaction). While in l_0 the value of variables evolves according to their flow, which is constant and equal to 1 for clock t, while is given by the current population of A for x_1 (therefore x_1 measures the integral of population A along the synchronising path). The synchronisation of a path σ with $\mathcal{A}$ works as follows: $\mathcal{A}$ stays in the initial location l_0 up until at $t = 4$ the autonomous transition the synchronisation with σ ends as soon as the *autonomous* transition $l_0 \xrightarrow{\sharp, t=4, \{x_1/=4\}} l_1$ becomes enabled (guard $t = 4$ gets true) hence is fired (by definition autonomous transitions have priority over *synchronised* transitions). As long as $t < 4$ the LHA is in l_0 where it synchronises with the occurrences of the reactions of the CRN model: on occurrence of R_1 the $\mathcal{M} \times \mathcal{A}$ transition $l_0 \xrightarrow{\{R_1\}, t<4, \{n_2{+}{+}\}} l_0$ (*synchronised* on event set $\{R_1\}$) is fired hence increasing the counter n_2, whereas on occurrence of any other reaction transition $l_0 \xrightarrow{ALL \setminus \{R_1\}, t<4, \emptyset} l_0$ (*synchronised* on event set $ALL \setminus \{R_1\}$, where ALL denotes all reactions of the CRN) fires without updating any variable. Finally on ending the synchronisation with σ variable x_1 is update to $x_1/4$ which corresponds to average population of A observed

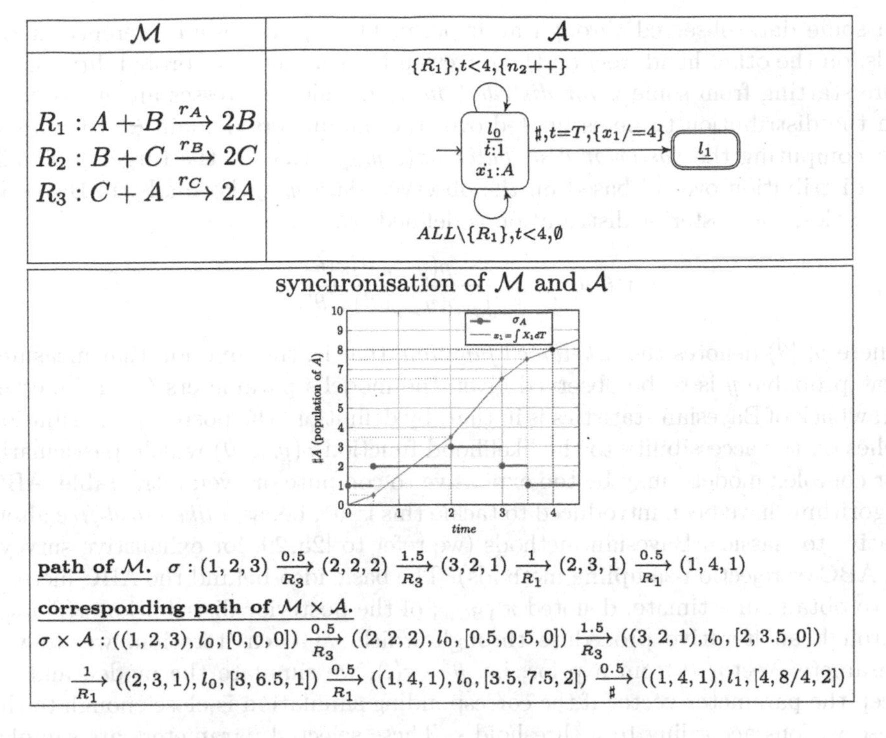

Fig. 2. Example of synchronisation of a CRN model with a LHA.

over the time interval $[0, 4]$. Such a LHA can therefore be used (through iterated synchronisation with a sufficiently large number of trajectories) for estimating the confidence interval of random variables such as the "average population of A" as well as the "number of R_1 occurrences" observed over time interval $[0, 4]$. An example of synchronisation between a path σ consisting of 2 occurrences of R_3 (at $t = 0.5$, $t = 2.0$ respectively) followed by 2 occurrences of R_1 (at $t = 3.0$ and $t = 3.5$ respectively) and for which $(A_0, B_0, C_0) = (1, 2, 3)$ is assumed as the initial state is depicted at bottom of Fig. 2. The synchronised path shows the combined evolution of the model's state, the LHA location and the value of LHA variables. Notice that when synchronisation ends (l_1 is reached) variable x_1 is assigned with $8/4$ which indeed is the mean population of A along σ until $t = 4$.

2.2 Approximate Bayesian Computation

The framework for tuning of stochastic oscillators we present in the remainder relies on the integration of HASL-based measurements within the class of *Bayesian inference* methods known as Approximate Bayesian Computation (ABC) [25,29]. Generally speaking *statistical inference* is interested with inferring properties of an underlying distribution of probability (in our case f_θ) based on some data observed through an experiment y_{exp}. Bayesian inference methods, on the other hand, rely on the Bayesian interpretation of probability, therefore starting from some *prior distribution* $\pi(.)$, which expresses an *initial belief* on the distribution to be estimated over the parameters domain Θ, they allow for computing the *posterior distribution* $\pi(\theta|y_{exp})$, that is, the target probability distribution over Θ based on the observed data y_{exp}. Formally, in Bayesian statistics, the posterior distribution is defined by:

$$\pi(\theta|y_{exp}) = \frac{p(y_{exp}|\theta)\pi(\theta)}{\int_{\theta'} p(y_{exp}|\theta')\pi(\theta')\,d\theta'}$$

where $p(.|\theta)$ denotes the *likelihood function*, that is, the function that measures how probable y is to be observed given the model's parameters θ. An inherent drawback of Bayesian statistics is in that, by definition, the posterior distribution relies on the accessibility to the likelihood function $p(y_{exp}|\theta)$ which, particularly for complex models, may be too expensive to compute or even intractable. ABC algorithms have been introduced to tackle this issue, i.e. as a *likelihood-free* alternative to classical Bayesian methods (we refer to [25,29] for exhaustive surveys of ABC or rejection-sampling methods). The basic idea behind the ABC method is to obtain an estimate, denoted $\pi_{ABC,\epsilon}$, of the posterior distribution $\pi(\theta|y_{exp})$ through an iterative procedure through which, at each iteration, we draw a parameter vector θ from a prior, i.e. $\theta \sim \pi(.)$, we simulate the model, and we keep the parameter vector if the corresponding simulation is close enough to the observations according to a threshold ϵ. These selected parameters are samples from $\pi_{ABC,\epsilon}$ and approximates the posterior distribution: the smaller the ϵ, the better the approximation. The chosen value of ϵ is crucial for the performance of ABC algorithm: a small ϵ is needed to achieve a good approximation, however this may result in high rejection rate leading to cumbersome computations. To overcome this issue, more elaborate algorithms were proposed, like ABC-SMC algorithms [9,20].

3 ABC-HASL Method for Tuning Oscillators

We introduce an approach for exploring the parameters space of stochastic oscillators so that a given oscillatory criteria, e.g., the mean duration of the oscillation period, is met. The approach is based on the Automata-ABC procedure described in Sect. 3.3. The overall idea is to provide one with the characterisation of some linear hybrid automaton capable of assessing oscillation related measures and to plug in the Automata-ABC scheme so that it can effectively be applied to the analysis the effect the model's parameters have on the oscillations. We start off with an overview of preliminary notions necessary for understanding the functioning of the automata for oscillation related measures.

3.1 Characterising of Noisy Periodicity

The mathematical notion of periodic function (i.e. a function $f : \mathbb{R}^+ \to \mathbb{R}$ for which $\exists t_p \in \mathbb{R}^+$ such that $\forall t \in \mathbb{R}^+$, $f(t) = f(t + t_p)$, with t_p being the period) is of little use in the context of stochastic models as paths of a stochastic oscillator are noisy by nature (e.g. Figure 3-left) hence will have (unless in degenerative cases) zero probability of matching such strict notion of periodicity.

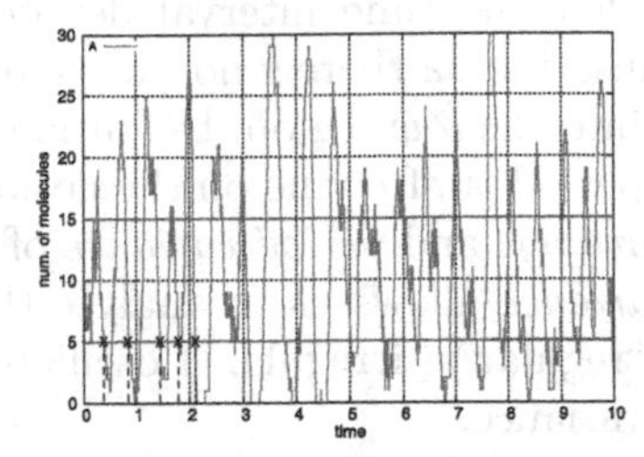

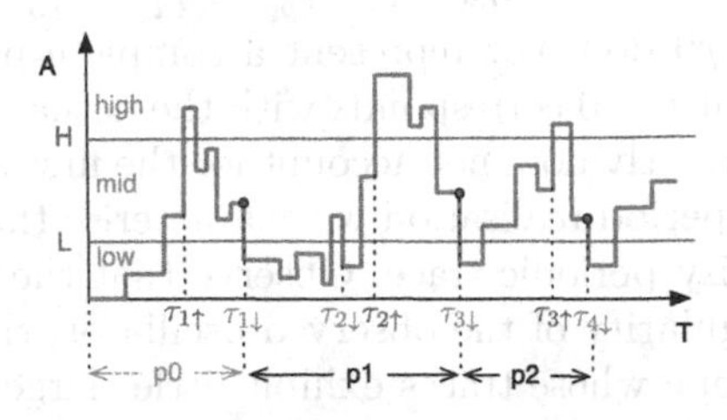

Fig. 3. A *noisy periodic* trajectory (left) and the corresponding *partition dependent* characterisation of noisy period realisations (right).

Therefore to take into account the noisy nature of stochastic oscillators we resort to a less strict notion of *noisy periodicity* [31].

Definition 1 (noisy periodic trajectory). *For $\mathcal{M}_\theta$ an n-dimensional DESP population model let $L, H \in \mathbb{N}$, $L < H$, be two levels establishing the partition $dom_i(\mathcal{M}_\theta) = low \cup mid \cup high$ with $low = [0, L)$, $mid = [L, H)$ and $high = [H, \infty)$. A path $\sigma \in Path(\mathcal{M}_\theta)$ is said* noisy periodic *w.r.t the ith dimension, and the considered L, H induced partition of $dom_i(\mathcal{M}_\theta)$ if the projection σ_i visits the intervals low, mid and high infinitely often.*

H/L-Crossing Points. Given a noisy periodic trace σ_A we denote $\tau_{j\downarrow}$ ($\tau_{j\uparrow}$), the instant of time when σ_A enters for the j-th time the *low* (*high*) region. $T_\downarrow = \cup_j \tau_{j\downarrow}$ (resp. $T_\uparrow = \cup_j \tau_{j\uparrow}$) is the set of all *low-crossing points* (reps. *high-crossing points*). Observe that $T_\downarrow$ and $T_\uparrow$ reciprocally induce a partition on each other. Specifically $T_\downarrow = \cup_k T_{k\downarrow}$ where $T_{k\downarrow}$ is the subset of $T_\downarrow$ containing the k-th sequence of contiguous *low-crossing points* not interleaved by any *high-crossing point*. Formally $T_{k\downarrow} = \{\tau_{i\downarrow}, \dots, \tau_{(i+h)\downarrow} | \exists k', \tau_{(i-1)\downarrow} < \tau_{k'\uparrow} < \tau_{i\downarrow}, \tau_{(i+h)\downarrow} < \tau_{(k'+1)\uparrow}\}$. Similarly $T_\uparrow$ is partitioned $T_\uparrow = \cup_k T_{k\uparrow}$ where $T_{k\uparrow}$ is the subset of $T_\uparrow$ containing the k-th sequence of contiguous *high-crossing points* not interleaved by any *low-crossing point*. For path σ_A in Fig. 3 (right) we have that $T_\downarrow = T_{1\downarrow} \cup T_{2\downarrow} \cup T_{3\downarrow} \dots$ with $T_{1\downarrow} = \{\tau_{1\downarrow}, \tau_{2\downarrow}\}$, $T_{2\downarrow} = \{\tau_{3\downarrow}\}$, $T_{3\downarrow} = \{\tau_{4\downarrow}\}$, while $T_\uparrow = T_{1\uparrow} \cup T_{2\uparrow} \cup T_{3\uparrow} \dots$ with $T_{1\uparrow} = \{\tau_{1\uparrow}\}$, $T_{2\uparrow} = \{\tau_{2\uparrow}\}$, $T_{3\uparrow} = \{\tau_{3\uparrow}\}$. Based on H/L crossing points we formalise the notion of *period realisation* for a noisy period path.

Definition 2 (k^{th}noisy period realisation). *For σ_A a noisy periodic trajectory with crossing point times $T_\downarrow = \cup_{k \geq 1} T_{k\downarrow}$, respectively $T_\uparrow = \cup_{k \geq 1} T_{k\uparrow}$, the realisation of the kth noisy period, denoted t_{p_k}, is defined as $t_{p_k} = min(T_{(k+1)\downarrow}) - min(T_{k\downarrow})$*[2].

[2] t_{p_k} could alternatively be defined as $t_{p_k} = min(T_{(k+1)\uparrow}) - min(T_{k\uparrow})$, that is, w.r.t. crossing into the *high* region, rather than into the *low* region. It is straightforward

Figure 3 (right) shows an example of *period realisations*: the first two period realisations, denoted $p1$ and $p2$, are delimited by the *mid*-to-*low* crossing points corresponding to the first entering of the *low* region which follows a previous sojourn in the *high* region and their duration (as per Definition 2) is $t_{p_1} = \tau_{3\downarrow} - \tau_{1\downarrow}$ respectively $t_{p_2} = \tau_{4\downarrow} - \tau_{3\downarrow}$. Notice that the time interval denoted as $p0$ does not represent a complete period realisation as there's no guarantee that $t = 0$ corresponds with the actual entering into the *low* region. Definition 2 correctly does not account for the first *spurious* period $p0$. Relying on the notion of period realisation we characterise the *period average* and *period variance* of a noisy periodic trace. Observe that the *period variance* allows us to analyse the regularity of the observed oscillator, that is, a "regular" ('irregular") oscillator is one whose traces exhibit little (large) period variance.

Definition 3 (period average). *For σ_A a noisy periodic trajectory the period average of the first $n \in \mathbb{N}$ period realisations, denoted $\bar{t}_p(n)$, is defined as $\bar{t}_p(n) = \frac{1}{n}\sum_{k=1}^{n} t_{p_k}$, where t_{p_k} is the k-th period realisation.*

Observe that for a sustained oscillator, the average value of the noisy-period, in the long run, corresponds to the limit $\bar{t}_p = \lim_{n\to\infty} \bar{t}_p(n)$.

Definition 4 (period variance). *For σ_A a noisy periodic trajectory the period variance of the first $n \in \mathbb{N}$ period realisations, denoted $s^2_{t_p}(n)$, is defined as $s^2_{t_p}(n) = \frac{1}{n-1}\sum_{k=1}^{n}(t_{p_k} - \bar{t}_p(n))^2$, where t_{p_k} is the k-th period realisation and $\bar{t}_p(n)$ is the period average for the first n period realisations.*

Based on the period average and variance we now introduce a notion of distance of noisy periodic path from a target mean period value. We will employ such distance in the HASL-based adaptation of the ABC method for inferring the parameters of an oscillator.

Definition 5 (distance from target period). *For σ_A a noisy periodic trajectory and $\bar{t_p}^{(obs)} \in \mathbb{R}_{>0}$ a target mean period duration we define the distance of σ_A from $\bar{t_p}^{(obs)}$ w.r.t. the first $n \in \mathbb{N}$ period realisations as*

$$dist(\sigma_A, n, \bar{t_p}^{(obs)}) = dist(\bar{t_p}(n), s^2_{t_p}(n), \bar{t_p}^{(obs)}) = \min\left(\frac{|\ \bar{t_p}(n) - \bar{t_p}^{(obs)}\ |}{\bar{t_p}^{(obs)}}, \frac{\sqrt{s^2_{t_p}(n)}}{\bar{t_p}^{(obs)}}\right) \quad (4)$$

where $\bar{t_p}(n)$ ($s^2_{t_p}(n)$) denotes the mean value (the variance) of the first n periods detected along σ_A (as per Definition 2 and Definition 4).

Notice that distance (4) establishes a form of multi-criteria selection of parameters as both the mean value and the variance of the detected periods are constrained. For example with a 10% tolerance (i.e. $\epsilon = 0.1$ in ABC terms) only the parameters $\theta \in \Theta$ that issue a relative error (w.r.t. the target mean period) not above 0.1 are selected.

to show that both definitions are semantically equivalent, i.e., the average value of t_{p_k} measured along a trace is equivalent with both definitions.

3.2 An Automaton for the Distance from a Target Period

We introduce a LHA named $\mathcal{A}_{per}^{\bar{t_p}^{(obs)}}$ (Fig. 4) for assessing the distance (as per Definition 5) between the mean period measured on paths of DESP oscillator $\mathcal{M}_\theta$ and a target period duration $\bar{t_p}^{(obs)}$. The automaton consists of three main locations **low**, **mid** and **high** (corresponding to the regions of the partition of A's domain induced by thresholds $L < H$). Its functioning is as follows: processing starts in either of the 3 state (which are all initial) depending on the initial state of the oscillator (population of oscillating species A is stored in n_A which is initialised through autonomous transitions before unfolding of path σ begins). Detection of one period realisation (as of Definition 2) correspond with the completion of a loop from **low** to **high** and back to **low** locations. The analysis of the simulated trajectory ends by entering location **end** as soon as the N-th period has been detected. Table 1 report about some of the variables

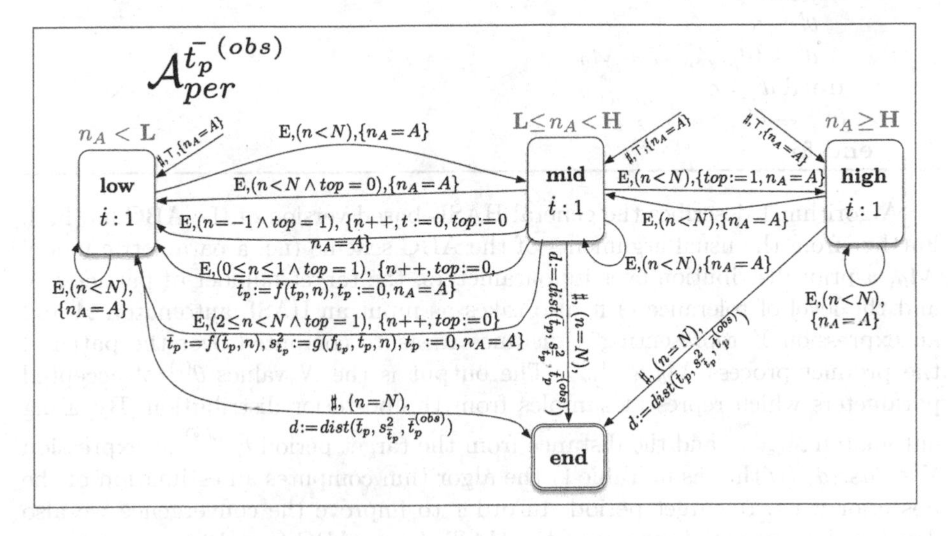

Fig. 4. $\mathcal{A}_{per}$: an LHA for selecting noisy periodic traces (with respect to an observed species A) related to partition $low = (-\infty, L]$, $mid = (L, H)$ and $high = [H, +\infty)$.

Table 1. Variables of the $\mathcal{A}_{per}^{\bar{t_p}^{(obs)}}$ automaton.

name	domain	update definition	description
n	$\mathbb{N}$	*increment*	counter of detected periods
t_p	$\mathbb{R}_{\geq 0}$	*reset*	duration last period
$\bar{t_p}$	$\mathbb{R}_{\geq 0}$	$f(\bar{t_p}, t_p, n) = \frac{\bar{t_{p_n}} \cdot n + t_p}{n+1}$	period mean
$s^2_{t_p}$	$\mathbb{R}_{\geq 0}$	$g(s^2_{t_p}, \bar{t_p}, t_p, n) = \frac{n-1}{n-2} \cdot s^2_{t_p} + \frac{(\bar{t_p}-t_p)^2}{n-1}$	period variance
d_p	$\mathbb{R}_{\geq 0}$	$\min(\frac{\bar{t_p}-\bar{t_p}^{(obs)}}{\bar{t_p}^{(obs)}}, \frac{\sqrt{s^2_{t_p}}}{\bar{t_p}^{(obs)}})$	distance from target period

(see [10] for the complete list) that $\mathcal{A}_{per}^{\bar{t_p}^{(obs)}}$ uses to store relevant statistics of the simulated path. Variable d is updated with the computed distance (as per Definition 4) on detection of the N-th period.

3.3 HASL-ABC Method for Tuning Oscillators

To calibrate the period of noisy oscillators we adapt the HASL-based version of the ABC method [12] so that it operates with $\mathcal{A}_{per}$ automaton.

Algorithm 1: Rejection sampling HASL-ABC Algorithm

Require: $(\mathcal{M}_\theta)_{\theta\in\Theta}$ parametric DESP, π prior,
N: number of particles, ϵ: tolerance level, $\mathcal{A}_{per}$ distance period LHA
Ensure: $(\theta^{(i)})_{1\leq i\leq N}$ drawn from π^{ϵ}_{ABC}
for $i = 1 : N$ **do**
  repeat
    $\theta' \sim \pi$
    $d' \sim (d_p, \mathcal{A}_{per}) \times \mathcal{M}_{\theta'}$
  until $d' \leq \epsilon$
  $\theta^{(i)} \leftarrow \theta'$
end for

Algorithm 1 describes the general HASL-based version of the ABC method. Further from the usual arguments of the ABC scheme (i.e. a parametric model $\mathcal{M}_\theta$, a prior distribution over its parameters, the target number of particles N and the level of tolerance ϵ) it also takes as input an HASL automaton $\mathcal{A}$ and an expression Y representing a distance measure computed over the paths of the product process $\mathcal{M}_\theta \times \mathcal{A}_{per}$. The output is the N values $\theta^{(i)}$ of accepted parameters which represent samples from the posterior distribution. By using automaton $\mathcal{A}_{per}^{\bar{t_p}^{(obs)}}$ and the distance from the target period $\bar{t_p}^{(obs)}$ as expression $Y \equiv last(d)$ (with d as in Table 1) the algorithm computes an estimation of the posterior w.r.t. to target period. In order to improve the convergence we also developed a sequential version of the HASL-based ABC (see [13]).

4 Case Studies

We demonstrate the HASL-ABC procedure for tuning stochastic oscillator on two examples of stochastic oscillators.

A Synthetic 3-Ways Oscillator. The CRN in (5) represents a model of synthetic sustained oscillator called *doping 3-way oscillator* [16].

$$\begin{array}{lll} R_1\!:\!A\!+\!B \xrightarrow{r_A} 2B & R_2\!:\!B\!+\!C \xrightarrow{r_B} 2C & R_3\!:\!C\!+\!A \xrightarrow{r_C} 2A \\ R_4\!:\!D_A\!+\!C \xrightarrow{r_C} D_A\!+\!A & R_5\!:\!D_B\!+\!A \xrightarrow{r_A} D_B\!+\!B & R_6\!:\!D_C\!+\!B \xrightarrow{r_B} D_C\!+\!C \end{array} \tag{5}$$

It consists of 3 main species A, B and C forming a positive feedback loop (through reactions R_1, R_2, R_3) plus 3 corresponding invariant (*doping*) species D_A, D_B, D_C whose goal is to avoid extinction of the main species (through reactions R_4, R_5, R_6). It can be shown that the total population is invariant and that the model yields sustained noisy oscillation for the 3 main species, whose period and amplitude depend on the model's parameters r_A, r_B and r_C (as well as on the total population). Figure 5 (left) depicts a simulated trajectory showing the oscillatory character of species A together with the period realisations induced by a given partition ($L = 300$ and $H = 360$). The corresponding location changes of the synchronised automaton $\mathcal{A}_{per}$ as the trajectory is unfolded are depicted in different colors.

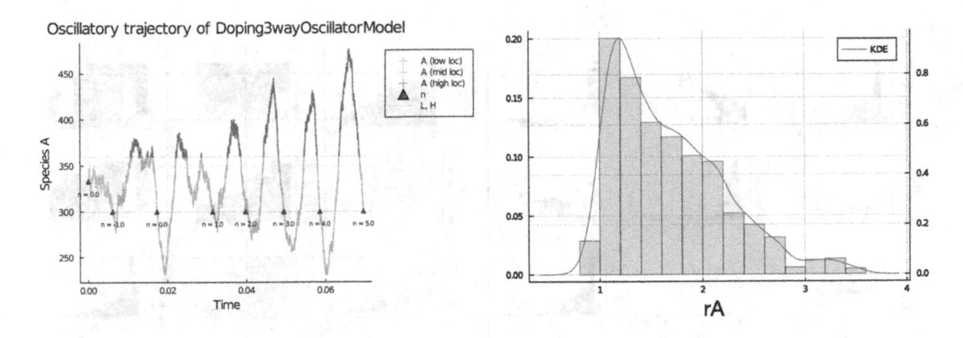

Fig. 5. The noisy periodic character of species A (left) and the posterior distribution for a single-parameter estimation experiment of the 3-ways oscillator.

Experiment 1. This is a 1-dimensional experiment in which we considered $\mathbf{s}_0 = (A_0, B_0, C_0, (D_A)_0, (D_B)_0, (D_C)_0) = (333, 333, 333, 10, 10, 10)$ as initial state, we fixed the rate constants $r_B = r_C = 1.0$, and estimated the posterior distribution for r_A considering a uniform $\mathcal{U}(0, 10)$ prior and a target mean period $\bar{t_p}^{(obs)} = 0.01$. For the automaton $\mathcal{A}_{per}$ the noisy-period dependent partition we considered is $L = 300$ and $H = 360$ while for each trajectory we observed $N = 4$ periods. For the ABC algorithm we used $N = 1000$ particles and considered a 20% tolerance ($\epsilon = 0.2$). Figure 5 (right) shows the resulting automaton-ABC posterior (histogram and Kernel density estimation). We observe that 1) the posterior support being included in the prior's $[0.0, 4.0] \subset [0, 10.0]$, we have reduced the parameter space to a subset where it is probable to obtain trajectories with a mean period of 0.01 (relative to a 20% tolerance) and 2) the posterior has only one mode, which is quite sensible, as having fixed r_B and r_C, one would expected the mean period duration being directly linked to the kinetics of reaction R_1, which is only parametrised by r_A. **Experiment 2.** This is a 3-dimensional version of the previous experiment in which we considered the same uniform prior for the 3 parameters $r_A, r_b, r_C \sim \mathcal{U}(0, 10)$. Figure 6 (left) shows the correlation plot matrix of the resulting automaton-ABC posterior, where each one-dimensional histogram on the diagonal of plot matrix represents the marginal distribution of each parameter, whereas the two-dimensional marginal distribution are given in the upper triangular part of the matrix (e.g., plot on position (1,3) refers to the marginal distribution $p(r_A, r_C|\bar{t_p}^{(obs)})$) and the scatter plots of the two-dimensional marginal distributions are in the lower triangular matrix. Based on the diagonal plots we observe that most of the parameters that results in a period close to the target one ($\bar{t_p}^{(obs)} = 20$) are within the support $[0,4] \times [0,3] \times [0,4]$ and, furthermore the particles form a 3D parabolic shape since the three two-dimensional projections have a parabolic shape, according to plots in the upper triangular part of Fig. 6. Also, one can notice that for each two-dimensional histogram, the area near the point $(1,1)$ is a high probability area, which is consistent with the previous one-dimensional experiment.

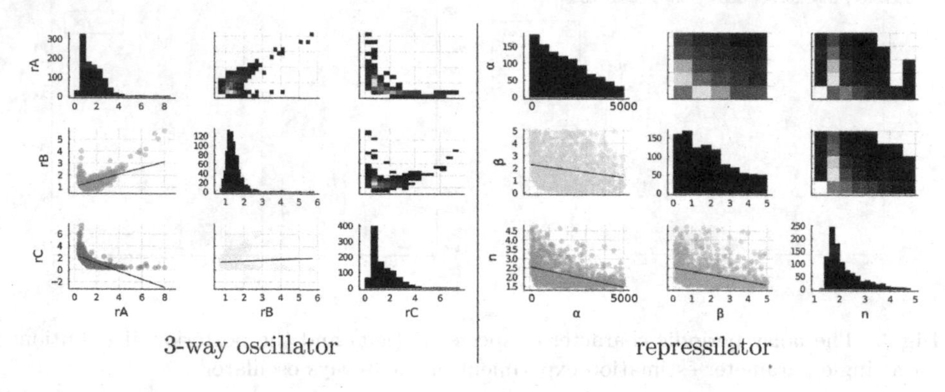

Fig. 6. Correlation plots of posterior distribution obtained through $\mathcal{A}per$-ABC for the 3D experiment of 3-way oscillator (left) and repressilator (right).

Repressilator. We consider an infinite-state model (6) of a synthetic genetic network known as Repressilator developed to reproduce oscillatory behaviours within a cell [21]. It consists of 3 proteins P_1, P_2, P_3 forming a negative feedback loop, with P_1 repressing P_2's transcription gene G_2, P_2 repressing P_3's transcription and so on.

$$\begin{array}{lll} R1: G_1 \xrightarrow{r_1} G_1 + M_1 & R2: G_2 \xrightarrow{r_2} G_2 + M_2 & R3: G_3 \xrightarrow{r_3} G_3 + M_3 \\ R4: M_1 \xrightarrow{\beta} M_1 + P_1 & R5: M_2 \xrightarrow{\beta} M_2 + P_2 & R6: M_3 \xrightarrow{\beta} M_3 + P_3 \\ R7: M_1 \xrightarrow{1} \emptyset & R8: M_2 \xrightarrow{1} \emptyset & R9: M_3 \xrightarrow{1} \emptyset \\ R10: P_1 \xrightarrow{1} \emptyset & R11: P_2 \xrightarrow{1} \emptyset & R12: P_3 \xrightarrow{1} \emptyset \end{array} \quad (6)$$

Following [21] we assumed mass-action dynamics with a common rate constant β for translation reactions (R_4, R_5, R_6) and 1 for species degradation (R_7 to R_{12}). Conversely transcription reactions ($R1, R_2, R_3$) are assumed to follow a Hill function dynamics given by the follow parameter definitions $r_1 = \frac{\alpha}{1+[P_3]^n} + \alpha_0$, $r_2 = \frac{\alpha}{1+[P_1]^n} + \alpha_0$ and $r_3 = \frac{\alpha}{1+[P_2]^n} + \alpha_0$, where n is the Hill coefficient,

α is related to transcription growth and α_0 is the parameter related to the minimum level of transcription growth. The parameter space is 4-dimensional with $\theta = (\alpha, \beta, n, \alpha_0) \in \mathbb{R}^4$. Each parameter affects the resulting oscillation **Experiment 1**. This is a 3-dimensional experiment in which we considered $\mathbf{s}_0 = ((M_1)_0, (M_2)_0, (M_3)_0, (P_1)_0, (P_2)_0, (P_3)_0) = (0, 0, 0, 5, 0, 15)$ as initial state, we fixed $\alpha_0 = 0$, and estimated the posterior distribution for the remaining parameters considering the following priors: $\alpha \sim \mathcal{U}(50, 5000)$, $\beta \sim \mathcal{U}(0.1, 5.0)$, $n \sim \mathcal{U}(0.5, 5.0)$. The target mean period was $\bar{t_p}^{(obs)} = 20$, while the noisy-period dependent partition was set at $L = 50$ and $H = 200$. For the ABC algorithm we used $N = 1000$ particles and considered a 10% tolerance ($\epsilon = 0.1$). Figure 6 (right) shows the correlation plot of the resulting automaton-ABC posterior. We observe that with this setting the Repressilator oscillations are most sensitive to parameter n as its marginal posterior is much narrower than that of α and β (i.e. varying n induces more instability than α and β). More experiments, including a 4 dimensional one, are illustrated in [10].

5 Conclusion

We introduced a methodology that given a parametric, discrete-state stochastic oscillator model, allows for inferring regions of the parameter space that exhibit a positive probability to match a desired mean oscillation period. Such framework relies on a formal characterisation of noisy periodicity which is assessed through a meter encoded by a hybrid automaton. Parameter inference is then obtained by plugging of such an automaton-meter within a ABC scheme. The added value of such a rather cumbersome formalism is in terms of automation, generality and separation of concerns. The period meter automaton, being completely configurable, can straightforwardly be generated automatically, therefore avoiding an annoying overhead to the end user. This combined with the fact that the framework inherently takes care of synchronising the model with the automaton results in a highly configurable and generic approach, one in which different oscillation tuning criteria can easily be taken into account as long as they can be encoded into a corresponding meter automaton. Notice that alternative approaches that are not based on formal methods, such as e.g., those based on auto-correlation analysis, although effective, are not easily adaptable as they require the implementation of a customised, hard-coded procedure where periodicity indicators are obtained by offline analysis of trajectories sampled from the model.

Future developments include the integration of oscillation-amplitude amongst the criteria for tuning oscillators through a *peak detector* automaton [6].

References

1. Andreychenko, A., Krüger, T., Spieler, D.: Analyzing oscillatory behavior with formal methods. In: Remke, A., Stoelinga, M. (eds.) Stochastic Model Checking. Rigorous Dependability Analysis Using Model Checking Techniques for Stochastic Systems. LNCS, vol. 8453, pp. 1–25. Springer, Heidelberg (2014). https://doi.org/10.1007/978-3-662-45489-3_1

2. Aziz, A., Sanwal, K., Singhal, V., Brayton, R.: Verifying continuous time Markov chains. In: Alur, R., Henzinger, T.A. (eds.) CAV 1996. LNCS, vol. 1102, pp. 269–276. Springer, Heidelberg (1996). https://doi.org/10.1007/3-540-61474-5_75
3. Baier, C., Haverkort, B., Hermanns, H., Katoen, J.-P.: Model-checking algorithms for continuous-time Markov chains. IEEE Trans. Software Eng. **29**, 524–541 (2003)
4. Ballarini, P., Barbot, B., Duflot, M., Haddad, S., Pekergin, N.: HASL: a new approach for performance evaluation and model checking from concepts to experimentation. Perform. Eval. **90**, 53–77 (2015)
5. Ballarini, P., Guerriero, M.L.: Query-based verification of qualitative trends and oscillations in biochemical systems. Theoret. Comput. Sci. **411**(20), 2019–2036 (2010)
6. Ballarini, P.: Analysing oscillatory trends of discrete-state stochastic processes through HASL statistical model checking. Int. J. Softw. Tools Technol. Transf. **17**(4), 505–526 (2015)
7. Ballarini, P., Duflot, M.: Applications of an expressive statistical model checking approach to the analysis of genetic circuits. Theor. Comput. Sci. **599**, 4–33 (2015)
8. Ballarini, P., Mardare, R., Mura, I.: Analysing biochemical oscillation through probabilistic model checking. Electron. Notes Theor. Comput. Sci. **229**(1), 3–19 (2009)
9. Beaumont, M.A., Cornuet, J.-M., Marin, J.-M., Robert, C.P.: Adaptive approximate Bayesian computation. Biometrika **96**(4), 983–990 (2009)
10. Bentriou, M.: Statistical inference and verification of chemical reaction networks. Ph.D. thesis, École doctorale Interfaces, University Paris Saclay (2021)
11. Bentriou, M., Ballarini, P., Cournède, P.-H.: Reachability design through approximate Bayesian computation. In: Bortolussi, L., Sanguinetti, G. (eds.) CMSB 2019. LNCS, vol. 11773, pp. 207–223. Springer, Cham (2019). https://doi.org/10.1007/978-3-030-31304-3_11
12. Bentriou, M., Ballarini, P., Cournède, P.-H.: Automaton-ABC: a statistical method to estimate the probability of spatio-temporal properties for parametric Markov population models. Theor. Comput. Sci. **893**, 191–219 (2021)
13. Bentriou, M., Boatto, S., Viaud, G., Bonnet, C., Cournède, P.-H.: Assimilation de données par filtrage particulaire régularisé dans un modèle d'épidémiologie, pp. 1–6 (2008)
14. Bortolussi, L., Milios, D., Sanguinetti, G.: Smoothed model checking for uncertain continuous-time Markov chains. Inf. Comput. **247**, 235–253 (2016)
15. Brim, L., Češka, M., Dražan, S., Šafránek, D.: Exploring parameter space of stochastic biochemical systems using quantitative model checking. In: Sharygina, N., Veith, H. (eds.) CAV 2013. LNCS, vol. 8044, pp. 107–123. Springer, Heidelberg (2013). https://doi.org/10.1007/978-3-642-39799-8_7
16. Cardelli, L.: Artificial biochemistry. In: Condon, A., Harel, D., Kok, J., Salomaa, A., Winfree, E. (eds.) Algorithmic Bioprocesses, pp. 429–462. Springer, Heidelberg (2009). https://doi.org/10.1007/978-3-540-88869-7_22
17. Češka, M., Dannenberg, F., Kwiatkowska, M., Paoletti, N.: Precise parameter synthesis for stochastic biochemical systems. In: Mendes, P., Dada, J.O., Smallbone, K. (eds.) CMSB 2014. LNCS, vol. 8859, pp. 86–98. Springer, Cham (2014). https://doi.org/10.1007/978-3-319-12982-2_7
18. Chabrier-Rivier, N., Chiaverini, M., Danos, V., Fages, F., Schächter, V.: Modeling and querying biomolecular interaction networks. Theor. Comput. Sci. **325**, 25–44 (2004)

19. Clarke, E.M., Emerson, E.A., Sistla, A.P.: Automatic verification of finite state concurrent systems using temporal logic specifications: a practical approach. In: Wright, J.R., Landweber, L., Demers, A.J., Teitelbaum, T. (eds.) Conference Record of the Tenth Annual ACM Symposium on Principles of Programming Languages, Austin, Texas, USA, January 1983, pp. 117–126. ACM Press (1983)
20. Del Moral, P., Doucet, A., Jasra, A.: An adaptive sequential Monte Carlo method for approximate Bayesian computation. Stat. Comput. **22**(5), 1009–1020 (2012)
21. Elowitz, M., Leibler, S.: A synthetic oscillatory network of transcriptional regulators. Nature **403**(335), 335–338 (2000)
22. Goldbeter, A.: Computational approaches to cellular rhythms. Nature **420**, 238–245 (2002)
23. Han, T., Katoen, J.P., Mereacre, A.: Approximate parameter synthesis for probabilistic time-bounded reachability. In: 2008 Real-Time Systems Symposium, pp. 173–182 (2008)
24. Thomas, P.J., Lindner, B., MacLaurin, J., Fellous, J.M.: Stochastic oscillators in biology: introduction to the special issue. Biol. Cybern. **116**(2), 119–120 (2022)
25. Marin, J.-M., Pudlo, P., Robert, C.P., Ryder, R.J.: Approximate Bayesian computational methods. Stat. Comput. **22**(6), 1167–1180 (2012)
26. Molyneux, G.W., Abate, A.: ABC(SMC)^2: simultaneous inference and model checking of chemical reaction networks. In: Abate, A., Petrov, T., Wolf, V. (eds.) CMSB 2020. LNCS, vol. 12314, pp. 255–279. Springer, Cham (2020). https://doi.org/10.1007/978-3-030-60327-4_14
27. Molyneux, G.W., Wijesuriya, V.B., Abate, A.: Bayesian verification of chemical reaction networks. In: Sekerinski, E., et al. (eds.) FM 2019. LNCS, vol. 12233, pp. 461–479. Springer, Cham (2020). https://doi.org/10.1007/978-3-030-54997-8_29
28. Pnueli, A.: The temporal logic of programs. In: 18th Annual Symposium on Foundations of Computer Science, Providence, Rhode Island, USA, 31 October–1 November 1977, pp. 46–57. IEEE Computer Society (1977)
29. Sisson, S.A., Fan, Y., Beaumont, M.: Handbook of Approximate Bayesian Computation. Chapman and Hall/CRC (2018)
30. Sneyd, J., Tsaneva-Atanasova, K., Reznikov, V., Bai, Y., Sanderson, M.J., Yule, D.I.: A method for determining the dependence of calcium oscillations on inositol trisphosphate oscillations. Proc. Natl. Acad. Sci. **103**(6), 1675–1680 (2006)
31. Spieler, D.: Characterizing oscillatory and noisy periodic behavior in Markov population models. In: Proceedings of QEST 2013 (2013)

Phenotype Control of Partially Specified Boolean Networks

Nikola Beneš[1], Luboš Brim[1], Samuel Pastva[2], David Šafránek[1], and Eva Šmijáková[1](✉)

[1] Systems Biology Laboratory, Masaryk University, Brno, Czech Republic
xsmijak1@fi.muni.cz

[2] Institute of Science and Technology Austria, Klosterneuburg, Austria

Abstract. Partially specified Boolean networks (PSBNs) represent a promising framework for the qualitative modelling of biological systems in which the logic of interactions is not completely known. Phenotype control aims to stabilise the network in states exhibiting specific traits.

In this paper, we define the phenotype control problem in the context of asynchronous PSBNs and propose a novel semi-symbolic algorithm for solving this problem with permanent variable perturbations.

1 Introduction

Boolean networks (BNs) are a widely used model to study dynamics of complex biological systems [2,4]. Recently, a new interesting variant of BN control problem called *phenotype* control was proposed [16,18]. The goal of the phenotype control is to stabilize the network in states exhibiting specific traits regardless of the source state. This approach does not limit the control target to a single state or a specific attractor but rather considers arbitrary combinations of traits (subspaces of BN states). To control the network, we use *variable perturbations* that fix variable values to specific constants.

The behavior of a BN is given by the Boolean update functions and the considered updating scheme. In this work, we specifically focus on *asynchronous* updates, where one update rule is triggered at a time, because they suitably capture the behaviour of biological systems [3]. However, it is important to note that in many cases, the exact Boolean functions of the model may not be precisely known due to various uncertainties such as insufficient experimental knowledge [22], inconsistent observations [21], genetic mutations [30], or any other ambiguities. To address this issue, we employ a framework of *partially specified* Boolean networks (PSBN), in which update functions can be defined using uninterpreted function symbols [8,9,11].

The most well-studied variant of BN control problem is *source-target* control in which both source and target states are specified. Generalized variants of this

This work was supported by the Czech Foundation grant No. GA22-10845S, Grant Agency of Masaryk University grant No. MUNI/G/1771/2020, and the European Union's Horizon 2020 research and innovation programme under the Marie Skłodowska-Curie Grant Agreement No. 101034413.

J. Pang and J. Niehren (Eds.): CMSB 2023, LNBI 14137, pp. 18–35, 2023.
https://doi.org/10.1007/978-3-031-42697-1_2

problem include target control [25], which aims to reach the target attractor regardless of the initial state, and full-control [20], which seeks a control strategy between all attractor pairs.

The control problems also differ significantly in the perturbations which they employ. The simplest perturbation variant is solved in the context of Boolean control networks (where only inputs are allowed to be set) [27]. In contrast, another line of research allows influence of any BN variable. In biology, such perturbations can be implemented in terms of gene knock-outs or over-expressions. Commonly used are *permanent* perturbations which consider variables being stabilized ad infinitum [33,34,40]. Other types of perturbations include one-step [5], temporary [28,36,38], and sequential dynamics [1,29,31,37]. It is worth noting that the source-target control was also studied in the context of PSBNs [11,12].

A common drawback of source-target, target and full control problems is that the target attractor must be fully known in advance in order to be provided as an input to the source-target control algorithm. However, the unknown parts might cause a PSBN to exhibit very similar attractors differing only in some negligible parts. Therefore, multiple attractors might be in fact a target of interest. A similar scenario can occur even for fully known networks, as the modeler's goal might be to stabilize just some subset of traits (phenotype) which are actually exhibited in several different attractors.

The phenotype control of BNs was previously solved using a method based on a reduction of the network into a layered network and converging trees [15]. However, the method is assuming synchronous update semantics. The phenotype control was later addressed also for asynchronous update semantics and solved using model checking methods [18]. A very similar problem, *marker control*, was also solved for the most permissive updating scheme [32].

Our contribution. In this work, we lift the problem of phenotype control from standard asynchronous BNs to partially specified BNs. We use permanent variable perturbations to achieve the control objective. We develop semi-symbolic method based on Binary Decision Diagrams (BDDs). This method allows us to find solutions for all PSBN interpretations in a single run, unlike brute-force methods, which need to compute solutions for each fully specified BN instance separately. Our method is based on a search of trap-sets, which makes its main idea transferable to other BN updating schemes as well. The method is conceptually related to [18], but aside from our novel incorporation of PSBNs, we also lift the perturbations into the symbolic domain (as opposed to the brute-force enumeration in [18]). We also demonstrate how this method can be applied to obtain interesting observations about PSBN models on real-world case studies.

2 Theory

This section presents a formal introduction to the topic of partially specified Boolean networks, including the notion of permanent perturbation, phenotype control and perturbation robustness.

Notation. Let us first note that we consider 0 and 1 as interchangeable with *false* and *true*, respectively. We then define $\mathbb{B} = \{0, 1\}$ to be the set of Boolean values and $\mathbb{B}_* = \{0, 1, *\}$ to be an extension of $\mathbb{B}$ which also admits a *free* value $*$ (i.e. neither *true* nor *false*). We write $\mathbb{B}^n$ to denote the set of all n-element vectors over $\mathbb{B}$. In the following, such n refers to the *size* of a Boolean network, while x represents its state (a configuration). For each $x \in \mathbb{B}^n$, x_i then denotes the i-th element of x. Finally, for $x \in \mathbb{B}^n$, index $i \in [1, n]$, and a Boolean value $b \in \mathbb{B}$, expression $x[i \mapsto b]$ denotes a substitution of the i-th element in x for the value b. Formally, the result is $x' = x[i \mapsto b]$ s.t. $x'_j = b$ for $j = i$, and $x'_j = x_j$ otherwise.

2.1 Boolean Networks

Before we define partially specified Boolean networks, we first introduce the classical (i.e. fully specified) Boolean network (BN) and other related concepts:

Definition 1. Let n be the number of system variables. A *Boolean network* is a collection $\mathbb{BN} = \{f_1, \ldots, f_n\}$ with each $f_i : \mathbb{B}^n \to \mathbb{B}$ being the Boolean *update function* (sometimes called *local function*) of the network's i-th variable.

In the context of a specific Boolean network, the set $\mathbb{B}^n$ is the network's *state space*, and the vectors $x \in \mathbb{B}^n$ are its *states*. Note that although the input of each f_i is a full state $x \in \mathbb{B}^n$, the output of f_i does not typically depend on all network variables, but rather on a smaller subset of variables which we say *regulate* the i-th variable. If a variable has no regulators (i.e. its update function is a constant), we also call it an *input* of $\mathbb{BN}$. Symmetrically, variable that does not regulate any other variable is called an *output*.

Additionally, we call the members of $\mathbb{B}^n_*$ the *subspaces* of $\mathbb{BN}$. Intuitively, each subspace $S \in \mathbb{B}^n_*$ describes a hypercube in the state space $\mathbb{B}^n$. This hypercube consists of states $x \in \mathbb{B}^n$ such that $x_i = S_i$ for all i where $S_i \in \mathbb{B}$. We can thus treat each subspace S as a set of states. Furthermore, to denote a specific subspace, we will often simply use a string of values from $\mathbb{B}_*$ instead of the full vector notation (e.g. $S = 11{*}0$ instead of $S = (1, 1, *, 0)$).

To formally reason about the evolution of the network's state, we consider its asynchronous state-transition graph:

Definition 2. For $\mathbb{BN} = \{f_1, \ldots, f_n\}$, the *state-transition graph* $\text{STG}(\mathbb{BN}) = (V, E)$ is a directed graph with $V = \mathbb{B}^n$ and $E \subseteq V \times V$ given as follows:

$$(u, v) \in E \Leftrightarrow (u \neq v \wedge \exists i \in [1, n].\ v = u[i \mapsto f_i(u)])$$

We can write $u \to v$ whenever $(u, v) \in E$. Observe that each transition within $\text{STG}(\mathbb{BN})$ always updates exactly one network variable. To then study the long-term behaviour of a network, we focus on the terms *trap set* and *attractor* [17,32]:

Definition 3. Let $\mathbb{BN} = \{f_1, \ldots, f_n\}$ be a Boolean network and $X \subseteq \mathbb{B}^n$ a set of network states. We say that X is a *trap set* when for all $x \in X$ and $y \in \mathbb{B}^n$ we have that $x \to y$ implies $y \in X$ (i.e. X cannot be escaped). We write that X is an *attractor* when X is strongly connected within $\text{STG}(\mathbb{BN})$.

Equivalently, we can also define attractors as exactly the inclusion-minimal trap sets of $\mathbb{BN}$. We write $\mathcal{A}(\mathrm{STG}(\mathbb{BN}))$ to denote the set of all attractors of $\mathbb{BN}$.

It has been shown that attractors are closely tied to the notion of biological phenotypes [18,26]. Each attractor represents a possible stable *outcome* achievable within a particular $\mathbb{BN}$. The combination of traits observable as part of this outcome then forms the actual phenotype. However, in many cases, there can be multiple attractors that exhibit the same set of traits, and thus the same biological phenotype. We formalise this concept as follows:

Definition 4. A *character* is a set of BN variables $\mathcal{U} \subseteq [1, n]$ which cover the observable real-world properties of the system. A *trait* T is then a valuation of these character variables: $T : \mathcal{U} \to \mathbb{B}$.

Such character variables $\mathcal{U}$ typically correspond to the network outputs, but this is not required. Each trait defines a subspace $S^T \in \mathbb{B}^n_*$ s.t. $S^T_i = T(i)$ for $i \in \mathcal{U}$, and $S^T_i = *$ otherwise. With a slight abuse of notation, we simply use T to also mean the subspace S^T when clear from context.

Definition 5. A *phenotype* is a set of states $P \subseteq \mathbb{B}^n$ described by an arbitrary combination of the BN traits. We say that phenotype P *exists* in $\mathbb{BN}$ when $A \subseteq P$ for some $A \in \mathcal{A}(\mathrm{STG}(\mathbb{BN}))$. We say that $\mathbb{BN}$ *exhibits* P when $A \subseteq P$ for all $A \in \mathcal{A}(\mathrm{STG}(\mathbb{BN}))$.

While a *trait* is always a subspace, a phenotype is an arbitrary combination of traits. For example, assuming $n = 4$ and $\mathcal{U} = \{1, 2\}$, $11{*}{*}$ and $00{*}{*}$ are two of the four admissible traits (or rather trait subspaces). Each of these traits can represent a single phenotype, but they can also represent a combined phenotype $00{*}{*} \cup 11{*}{*}$. We typically assume that if a network admits more than one phenotype, these are mutually disjoint. Finally, note that not all traits have to belong to some phenotype (e.g. if a trait is not biologically viable).

2.2 Partially Specified Boolean Networks

A shortcoming of classical BNs as defined above is that to study network dynamics, all update functions must be fully known. However, this is often not realistic for large-scale systems. To address this problem, we consider the notion of *Partially Specified Boolean Networks* (PSBNs) [8] (also termed *Coloured Boolean Networks*).

In a PSBN, we can use *uninterpreted functions* as stand-ins for unknown (fixed but arbitrary) parts of the network's dynamics. Each uninterpreted function is then denoted by its *symbol* (a name) and input arguments.

Definition 6. Let n be the number of system variables, and $\mathfrak{F}$ a set of *uninterpreted function symbols*. A *partially specified Boolean network* $\mathbb{PSBN} = \{P_1, \ldots, P_n\}$ consists of expressions P_i given by the following grammar:

$$E ::= 0 \mid 1 \mid x \mid \neg E \mid E \wedge E \mid E \vee E \mid \mathcal{F}^{(a)}(E, \ldots, E)$$

Here, x ranges over the network variables and $\mathcal{F}$ over the uninterpreted functions of $\mathfrak{F}$ (superscript $a \in \mathbb{N}_0$ denotes the arity of $\mathcal{F}$). Other Boolean operators (e.g. $\Rightarrow$ or $\Leftrightarrow$) can be implemented as syntactic abbreviations using $\vee$, $\wedge$, and $\neg$.

In other words, a PSBN is defined using standard Boolean constants (0 and 1), variable state propositions (x) and Boolean connectives ($\neg$, $\wedge$, $\vee$), but it can also use uninterpreted functions from $\mathfrak{F}$ as a way of incorporating unknown behaviour. This makes it possible to idiomatically describe systems whose dynamics are not fully known.

Note that this definition also allows zero-arity uninterpreted functions (e.g. $\mathcal{F}^{(0)} \in \mathfrak{F}$). These are effectively unknown Boolean constants. As such, they are functionally equivalent to network inputs with an unknown value. To distinguish them, we sometimes call these uninterpreted functions *logical parameters.*

To assign meaning to a particular $\mathbb{PSBN}$, we rely on the term *interpretation.* An interpretation $\mathcal{I}$ is a function which assigns each symbol from $\mathfrak{F}$ a Boolean function of the corresponding arity. By substituting each $\mathcal{F} \in \mathfrak{F}$ in expressions $P_1, \ldots, P_n$ for its corresponding $\mathcal{I}(\mathcal{F})$ (written $P_i(\mathcal{I})$), we obtain a classical fully specified Boolean network which we denote $\mathbb{PSBN}(\mathcal{I}) = \{P_1(\mathcal{I}), \ldots, P_n(\mathcal{I})\}$.

Definitions of trap set and attractor for fully specified BNs then extend to PSBNs naturally per individual interpretations. For example, we write that a set $A \subseteq \mathbb{B}^n$ is an attractor of $\mathbb{PSBN}$ *for interpretation* $\mathcal{I}$ when it is an attractor of $\mathbb{PSBN}(\mathcal{I})$ (i.e. there are no attractors of $\mathbb{PSBN}$, only attractors of its interpretations). The definitions of character, trait and phenotype do not depend on the actual dynamics of the network (only on its variables). As such, these are identical for both BNs and PSBNs.

Note that for any given $\mathbb{PSBN}$ based on uninterpreted functions $\mathfrak{F}$, there is a logically equivalent *normalised* $\widehat{\mathbb{PSBN}}$ based on $\widehat{\mathfrak{F}}$, such that $\widehat{\mathfrak{F}}$ only admits zero-arity uninterpreted functions. The details of this conversion can be found in [11]. However, in general, the size of $\widehat{\mathfrak{F}}$ is exponential w.r.t. the arity of functions in $\mathfrak{F}$. Considering that $\widehat{\mathfrak{F}}$ only admits zero-arity logical parameters, the possible valuations of these parameters (which we also call *colours*) can be encoded as vectors $c \in \mathbb{B}^m$ where $m = |\widehat{\mathfrak{F}}|$. We then write $\mathcal{I}_c$ to denote an interpretation of $\mathbb{PSBN}$ that is encoded by a particular colour $c \in \mathbb{B}^m$.

Finally, to algorithmically study the dynamics of possible PSBN interpretations, we rely on the term *coloured state-transition graph.*

Definition 7. Let $\mathbb{PSBN} = \{P_1, \ldots, P_n\}$ be a partially specified Boolean network such that $\widehat{\mathbb{PSBN}}$ admits m logical parameters. The *coloured state-transition graph* $\mathrm{STG}(\mathbb{PSBN}) = (V, C, E)$ is a directed graph where $V = \mathbb{B}^n$, $C = \mathbb{B}^m$ and $E \subseteq V \times C \times V$ is given as $(u, c, v) \in E \Leftrightarrow (u, v) \in E(\mathrm{STG}(\mathbb{PSBN}(\mathcal{I}_c)))$.

In other words, the coloured state-transition graph is a unifying structure which incorporates the STGs of all interpretations of $\mathbb{PSBN}$: A transition from a state u to state v is enabled for colour c if the same transition appears in the STG of $\mathbb{PSBN}(\mathcal{I}_c)$. An example of such an STG is depicted in Fig. 1a. We further explore the utility of coloured STGs in the Methods section, where we show how these can be efficiently manipulated *symbolically.*

Example 1. Consider a PSBN with $\mathfrak{F} = \{\mathcal{F}^{(2)}\}$ and $P_1 \equiv \mathcal{F}(x_1, x_2)$; $P_2 \equiv x_1 \wedge x_3$; $P_3 \equiv x_1 \vee \neg x_3$. After normalisation, we have $\widehat{\mathfrak{F}} = \{\mathcal{F}_{00}, \mathcal{F}_{01}, \mathcal{F}_{10}, \mathcal{F}_{11}\}$ and $\widehat{P}_1 \equiv ((\neg x_1 \wedge \neg x_2) \Rightarrow \mathcal{F}_{00}) \wedge ((\neg x_1 \wedge x_2) \Rightarrow \mathcal{F}_{01}) \wedge ((x_1 \wedge \neg x_2) \Rightarrow \mathcal{F}_{10}) \wedge ((x_1 \wedge x_2) \Rightarrow \mathcal{F}_{11})$.

The coloured STG of this PSBN is shown in Fig. 1a. Each edge label describes a subspace of the colour set $\mathbb{B}^m$ which enables said edge (edges without labels are enabled for all colours). Note that the presence of each edge always depends on a single logical parameter. Figure 1b then shows STGs of two specific PSBN interpretations: $\mathcal{I}_{0101}$ and $\mathcal{I}_{0110}$ (i.e. $\mathcal{F}(x_1, x_2) = x_2$ and $\mathcal{F}(x_1, x_2) = x_1 \not\Leftrightarrow x_2$).

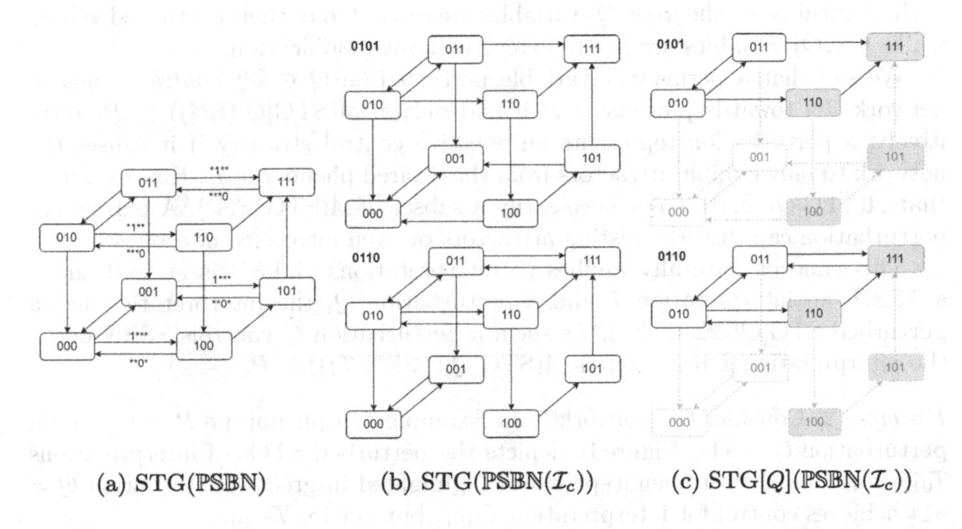

(a) STG($\mathbb{PSBN}$) (b) STG($\mathbb{PSBN}(\mathcal{I}_c)$) (c) STG[$Q$]($\mathbb{PSBN}(\mathcal{I}_c)$)

Fig. 1. STGs of the PSBN from Example 1: (a) The full coloured STG; (b) STGs for interpretations $\mathcal{I}_{0101}$ and $\mathcal{I}_{0110}$; (c) Perturbed STGs for $Q = {*}1{*}$. Green states represent phenotype $P = 1{*}{*}$.

2.3 Phenotype Control of Partially Specified Boolean Networks

Finally, we can discuss the topic of phenotype control for PSBNs: we assume a set of phenotype states $P \subseteq \mathbb{B}^n$ and the goal is to perturb the network such that it *exhibits* this phenotype. In other words, after the perturbation is applied to the network, the network stabilizes in some attractor A such that $A \subseteq P$. However, in the presence of partially unknown dynamics, this may not be achievable using a sufficiently small perturbation. In such cases, we rely on control *robustness* to assess viability of possible perturbations.

As before, for simplicity, we first introduce some of the concepts for classical BNs and then expand these to PSBNs. First, to control the behaviour of a BN, we use the notion of variable perturbation:

Definition 8. A *variable perturbation* is a vector $Q \in \mathbb{B}^n_*$. For every $Q_i = *$, we say that the i-th variable is *unperturbed*, whereas for $Q_i = 0$ or $Q_i = 1$, we say that the variable is *perturbed* to either 0 or 1.

We write *fixed*(Q) and *free*(Q) to denote the subset of perturbed and unperturbed variables, respectively. The *size* of Q is the size of the *fixed*(Q) set. Furthermore, a perturbation can again be seen as a *subspace* of $\mathbb{B}^n$. We then write that the states in this subspace are *compliant* with Q.

For a particular $\mathbb{BN}$ and Q, we consider a *perturbed* state-transition graph $\mathrm{STG}[Q](\mathbb{BN})$, which is a sub-graph of $\mathrm{STG}(\mathbb{BN})$ *induced* by the states compliant with Q. Intuitively, the *fixed*(Q) variables are restricted to their perturbed values, while *free*(Q) variables are left to evolve without modification.

We say that a permanent variable perturbation $Q \in \mathbb{B}^n_*$ *controls* Boolean network $\mathbb{BN}$ towards phenotype P if and only if $\mathcal{A}(\mathrm{STG}[Q](\mathbb{BN})) \subseteq P$. Intuitively, a perturbation represents an effective control strategy if it causes the network to only exhibit attractors from the desired phenotype P. However, note that $\mathcal{A}(\mathrm{STG}[Q](\mathbb{BN}))$ is not necessarily a subset of $\mathcal{A}(\mathrm{STG}(\mathbb{BN}))$. A permanent perturbation can disturb existing attractors or even introduce new ones.

This concept naturally applies to interpretations of PSBNs as well: given a $\mathbb{PSBN}$, an interpretation $\mathcal{I}$ and a perturbation Q, the interpretation has a perturbed $\mathrm{STG}[Q](\mathbb{PSBN}(\mathcal{I}))$. As such, a perturbation Q controls $\mathbb{PSBN}$ under the interpretation $\mathcal{I}$ if it ensures $\mathcal{A}(\mathrm{STG}[Q](\mathbb{PSBN}(\mathcal{I}))) \subseteq P$.

Example 2. Consider the network from Example 1, a phenotype $P = 1{*}{*}$ and a perturbation $Q = {*}1{*}$. Figure 1c depicts the perturbed STGs of interpretations $\mathcal{I}_{0101}$ and $\mathcal{I}_{0110}$, with phenotype states highlighted in green. Perturbation $Q = {*}1{*}$ achieves control for interpretation $\mathcal{I}_{0101}$, but not for $\mathcal{I}_{0110}$.

Definition 9. Assume a partially specified network $\mathbb{PSBN}$, a phenotype $P \subseteq \mathbb{B}^n$ and a set of admissible perturbations $\mathcal{Q} \subseteq \mathbb{B}^n_*$. The goal of the *complete* phenotype control is to compute all pairs $(c, Q) \in \mathbb{B}^m \times \mathbb{B}^n_*$ such that $Q \in \mathcal{Q}$ and Q controls $\mathbb{PSBN}(\mathcal{I}_c)$ towards phenotype P.

Intuitively, the result of PSBN phenotype control are all combinations of colours and perturbations for which the perturbation necessarily stabilizes the associated network interpretation in the given phenotype. The set $\mathcal{Q}$ is in practice used to restrict the problem setting to perturbations of a certain size or to otherwise biologically feasible perturbations.

Perturbation size and robustness In practice, it is often unnecessary to compute the complete set of colour-perturbation pairs. The goal is instead to find the smallest functioning perturbation [37,39]. However, in partially specified networks, such minimal perturbation typically only works for a small subset of interpretations, making it potentially unreliable in practice. To address this, we consider robust phenotype control:

Definition 10. The *robustness* ρ of a perturbation $Q \in \mathbb{B}_*^n$ is defined as:

$$\rho(Q) = \frac{|\{c \in \mathbb{B}^m \mid Q \text{ controls } \mathbb{PSBN}(\mathcal{I}_c)\}|}{|\mathbb{B}^m|}$$

Given a robustness threshold $r \in (0,1]$, the goal of *robust* phenotype control is to compute the smallest perturbation (or perturbations) Q s.t. $\rho(Q) \geq r$.

Ideally, a suitable perturbation Q with $\rho(Q) = 1$ achieves control for all interpretations of $\mathbb{PSBN}$. However, if no such perturbation exists, the parameter r can be tuned to achieve a trade off between perturbation size and robustness.

3 Methods

Partially specified Boolean networks suffer from both state and parameter space explosion, which in practice makes them hard to analyse exhaustively. In the context of control, this problem is further complicated by perturbation space explosion. To mitigate this, we employ symbolic state space exploration which allows us to analyse the full set of possible perturbed STGs within a single algorithm pass. This technique exploits similarities between the STGs of various network interpretations (and perturbations) to significantly speed up the exploration process compared to the naive enumeration [6,12].

3.1 Symbolic Computation Model

A binary decision diagram (BDD) [13] is a directed acyclic graph which represents a Boolean function. A BDD-encoded function $f : \mathbb{B}^k \to \mathbb{B}$ can be used to encode a set (or relation) of Boolean vectors $X \subseteq \mathbb{B}^k$, such that $f(x) = 1 \Leftrightarrow x \in X$. Since both network states ($\mathbb{B}^n$) and interpretations ($\mathbb{B}^m$ after normalisation) correspond to Boolean vectors, such sets have a direct BDD representation.

For sets of perturbations, a naive approach requires two Boolean variables per component, since their domain is $\mathbb{B}_*$ instead of $\mathbb{B}$. However, in [11], we introduce an alternative encoding suitable for representing perturbed state-transition systems that only requires one variable. This significantly improves the efficiency of the encoding, since fewer symbolic variables typically result in smaller BDDs.

Perturbed STG encoding Observe that in the context of a state $x \in \mathbb{B}^n$ which is *compliant* with a perturbation Q, it is sufficient to know the set *fixed*(Q) to fully reconstruct Q: we have $Q_i = x_i$ for every $i \in$ *fixed*(Q) and $Q_i = *$ otherwise.

Consequently, a relation over states x, interpretations $\mathcal{I}_c$ and perturbations Q where all states are compliant with their respective perturbations can be encoded as a relation over $\mathbb{B}^n \times \mathbb{B}^m \times \mathbb{B}^n$. Here, the last component encodes the possible sets *fixed*(Q). Since each perturbed STG$[Q](\mathbb{BN})$ only admits states compliant with Q, such representation is suitable for collectively encoding dynamics of such perturbed STGs for different perturbations.

To avoid confusion, we write $\mathcal{V} \equiv \mathbb{B}^n$ to mean the set of network states, $\mathcal{C} \equiv \mathbb{B}^m$ to mean the set of network interpretations (colours), and $\mathcal{L} \equiv \mathbb{B}^n$ to mean the set of sets of perturbed variables *fixed*(Q), collectively $\mathcal{S} = \mathcal{V} \times \mathcal{C} \times \mathcal{L}$. For $(x, l) \in \mathcal{V} \times \mathcal{L}$, we also write that $Q \equiv (x, l)$ when $Q_i = x_i$ for all $l_i = 1$ and $Q_i = *$ when $l_i = 0$ (i.e. Q can be reconstructed based on the state x and the set of perturbed variables l).

When the goal is to symbolically represent a subset of perturbations $\mathcal{Q} \subseteq \mathbb{B}^n_*$, we do this through a relation $S_{\mathcal{Q}} \subseteq \mathcal{V} \times \mathcal{L}$ where $S_{\mathcal{Q}}$ contains exactly *all* states compliant with every $Q \in S$. Finally, in some cases, we may need to directly reference the Boolean BDD variables used in our encoding. We then use the notation $\mathcal{V}_i$, $\mathcal{C}_i$, resp. $\mathcal{L}_i$ to denote the i-th BDD variable that encodes the state, colour or perturbation component of $\mathcal{S}$.

Symbolic operations Set operations ($\cap, \cup, \setminus$, etc.) on symbolic sets (or relations) are implemented through Boolean logical operators ($\wedge, \vee, \neg$, etc.) as is customary for BDDs. Furthermore, the following standard BDD operations are used:

$$\begin{aligned}
\textsc{Project}_{\mathcal{X}}(X \subseteq \mathbb{B}^k) &= \{x \in \mathbb{B}^k \mid \exists b \in \mathbb{B}.\ x[\mathcal{X} \mapsto b] \in X\} \\
\textsc{Select}_{\mathcal{X}=b}(X \subseteq \mathbb{B}^k) &= \{x \in \mathbb{B}^k \mid x \in X \wedge x[\mathcal{X}] = b\} \\
\textsc{Restrict}_{\mathcal{X}=b}(X \subseteq \mathbb{B}^k) &= \textsc{Project}_{\mathcal{X}}(\textsc{Select}_{\mathcal{X}=b}(X))
\end{aligned}$$

Here, $\mathcal{X}$ denotes a variable of the BDD encoding (e.g. $\mathcal{V}_3$). We can also use a list of conditions in the subscript of the method as a shorthand for multiple nested calls to the same method. Intuitively, Project is equivalent to existential quantification, Select is implemented through conjunction, and Restrict is a combination of both. To explore the perturbed STGs, we use:

$$\begin{aligned}
\textsc{Pre}(X \subseteq \mathcal{S}) &= \{\ (s, c, l) \mid \exists (t, c, l) \in X.\ s \to_{c,Q} t \text{ for } Q \equiv (s, l)\ \} \\
\textsc{Post}(X \subseteq \mathcal{S}) &= \{\ (t, c, l) \mid \exists (s, c, l) \in X.\ s \to_{c,Q} t \text{ for } Q \equiv (s, l)\ \}
\end{aligned}$$

Here, $s \to_{c,Q} t$ denotes that $s \to t$ within the graph $\mathrm{STG}[Q](\mathbb{PSBN}(\mathcal{I}_c))$. Intuitively, Pre and Post compute the sets of predecessors and successors of the states in X within $\mathbb{PSBN}$ under their respective interpretations $\mathcal{I}_c$ and perturbations Q. The implementation details of these operations can be found in [11].

We then rely on a method $\textsc{Trap}(X \subseteq \mathcal{S})$ which computes the *maximal trap set* within the given set X. Naively, this method can be implemented as the greatest fixed point of $\textsc{Trap}(X) = X \cap \textsc{Trap}(X \setminus \textsc{Pre}(\textsc{Post}(X)))$. However, we use a more efficient implementation based on the technique called *saturation* [8].

Finally, we use $\mathbb{VAL}^n_k$ to denote a BDD which encodes a function of n inputs that is *true* iff exactly k inputs are *true*. To construct such BDD efficiently, we observe that $\mathbb{VAL}^n_0 = \bigwedge_{i \in [1,n]} \neg x_i$, and that $\mathbb{VAL}^n_k = \bigvee_{i \in [1,n]} (\mathbb{VAL}^n_{k-1} \vee x_i)$.

3.2 Control Algorithm

Our control approach is based on Algorithm 1 where we describe:

- CompletePhenotypeControl: Core algorithm that iteratively computes the *complete* control map for an admissible set $\mathcal{Q}$.
- RobustPhenotypeControl: A wrapper for the core algorithm that facilitates minimal control under the desired *robustness*.
- Enumerate: An auxiliary method to enumerate all *working* perturbations and find the maximum robustness.

Algorithm 1: Symbolic permanent phenotype control of PSBNs.

```
Fn CompletePhenotypeControl(P ⊆ V, Q ⊆ V × L (encodes B^n_*))
    universe ← { (x, c, l) ∈ S | (x, l) ∈ Q };
    phenotype ← universe ∩ (P × C × L);
    phenotype_trap ← Trap(phenotype);
    non_phenotype ← universe \ phenotype_trap;
    cannot_control ← Trap(non_phenotype);
    for i ∈ [1, n] do
        not_perturbed ← Project_{V_i}(Select_{L_i=0}(cannot_control));
        cannot_control ← not_perturbed ∪ Select_{L_i=1}(cannot_control);
    control_map ← universe \ cannot_control;
    return control_map;
Fn RobustPhenotypeControl(r, P ⊆ V, Q ⊆ V × L)
    for k ∈ [0, n] do
        Q_k ← Q ∩ (V × VAL^n_k);
        control_map ← CompletePhenotypeControl(P, Q_k);
        ρ_best ← Enumerate(1, control_map);
        if ρ_best ≥ r then return;
Fn Enumerate(i ∈ N, control_map ⊆ S (encodes B^m × B^n_*))
    if control_map = ∅ then return 0;
    if i > n then return |{c ∈ C | ∃(x, c, l) ∈ S.(x, c, l) ∈ control_map}| / |C|;
    not_controlled ← Restrict_{L_i=0}(control_map);
    controlled_true ← Restrict_{L_i=1, V_i=1}(control_map);
    controlled_false ← Restrict_{L_i=1, V_i=0}(control_map);
    best ← Enumerate(i + 1, not_controlled);
    best ← max(best, Enumerate(i + 1, controlled_true));
    best ← max(best, Enumerate(i + 1, controlled_false));
    return best;
```

Complete control. The main idea behind Algorithm 1 is the following: A perturbation achieves phenotype control if and only if all attractors in the perturbed STG are within the phenotype subset P. Consequently, we want to *discard* any perturbation that provably retains some attractor not contained in P.

Therefore, we compute a non-phenotype trap set that is guaranteed to contain all non-phenotype attractor states in it. First, we compute a similar trap subset of the phenotype P which is guaranteed to contain all phenotype attractors. Then, we invert this set and repeat the operation to obtain a trap which is a superset of all non-phenotype attractors. Note that simply computing the largest trap set within $\mathcal{V} \setminus P$ would only cover attractors that are completely within $\mathcal{V} \setminus P$. The above-described process is necessary to also cover attractors that intersect P but are not subsets of P.

The algorithm then iterates over all network variables and performs projection in cases where the variable is *not perturbed.* Initially, the `cannot_control` set contains at least one state of the perturbed STG of each $Q \in \mathcal{Q}$ that admits a non-phenotype attractor. After this operation, `cannot_control` contains *all* states of such perturbed STG. In other words: the resulting set can depend on variable $\mathcal{V}_i$ only if $\mathcal{L}_i = 1$ (i.e. the variable is perturbed). In such cases, the role of $\mathcal{V}_i$ is to encode the actual perturbed value of the i-th network variable.

Finally, we invert the `cannot_control` set to only retain perturbations where no non-phenotype attractor state exists.

Robust control To extend this algorithm to robust control, we test the perturbations of increasing size (using the $\mathbb{VAL}_k^n$ BDD) and use a recursive ENUMERATE method to find perturbations with maximal robustness. Notice that the necessity of this step makes our algorithm *semi-symbolic.*

Instead of iterating through all possible control strategies of size k, procedure ENUMERATE recursively branches into three cases for each variable i: i is not perturbed, i is perturbed to *true*, and i is perturbed to *false*. Note that each call completely eliminates both $\mathcal{L}_i$ and $\mathcal{V}_i$ from `control_map` (for the case of $\mathcal{L}_i = 0$, $\mathcal{V}_i$ was already eliminated by the core algorithm). As such, once $i > n$, the resulting `control_map` only depends on variables $\mathcal{C}_i$ and we can use it to compute the robustness.

Note that for simplicity, we do not store the actual perturbations with maximal robustness explicitly in the algorithm. However, these can be easily reconstructed from the recursion path in the ENUMERATE algorithm. Furthermore, note that the recursion in the ENUMERATE algorithm can be replaced using *projected iteration*, where we first project the `control_map` to the admissible $l \in \mathcal{L}$, and then project only to the state variables perturbed within each such l. This approach can be faster as it uses fewer symbolic steps. However, it is also highly specific to each BDD library. As such, we chose to present the more widely applicable algorithm.

4 Evaluation

In this section, we evaluate our proposed semi-symbolic method. For the implementation, we use the Rust language and libraries of the AEON tool [7]. This prototype implementation is available as an open-source GitHub repository[1]. The repository also includes auxiliary scripts we used for the measurements and the raw results. All experiments were performed using a computer with AMD Ryzen Threadripper 2990WX 32-Core Processor and 64GB of memory.

[1] https://github.com/sybila/biodivine-pbn-control/.

Table 1. Comprehensive overview of the tested real-world Boolean networks. The first four columns contain the model name and the counts of network inputs, all variables, and perturbable variables, respectively. Column $\mathcal{Q}_{\leq 3}$ represents the number of all admissible perturbations of size up to three. Sixth column gives a reference to the literature that describes each model. Lastly, the seventh column contains variables that we do not allow to be perturbed.

Model	Ins	Vars	Per	$\mathcal{Q}_{\leq 3}$	Ref.	Uncontrollable vars
Cardiac	2	13	11	6,252	[23]	`Tbx1`, `Tbx5`
Red. MAPK	4	14	11	25,008	[22]	`Apoptosis`, `Growth_Arrest`, `Proliferation`
ERBB	1	19	18	14,354	[24]	`pRB1`
Tumour	2	30	24	69,380	[19]	`Apoptosis`, `Metastasis`, `Invasion`, `Migration`, `EMT`, `CellCycleArrest`
Cell Fate	2	31	26	88,612	[14]	`Apoptosis`, `Survival`, `Death`, `Division`, `NonACD`
Full MAPK	2	49	46	2,010,768	[22]	`Apoptosis`, `Growth_Arrest`, `Proliferation`

4.1 Performance

Benchmark model set We use real-world Boolean networks to evaluate our method. All tested networks contain input nodes which can be viewed as functionally equivalent to zero-arity uninterpreted functions in PSBNs. We thus treat these inputs as unknown external signal beyond our control.

Table 1 lists all models and their relevant characteristics, including number of inputs, variables, perturbable variables, and reference to the original publication. We also state the number of admissible perturbations of size up to three. Previous work shows that such relatively small size is both realistic to implement in practice [10,18,38] and robust enough in the presence of partially unknown dynamics [11]. This number therefore represents how many models would need to be explored in a brute-force based methods for the models of the given size. The table also lists variables which we explicitly do not allow to be perturbed. These variables are mostly outputs and they represent traits which induce individual phenotypes. Therefore, perturbing these variables would lead to trivial control which is neither viable nor interesting.

The first model illustrates cardiac progenitor cells differentiation into the first heart field (FHF) or second heart field (SHF) [23]. The second and sixth models represent Mitogen-Activated Protein Kinase (MAPK) network representing signalling pathways involved in diverse cellular processes including cancer deregulations. We use both the full and reduced versions of this model as stated in [22]. The third model depicts the key event preceding breast cancer cells proliferation which is the hyper-phosphorylation and subsequent lack of pRB. This process is regulated by ERBB kinase, the lack of which is considered a breast

Table 2. Performance of PSBN control. The first two columns contain model and phenotype names. Then, for perturbations of the size up to three, we list the computation time and maximal robustness found in perturbations of this size. Last column gives the number of minimal perturbations with $\rho = 1.0$.

Model	Phenotype	Size 1		Size 2		Size 3		# Min. per.
		Time	ρ	Time	ρ	Time	ρ	($\rho = 1.0$)
Cardiac	FHF	<1s	0.5	<1s	1.0	<1s	1.0	4
	SHF	<1s	0.5	<1s	1.0	<1s	1.0	2
	No mesoderm	<1s	0.5	<1s	1.0	<1s	1.0	3
Red. MAPK	Apoptosis	<1s	1.0	<1s	1.0	<1s	1.0	1
	Growth arrest	<1s	0.75	<1s	1.0	<1s	1.0	1
	No decision	<1s	1.0	<1s	1.0	<1s	1.0	1
	Proliferation	<1s	0.25	<1s	1.0	<1s	1.0	3
ERBB	Phosphor	<1s	1.0	<1s	1.0	<1s	1.0	7
	Non-phospor	<1s	1.0	<1s	1.0	<1s	1.0	8
Tumour	Apoptosis	2s	1.0	8s	1.0	23s	1.0	2
	EMT	<1s	0.5	3s	1.0	11s	1.0	15
	Hybrid	<1s	0.25	5s	1.0	24s	1.0	3
	Metastasis	<1s	0.5	<1s	1.0	1s	1.0	6
Cell Fate	Apoptosis	<1s	0	4s	1.0	58s	1.0	24
	Naive	<1s	0	2s	1.0	29s	1.0	8
	Necrosis	<1s	1.0	<1s	1.0	11s	1.0	1
	Survival	<1s	1.0	2s	1.0	21s	1.0	1
Full MAPK	Apoptosis	<1s	1.0	7s	1.0	14min	1.0	6
	Growth arrest	4s	0.75	4s	1.0	22min	1.0	47
	No decision	3s	0.81	107s	1.0	22min	1.0	45
	Proliferation	3s	0.25	3s	1.0	20min	1.0	8

cancer marker [35]. The fourth model focuses on specific conditions which lead to a metastatic tumour [19]. Finally, the cell fate model provides a high-level understanding of the interplays between pro-survival, necrosis, and apoptosis pathways in response to death receptor-mediated signals [14].

Performance evaluation on real-world models In Table 2, we show results of computing phenotype control on all our benchmark models from Table 1. We use real-world phenotypes as described in the source literature. These are typically subspaces obtained by fixing network outputs to the desired values. The details of exact phenotypes can be found in the git repository.

For each model and its phenotype, we computed all working perturbations of size up to three. We then show the times needed for individual computations (including enumeration and robustness calculation), the highest robustness achieved for each case, and the number of minimal perturbations with 100% robustness.

Table 3. Minimal perturbations for reduced MAPK model. The first column is target phenotype while other columns contain minimal perturbations for unperturbed model, model with over-expressed EGFR, and model with FGFR3 gain-of-function. Variables divided by / stand for perturbing either of them. ∅ is used when no perturbation is necessary.

Phenot.	**Unperturbed**	**Perturbed** EGFR=1	**Perturbed** FGFR3=1
Apoptosis	DNA_dmg=1, TGFBR_st=1, FRS2=1	DNA_dmg=1, TGFBR_st=1, ERK=0, p53=1	DNA_dmg=1, TGFBR_st=1, ERK=0, p53=1
¬Apoptosis	∅	AKT=1, ERK=1, MSK=0, PTEN=0, p14=0, p53=0	AKT=1, ERK=1, MSK=0, PTEN=0, p14=0, p53=0
Prolif.	ERK=1	p14=0, p53=0	{p14/p53=0, FRS2=1}, {p14/p53=0, PI3K=1}, {p14/p53=0, EGFR=1}
¬Prolif.	∅	DNA_dmg=1, TGFBR_st=1, AKT=0, ERK=0, MSK=0, PI3K=0, PTEN=1, p53=1	DNA_dmg=1, TGFBR_st=1, AKT=0, ERK=0, MSK=0, PI3K=0, PTEN=1, p53=1
No decis.	∅	MSK=0	MSK=0
¬No decis.	DNA_dmg=1, TGFBR_st=1 EGFR=1, ERK=1, FRS2=1, p53=1	∅	DNA_dmg=1, TGFBR_st=1 ERK=1, FRS2=1, p53=1, PI3K=1

The performance of our method is almost instant for small-sized models which is very convenient for practical use. As for the bigger models, the method also performs sufficiently, nonetheless, due to complex unpredictable nature of PSBN dynamics, its performance may vary significantly from case to case. For example, even though Cell Fate model has only one more perturbable variable than the Tumour model, we can notice that perturbations of size three take on average twice longer to compute. The differences in PSBN dynamics can also have big impact within the same model for different phenotypes. Even though the phenotypes within the same model are all of the same size, we can notice that in any of the bigger models and perturbations of size three, the control takes much longer to compute. There are many factors which can contribute to these big performance differences such as fixed-point nature of the trap set algorithm (which might require numerous iterations in case of long paths with few neighbors) or heuristic BDD representation.

4.2 MAPK Case Study

Now we demonstrate how phenotype control can be used to replicate observations conducted in [22] and even strengthen these results with a more robust assessment of the model. In [22], various perturbations of a reduced MAPK model are simulated and their effect on phenotypes is observed. Specifically, networks with all inputs set to *false* and with EGFR or FGFR3 over-expressed (gain-of-function) are exposed to a set of further perturbations. The model has three outputs (Apoptosis, Growth_arrest, Proliferation) and three attractors are observed:

Table 4. Full MAPK model with uninterpreted functions controlled to the apoptosis phenotype. Each of the three sections covers a different set of uninterpreted functions (increasing in size). For each case, we state: maximal robustness, count of perturbations having such robustness (- for no perturbation) and the computation time. N/A denotes a 24h timeout.

Uninter. functions	EGFR, FGFR3, p53, p14			EGFR, FGFR3, p53, p14, PTEN			EGFR, FGFR3, p53, p14, PTEN, PI3K, AKT		
Colours	806,400			2,419,200			1,161,216,000		
Metrics	ρ	#	Time	ρ	#	Time	ρ	#	Time
$\emptyset$	0.58	-	4s	0.39	-	5s	0.24	-	22s
Size 1	0.87	1	74s	0.58	4	2min	0.59	1	37min
Size 2	1.0	17	27min	0.87	4	38min	0.87	1	20.2hrs
Size 3	1.0	1165	5.6hrs	1.0	68	8.2hrs	N/A	N/A	N/A

apoptosis (`Apoptosis=Growth_arrest=1`), proliferation (`Proliferation=1`) and no decision (all outputs set to false).

We first compute phenotype control on the fully specified but reduced model. We list discovered minimal perturbations for network variants of interest in Table 3. Here, the perturbations which were also discovered in [22] are shown as green. In the enumeration, where appropriate, we only considered perturbations with over-expressed `EGFR` or `FGFR3`, as in the original paper. We also list the minimal perturbations working for the unperturbed network. This way we can compare such perturbations with the `EGFR` and `FGFR3` over-expressed variants. We can see that with our method we were able to not only replicate all solutions from [22], but also conveniently obtain more perturbation options, including the truly minimal controls (if we consider over-expressions of `EGFR` or `FGFR3` as perturbations, the further perturbations to these networks lead to a non-minimal perturbations in most of the cases).

The interest of the original MAPK study [22] is also to observe the impact on phenotypes caused by various gain- or loss-of-function mutations. Here, authors replace such functions with constants to simulate these effects. Nonetheless, such an approach could be too restrictive: a mutation could alter the function in unpredictable ways instead of knocking-out (resp. over-expressing) the variable permanently. To model such mutations, we employ the uninterpreted functions of the PSBN framework. This application is demonstrated in Table 4. Here, we selected apoptosis phenotype as the phenotype of interest (the "healthy" phenotype preserving non-cancerous cell behaviour). We then replace the dynamics of variables studied in [22] with uninterpreted functions in the *full* MAPK model.

Our method performs well in spite of the significant amount of colours introduced by the model uncertainty. Moreover, we see that perturbations with relatively small size are still capable of successful control. The obtained observations can be for example used to refute hypotheses about model's update functions. If a candidate perturbation is shown as non-viable, this indicates that the colours where such perturbation works do not represent the true dynamics of the system. This can guide further refinements of the partially specified model.

5 Conclusion

In this work, we presented a novel problem of phenotype control for partially specified Boolean networks. We proposed a semi-symbolic method using permanent variable perturbations to solve this problem. Our approach offers a practical and flexible solution for stabilizing a network at specific collection of traits, regardless of its initial state. We have also demonstrated the applicability of our method to real-world networks. In future work, we would like to further optimize our method, investigate other types of perturbations for phenotype control and further explore applicability of these methods.

References

1. Abou-Jaoudé, W., et al.: Logical modeling and dynamical analysis of cellular networks. Front. Genet. **7**, 94 (2016)
2. Albert, R.: Boolean modeling of genetic regulatory networks. In: Complex Networks, pp. 459–481. Springer (2004)
3. Aracena, J., Goles, E., Moreira, A., Salinas, L.: On the robustness of update schedules in Boolean networks. Biosystems **97**(1), 1–8 (2009)
4. Barbuti, R., Gori, R., Milazzo, P., Nasti, L.: A survey of gene regulatory networks modelling methods: from differential equations, to Boolean and qualitative bioinspired models. J. Membr. Comput. **2**(3), 207–226 (2020)
5. Baudin, A., Paul, S., Su, C., Pang, J.: Controlling large Boolean networks with single-step perturbations. Bioinformatics **35**(14), i558–i567 (2019)
6. Beneš, N., et al.: Aeon. py: Python library for attractor analysis in asynchronous boolean networks. Bioinformatics **38**(21), 4978–4980 (2022)
7. Beneš, N., Brim, L., Kadlecaj, J., Pastva, S., Šafránek, D.: AEON: Attractor Bifurcation Analysis of Parametrised Boolean Networks. In: Lahiri, S.K., Wang, C. (eds.) CAV 2020. LNCS, vol. 12224, pp. 569–581. Springer, Cham (2020). https://doi.org/10.1007/978-3-030-53288-8_28
8. Beneš, N., Brim, L., Pastva, S., Šafránek, D.: BDD-based algorithm for scc decomposition of edge-coloured graphs. Logical Methods Comput. Sci. **18** (2022)
9. Beneš, N., Brim, L., Huvar, O., Pastva, S., Šafránek, D.: Boolean network sketches: a unifying framework for logical model inference. Bioinformatics **39**, btad158 (2023)
10. Borriello, E., Daniels, B.C.: The basis of easy controllability in Boolean networks. Nature Commun. **12**(1), 1–15 (2021)
11. Brim, L., Pastva, S., Šafránek, D., Šmijáková, E.: Temporary and permanent control of partially specified boolean networks. Biosystems **223**, 104795 (2023)
12. Brim, L., Pastva, S., Šafránek, D., Šmijáková, E.: Parallel one-step control of parametrised Boolean networks. Mathematics **9**(5), 560 (2021)
13. Bryant, R.E.: Graph-based algorithms for Boolean function manipulation. IEEE Trans. Comput. **35**(8), 677–691 (1986)
14. Calzone, L., et al.: Mathematical modelling of cell-fate decision in response to death receptor engagement. PLOS Comput. Bio. **6**(3), 1–15 (2010)
15. Choo, S.M., Ban, B., Joo, J.I., Cho, K.H.: The phenotype control kernel of a biomolecular regulatory network. BMC Syst. Biol. **12**(1), 1–15 (2018)
16. Choo, S.M., Cho, K.H.: An efficient algorithm for identifying primary phenotype attractors of a large-scale Boolean network. BMC Syst. Bio. **10**(1), 1–14 (2016)

17. Cifuentes Fontanals, L., Tonello, E., Siebert, H.: Control strategy identification via trap spaces in Boolean networks. In: Computational Methods in Systems Biology. Lecture Notes in Computer Science, vol. 12314, pp. 159–175. Springer (2020)
18. Cifuentes Fontanals, L., Tonello, E., Siebert, H.: Control in Boolean networks with model checking. Front. Appl. Math. Stat. **8**, 838546 (04 2022)
19. Cohen, D.P., Martignetti, L., Robine, S., Barillot, E., Zinovyev, A., Calzone, L.: Mathematical modelling of molecular pathways enabling tumour cell invasion and migration. PLOS Comput. Bio. **11**(11), 1–29 (2015)
20. Fiedler, B., Mochizuki, A., Kurosawa, G., Saito, D.: Dynamics and control at feedback vertex sets. I: Informative and determining nodes in regulatory networks. J. Dyn. Diff. Equat. **25**(3), 563–604 (2013)
21. Geris, L., Gomez-Cabrero, D.: An Introduction to Uncertainty in the Development of Computational Models of Biological Processes. In: Geris, L., Gomez-Cabrero, D. (eds.) Uncertainty in Biology. SMTEB, vol. 17, pp. 3–11. Springer, Cham (2016). https://doi.org/10.1007/978-3-319-21296-8_1
22. Grieco, L., Calzone, L., Bernard-Pierrot, I., Radvanyi, F., Kahn-Perles, B., Thieffry, D.: Integrative modelling of the influence of MAPK network on cancer cell fate decision. PLOS Comput. Bio. **9**(10), e1003286 (2013)
23. Herrmann, F., Groß, A., Zhou, D., Kestler, H.A., Kühl, M.: A boolean model of the cardiac gene regulatory network determining first and second heart field identity. PloS one **7**(10), e46798 (2012)
24. Ito, N., Kuwahara, G., Sukehiro, Y., Teratani, H.: Segmental arterial mediolysis accompanied by renal infarction and pancreatic enlargement: a case report. J. Med. Case Rep. **6**(1), 1–5 (2012)
25. Kim, J., Park, S.M., Cho, K.H.: Discovery of a kernel for controlling biomolecular regulatory networks. Sci. Rep. **3**(1), 1–9 (2013)
26. Klarner, H., Heinitz, F., Nee, S., Siebert, H.: Basins of attraction, commitment sets, and phenotypes of Boolean networks. IEEE/ACM Trans. Comput. Bio. Bioinf. **17**(4), 1115–1124 (2018)
27. Kobayashi, K., Hiraishi, K.: Optimal control of asynchronous Boolean networks modeled by Petri nets. In: Biological Process & Petri Nets. pp. 7–20. CEUR-WS (2011)
28. Mandon, H., Haar, S., Paulevé, L.: Temporal Reprogramming of Boolean Networks. In: Feret, J., Koeppl, H. (eds.) CMSB 2017. LNCS, vol. 10545, pp. 179–195. Springer, Cham (2017). https://doi.org/10.1007/978-3-319-67471-1_11
29. Mandon, H., Su, C., Haar, S., Pang, J., Paulevé, L.: Sequential Reprogramming of Boolean Networks Made Practical. In: Bortolussi, L., Sanguinetti, G. (eds.) CMSB 2019. LNCS, vol. 11773, pp. 3–19. Springer, Cham (2019). https://doi.org/10.1007/978-3-030-31304-3_1
30. Martin, A.J., Dominguez, C., Contreras-Riquelme, S., Holmes, D.S., Perez-Acle, T.: Graphlet based metrics for the comparison of gene regulatory networks. PLOS ONE **11**(10), e0163497e (2016)
31. Pardo, J., Ivanov, S., Delaplace, F.: Sequential Reprogramming of Biological Network Fate. In: Bortolussi, L., Sanguinetti, G. (eds.) CMSB 2019. LNCS, vol. 11773, pp. 20–41. Springer, Cham (2019). https://doi.org/10.1007/978-3-030-31304-3_2
32. Paulevé, L.: Marker and source-marker reprogramming of most permissive Boolean networks and ensembles with BoNesis. Peer Commun. J. (2023)
33. Rozum, J.C., Deritei, D., Park, K.H., Gómez Tejeda Zañudo, J., Albert, R.: pystablemotifs: python library for attractor identification and control in Boolean networks. Bioinformatics **38**(5), 1465–1466 (2022)

34. Rozum, J.C., Gómez Tejeda Zañudo, J., Gan, X., Deritei, D., Albert, R.: Parity and time reversal elucidate both decision-making in empirical models and attractor scaling in critical Boolean networks. Sci. Adv. **7**(29), eabf8124 (2021)
35. Sahin, Ö., et al.: Modeling ERBB receptor-regulated G1/S transition to find novel targets for de novo trastuzumab resistance. BMC Syst. Bio. **3**(1), 1–20 (2009)
36. Su, C., Pang, J.: A dynamics-based approach for the target control of Boolean networks. In: ACM International Conference on Bioinformatics, Computational Biology and Health Informatics. pp. 1–8. Association for Computing Machinery (2020)
37. Su, C., Pang, J.: Sequential Temporary and Permanent Control of Boolean Networks. In: Abate, A., Petrov, T., Wolf, V. (eds.) CMSB 2020. LNCS, vol. 12314, pp. 234–251. Springer, Cham (2020). https://doi.org/10.1007/978-3-030-60327-4_13
38. Su, C., Paul, S., Pang, J.: Controlling Large Boolean Networks with Temporary and Permanent Perturbations. In: ter Beek, M.H., McIver, A., Oliveira, J.N. (eds.) FM 2019. LNCS, vol. 11800, pp. 707–724. Springer, Cham (2019). https://doi.org/10.1007/978-3-030-30942-8_41
39. Su, C., Paul, S., Pang, J.: Scalable control of asynchronous Boolean networks. In: Computational Methods in Systems Biology. Lecture Notes in Computer Science, vol. 11773, pp. 364–367. Springer (2019)
40. Zañudo, J.G.T., Albert, R.: Cell fate reprogramming by control of intracellular network dynamics. PLOS Comput. Bio. **11**(4), 1–24 (2015)

A More Expressive Spline Representation for SBML Models Improves Code Generation Performance in AMICI

Lorenzo Contento[1](✉), Paul Stapor[2,3], Daniel Weindl[2], and Jan Hasenauer[1,2,3]

[1] Faculty of Mathematics and Natural Sciences, University of Bonn, Bonn, Germany
lorenzo.contento@uni-bonn.de
[2] Computational Health Center, Helmholtz Zentrum München, München, Germany
[3] Center for Mathematics, Technische Universität München, München, Germany

Abstract. Spline interpolants are commonly used for discretizing and estimating functions in mathematical models. While splines can be encoded in the Systems Biology Markup Language (SBML) using piecewise functions, the resulting formulas are very complex and difficult to derive by hand. Tools to create such formulas exist but only deal with numeric data and thus cannot be used for function estimation. Similarly, simulation tools suffer from several limitations when handling splines. For example, in the AMICI library splines with large numbers of nodes lead to long model import times.

We have developed a set of SBML annotations to mark assignment rules as spline formulas. These compact representations are human-readable and easy to edit, in contrast to the piecewise representation. Different boundary conditions and extrapolation methods can also be specified. By extending AMICI to create and recognize these annotations, model import can be sped up significantly. This allows practitioners to increase the expressivity of their models.

While the performance improvement is limited to AMICI, our tools for creating spline formulas can be used for other tools as well and our syntax for compact spline representation may be a starting point for an SBML-native way to represent spline interpolants.

Keywords: dynamic modeling · splines · systems biology · parameter estimation

1 Background

Mathematical modelling has always been an essential component of the scientific method and, thanks to an increase in available computational power, the complexity of the models has been growing rapidly in the last years [3]. Models are defined in terms of a mathematical function that maps a set of parameters,

Lorenzo Contento and Paul Stapor: equal contribution first authors.

J. Pang and J. Niehren (Eds.): CMSB 2023, LNBI 14137, pp. 36–43, 2023.
https://doi.org/10.1007/978-3-031-42697-1_3

some of them known and some of them unknown, to an observation (possibly stochastic in nature). For example, such mapping may be defined by a system of differential equations representing a reaction network in systems biology.

The parameters of a mathematical model are usually real numbers, but they can also be infinite-dimensional objects such as functions [6]. A common use case is that of an input function whose governing law is not known and thus cannot be modelled directly, but which is otherwise necessary to predict the evolution of the other components of the system. When dealing with infinite-dimensional objects computationally, it is necessary to employ discretization methods that replace the original set of possible values with a finite-dimensional one.

A popular class of discretization methods for univariate functions consists of interpolation methods. A finite set of points, called nodes, is chosen in the function's domain and the function is parameterized by its values (and possibly some of its derivatives) at the nodes. The value at a point between nodes is obtained by combining the known values at nearby nodes, usually in a way that preserves the continuity of the function (and of some of its derivatives). A very popular and effective approach is cubic spline interpolation [14], which, in virtue of being continuous up to the first derivative, results in natural-looking and plausible functions. For example, splines are commonly used to model and estimate metabolic fluxes [12,15,19].

Systems biology models are often encoded using the Systems Biology Markup Language (SBML) [9], which allows for a tool-agnostic declarative formulation of the system. The SBML format does not have native support for spline interpolation, and neither do other commonly used alternatives such as CellML [1] and BNGL [7]. However, since arbitrary piecewise functions can be defined in MathML, encoding a spline function is possible by calculating its constituting polynomial segments. Software tools that support the SBML specification can then read and simulate such models. In practice, existing software suffers from several limitations when dealing with complex piecewise functions. For example, COPASI [8] succeeds in computing the sensitivities but is not able to import the SBML file when the spline has more than roughly 40 nodes. On the other hand, libRoadRunner [17,20] simulates the model correctly but does not support adjoint sensitivity analysis, making it less suitable for estimating large numbers of parameters. Finally, the AMICI library for simulation and sensitivity analysis [4] supports adjoint sensitivities, but in presence of piecewise functions with many segments it suffers from extremely long code generation times.

In this paper, we will present an extension to AMICI 0.18 [5] that allows it to efficiently deal with SBML models containing splines. The workflow for creating and simulating SBML models with splines (summarized in Fig. 1) is as follows. We encode the spline in an SBML model using an assignment rule which, in addition to the MathML piecewise formula for the spline, contains an annotation describing the interpolator in a compact form (Sect. 2). Such annotated assignment rules can be created using the AMICI Python library from a high-level description of the spline, without requiring the user to derive the piecewise formula for it (Sect. 3). Finally, models can be imported and simulated

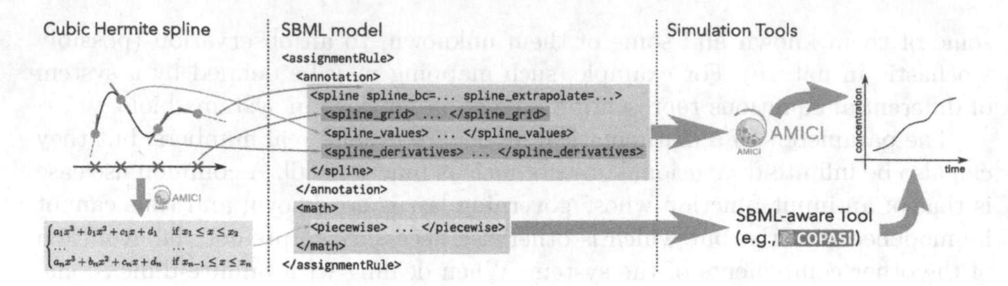

Fig. 1. Workflow for adding splines to an SBML model. (i) The user provides a description of a cubic Hermite spline using the nodes positions, the values and derivatives of the spline at the nodes, the boundary conditions and the extrapolation method. (ii) Using the AMICI Python library, an SBML assignment rule for the spline is generated and added to a pre-existing SBML model (the piecewise MathML expression is automatically generated). (iii) The resulting model can be imported and simulated either by AMICI (which parses the annotation) or by other SBML-aware tools (which use the MathML formula).

using any SBML-compliant tool (Sect. 4). AMICI parses the annotation and uses it to quickly generate the required C++ code, while other tools will ignore it and fall back to the MathML expression. We conclude the paper with a real-world application of this workflow (Sect. 5).

2 Annotation-Based Compact Representation for Splines

In our extension to AMICI we use cubic Hermite splines [14] in order to model time-dependent parameters. Such splines are piecewise cubic polynomials continuous up to the first derivative which are parameterized by the values of the spline and of its first derivative at its nodes. Often derivatives are computed from the values at the nodes by finite differencing, taking into account the choice of boundary conditions (e.g., enforcing the first or second derivative to be zero). Boundary conditions are deeply linked with the choice of the extrapolation method to use before the first node and/or after the last node. For example, constant extrapolation requires a zero derivative boundary condition to preserve continuity of the first derivative at the boundary. Periodic boundary conditions are also possible.

The native way to represent a cubic Hermite spline in an SBML model is to use a MathML piecewise construct containing the equations of all polynomial segments of the spline. The resulting formula is extremely long and difficult to understand and/or modify. Thus, we propose a higher-level compact XML syntax to specify such splines using their natural parameters: node positions, spline values at the nodes, and either derivatives at the nodes or boundary conditions in case finite differences are to be used.

3 User-Friendly Insertion of Splines into SBML Models

A MathML spline representation can be inserted into any other MathML formula appearing in an SBML model. However, for readability's sake, one may wish to create a new non-constant parameter for the spline value and create an assignment rule for it, thus avoiding duplication of the very long MathML piecewise formula. This second approach also offers a natural place to store our compact spline representation: the annotation element of the assignment rule.

Since MathML representations of splines are too cumbersome to write manually, they are usually generated with dedicated software tools, such as sbmlutils [10] and SBMLDataTools [13]. However, such tools require the data to be interpolated to be numeric. While very useful for interpolating data series [11], they cannot be used for parameter inference on functions. We have thus added to the AMICI Python library some convenience functions to generate the above mentioned assignment rules, containing both the AMICI-specific compact representation and the usual MathML formula, also covering the case where the spline values at the nodes (or its derivatives) depend on the model parameters. We point the interested reader to the AMICI documentation for details and examples[1].

4 Parsing and Simulating Splines from an SBML Model

A model containing an assignment rule of the type presented in Sect. 3 is a valid SBML model. This means SBML-aware simulation tools can read it, ignore the annotations, parse the piecewise MathML expression and simulate the model. While AMICI is able to handle piecewise formulas correctly too, due to the way symbolic computation is carried out when AMICI processes the SBML file, the resulting code generation times are extremely long, making the current version of AMICI unsuited to run such models. This problem is solved when a compact XML representation of the spline is contained in the annotation to the assignment rule: AMICI can read the annotation, and efficiently generate the spline C++ code from it, without having to parse the MathML expression.

To quantify the resulting speed-up, we consider a simple SBML model containing a spline with n nodes. We study, as n changes, the time required by AMICI to read the model and compile it to native code. Results are displayed in Fig. 2, showing an exponential dependence of the import time on n, independently of whether MathML piecewise expressions or AMICI-specific spline annotations are used. However, the growth rate in the case of piecewise expressions is much larger, making it clear that splines with more than a few dozens of nodes are unfeasible, especially in the case of iterative model development. Model simulations, on the other hand, take similar time and have similar accuracy for both methods. This is expected, since the the underlying mathematical expressions are independent of how the spline is parsed from the SBML file. The

[1] https://github.com/AMICI-dev/AMICI/blob/0fc51c0ddcac0661a83931e8b1c4a59f0c6db85c/python/examples/example_splines/ExampleSplines.ipynb.

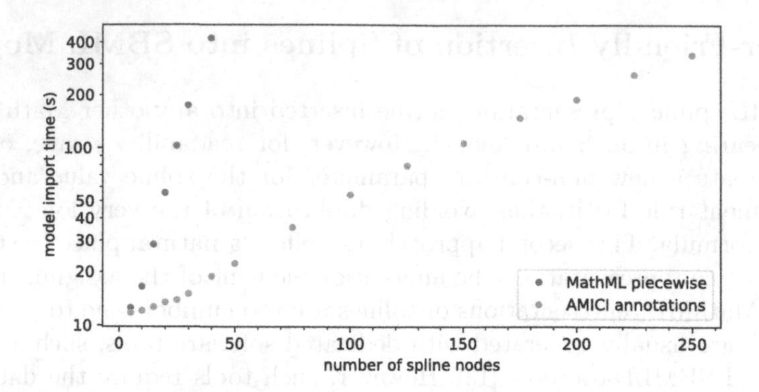

Fig. 2. Dependence of model import time on the number of spline nodes. Timing results using MathML expressions and AMICI annotations are both shown.

C++ code generated from MathML is longer and more repetitive than the code generated from the XML annotation, but execution time is essentially the same. We suspect this is due to modern compilers being very good at optimizing code.

5 A Real-World Example

As a real-world example[2], we consider a chemical reaction network model for the JAK2-STAT5 signaling pathway proposed by [18], in which the dynamics of the system depend on a continuous input function. This input function is only measured (with noise) at a finite number of time points, but its value at all intermediate time points is required to simulate the model. In [16] a 5-node interpolating cubic spline was used to model the logarithm of the input function (the logarithm is used in order to ensure positivity). We follow the same approach, except that we employ a cubic Hermite spline.

While in [16] the spline had only a few degrees of freedom, in order to show what can be achieved with our implementation we consider two different families of splines with higher expressivity: (i) a spline with 15 nodes where derivatives are estimated by finite differences; (ii) a spline with 5 nodes with derivatives fitted alongside the values at the nodes. In both cases, the derivative at the rightmost node is set to zero since the input function appears to have reached a steady state. In order to reduce overfitting, we follow [16] and add a regularization term consisting of the squared L2 norm of the curvature of the spline, which can be easily computed using the AMICI Python library. Figure 3 shows the best fits for different values of the regularization strength λ. Then, the most appropriate value for λ can be chosen by comparing the sum of the squared normalized residuals (which is approximately χ^2-distributed) with its expected value [16].

[2] https://github.com/AMICI-dev/AMICI/blob/0fc51c0ddcac0661a83931e8b1c4a59f0c6db85c/python/examples/example_splines_swameye/ExampleSplinesSwameye2003.ipynb.

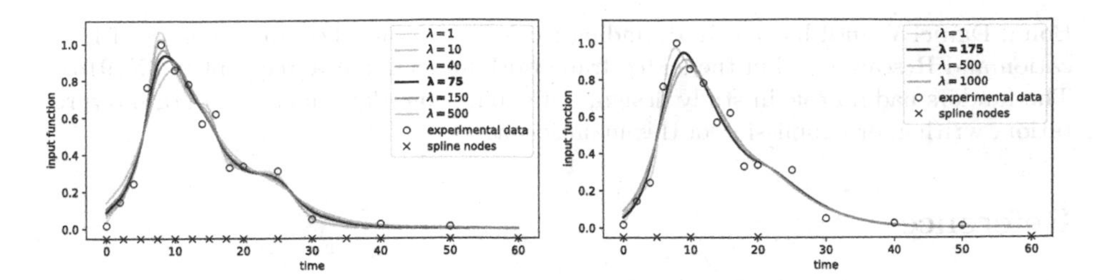

Fig. 3. Estimated input function of the JAK2-STAT5 model. The left panel shows the results for a 15-node cubic Hermite spline with derivatives estimated by finite differences, while the right panel uses a 5-node spline with derivatives fitted alongside the other parameters. The squared L2 norm of the input function curvature is used as a regularizer and the estimates for different values of the regularization strength λ are plotted. The value of λ that results in the sum of the squared normalized residuals being nearest to its expected value is highlighted in bold in the legend.

6 Discussion

In conclusion, the new annotation-based spline implementation in AMICI allows to integrate complex spline functions in SBML models without having to spend an inordinate amount of time on code creation. While other tools do not directly benefit from the AMICI-specific annotations, being AMICI one of the few available tools implementing adjoint sensitivity analysis coupled with extensive support of SBML features, we believe this addition will be of use to many members of our community. However, even users of other simulation tools can benefit from using our utility functions to create spline formulas and add them to SBML models, since existing methods only work with numeric data and support fewer spline types. Finally, we believe the SBML standard would greatly benefit from integrating native support for splines and other interpolating functions using a human-readable and easy-to-edit syntax; our XML representation could be a starting point for a discussion among the stakeholders of the SBML ecosystem.

Finally, we remark that, while the main focus of the SBML modelling language is systems biology, it can be used to describe arbitrary systems of differential equations. This allows for the spline implementation described in this paper to be used in other application domains too. For example, AMICI splines have already been used in the epidemiological model proposed by [2] to model the time-dependent effect of non-pharmaceutical interventions on the transmission rate of COVID-19. Thus, we believe this feature will be very useful to mathematical modellers in general, enabling them to increase the expressivity of their models through highly-resolved discretizations of input functions.

Acknowledgments. This study was funded by the German Ministry for Education and Research (MoKoCo19: 01KI20271; INSIDe: 031L0297A), German Research Foundation (SEPAN: 458597554; SFB 1454: 432325352; AMICI: 443187771; Germany's Excellence Strategy EXC 2047/1 - 390685813 and EXC 2151 - 390873048). Jan Hasenauer acknowledges financial support via a Schlegel Professorship at the University of

Bonn. Daniel Weindl has received funding from the German Federal Ministry of Education and Research within the e:Med framework under grant agreement 01ZX1916A. The funders had no role in study design, data collection, data analyses, data interpretation, writing, or submission of this manuscript.

References

1. Clerx, M., et al.: CellML 2.0. J. Integr. Bioinf. **17**(2–3), 20200021 (2020). https://doi.org/10.1515/jib-2020-0021
2. Contento, L., et al.: Integrative modelling of reported case numbers and seroprevalence reveals time-dependent test efficiency and infectious contacts. Epidemics **43**, 100681 (2023). https://doi.org/10.1016/j.epidem.2023.100681
3. Fröhlich, F., Gerosa, L., Muhlich, J., Sorger, P.K.: Mechanistic model of MAPK signaling reveals how allostery and rewiring contribute to drug resistance. Mol. Syst. Bio. **19**(2), e10988 (2023). https://doi.org/10.15252/msb.202210988
4. Fröhlich, F., Weindl, D., Schälte, Y., Pathirana, D., Paszkowski, Ł, Lines, G.T., Stapor, P., Hasenauer, J.: AMICI: high-performance sensitivity analysis for large ordinary differential equation models. Bioinformatics **37**(20), 3676–3677 (2021). https://doi.org/10.1093/bioinformatics/btab227
5. Fröhlich, F.,et al.: AMICI: high-performance sensitivity analysis for large ordinary differential equation models (2023). https://doi.org/10.5281/zenodo.7974682
6. Giné, E., Nickl, R.: Mathematical foundations of infinite-dimensional statistical models. Cambridge University Press (2015). https://doi.org/10.1017/CBO9781107337862
7. Harris, L.A., et al.: BioNetGen 2.2: advances in rule-based modeling. Bioinformatics **32**(21), 3366–3368 (2016). https://doi.org/10.1093/bioinformatics/btw469
8. Hoops, S., et al.: COPASI-a Complex Pathway Simulator. Bioinformatics **22**(24), 3067–3074 (2006). https://doi.org/10.1093/bioinformatics/btl485
9. Hucka, M., et al.: The systems biology markup language (SBML): language specification for level 3 version 2 core release 2. J. Integr. Bioinf. **16**(2), 20190021 (2019). https://doi.org/10.1515/jib-2019-0021
10. König, M.: sbmlutils: Python utilities for SBML (2022). https://doi.org/10.5281/zenodo.7462781
11. Maheshvare, M.D., Raha, S., König, M., Pal, D.: A Consensus Model of Glucose-Stimulated Insulin Secretion in the Pancreatic β-Cell. bioRxiv (2023). https://doi.org/10.1101/2023.03.10.532028
12. Martínez, V.S., Buchsteiner, M., Gray, P., Nielsen, L.K., Quek, L.E.: Dynamic metabolic flux analysis using B-splines to study the effects of temperature shift on CHO cell metabolism. Metabolic Eng. Commun. **2**, 46–57 (2015). https://doi.org/10.1016/j.meteno.2015.06.001
13. Millar, A.J., et al.: Practical steps to digital organism models, from laboratory model species to 'Crops in silico'. J. Exp. Bot. **70**(9), 2403–2418 (2019). https://doi.org/10.1093/jxb/ery435
14. Quarteroni, A., Saleri, F., Gervasio, P.: Scientific Computing with MATLAB and Octave. TCSE, vol. 2. Springer, Heidelberg (2014). https://doi.org/10.1007/978-3-642-45367-0
15. Quek, L.E., et al.: Dynamic ^{13}C Flux Analysis Captures the Reorganization of Adipocyte Glucose Metabolism in Response to Insulin. iScience **23**(2), 100855 (2020). https://doi.org/10.1016/j.isci.2020.100855

16. Schelker, M., Raue, A., Timmer, J., Kreutz, C.: Comprehensive estimation of input signals and dynamics in biochemical reaction networks. Bioinformatics **28**(18), i529–i534 (2012). https://doi.org/10.1093/bioinformatics/bts393
17. Somogyi, E.T., Bouteiller, J.M., Glazier, J.A., König, M., Medley, J.K., Swat, M.H., Sauro, H.M.: libRoadRunner: a high performance SBML simulation and analysis library. Bioinformatics **31**(20), 3315–3321 (2015). https://doi.org/10.1093/bioinformatics/btv363
18. Swameye, I., Müller, T.G., Timmer, J., O.Sandra, Klingmüller, U.: Identification of nucleocytoplasmic cycling as a remote sensor in cellular signaling by databased modeling. Proc. National Acad. Sci. **100**(3), 1028–1033 (2003). https://doi.org/10.1073/pnas.0237333100
19. Vercammen, D., Logist, F., Impe, J.V.: Dynamic estimation of specific fluxes in metabolic networks using non-linear dynamic optimization. BMC Syst. Bio. **8**, 132 (2014). https://doi.org/10.1186/s12918-014-0132-0
20. Welsh, C., Xu, J., Smith, L., König, M., Choi, K., Sauro, H.M.: libRoadRunner 2.0: a high performance SBML simulation and analysis library. Bioinformatics **39**(1) (2022). https://doi.org/10.1093/bioinformatics/btac770

Intuitive Modelling and Formal Analysis of Collective Behaviour in Foraging Ants

Rocco De Nicola[1], Luca Di Stefano[2], Omar Inverso[3], and Serenella Valiani[1(✉)]

[1] IMT School of Advanced Studies, Lucca, Italy
serenella.valiani@imtlucca.it
[2] University of Gothenburg, Gothenburg, Sweden
[3] Gran Sasso Science Institute (GSSI), L'Aquila, Italy

Abstract. We demonstrate a novel methodology that integrates intuitive modelling, simulation, and formal verification of collective behaviour in biological systems. To that end, we consider the case of a colony of foraging ants, where, for the combined effect of known biological mechanisms such as stigmergic interaction, pheromone release, and path integration, the ants will progressively work out the shortest path to move back and forth between their nest and a hypothetical food repository. Starting from an informal description in natural language, we show how to devise intuitive specifications for such scenario in a formal language. We then make use of a prototype software tool to formally assess whether such specifications would indeed replicate the expected collective behaviour of the colony as a whole.

Keywords: Agent-based models · Collective behaviour · Foraging · Ant colonies · Simulation · Formal verification

1 Introduction

Researchers in systems biology, ecology, and countless other disciplines have long since been interested in computational methods to rapidly explore promising ideas and hypotheses without having to resort to controlled real-world experiments or field studies. More specifically, agent- or individual- based models have been suggested as an effective research aid across several research areas [5,19]. In these models, the system is represented as a collection of autonomous, interacting components (or *agents*), whose interaction both lead to, and may be influenced by, the *emergence* of collective phenomena.

This form of modelling is becoming increasingly popular, owing to the continuous growth in available computational power and its ability to faithfully reproduce non-linear and complex dynamics [27,33]. A common approach to so-called agent-based modelling [1,12,23,39] is to describe the individual agents

Work partially funded by MIUR project PRIN 2017FTXR7S *IT MATTERS* (Methods and Tools for Trustworthy Smart Systems), ERC consolidator grant no. 772459 *D-SynMA* (Distributed Synthesis: from Single to Multiple Agents), and PRO3 MUR project *Software Quality*.

J. Pang and J. Niehren (Eds.): CMSB 2023, LNBI 14137, pp. 44–61, 2023.
https://doi.org/10.1007/978-3-031-42697-1_4

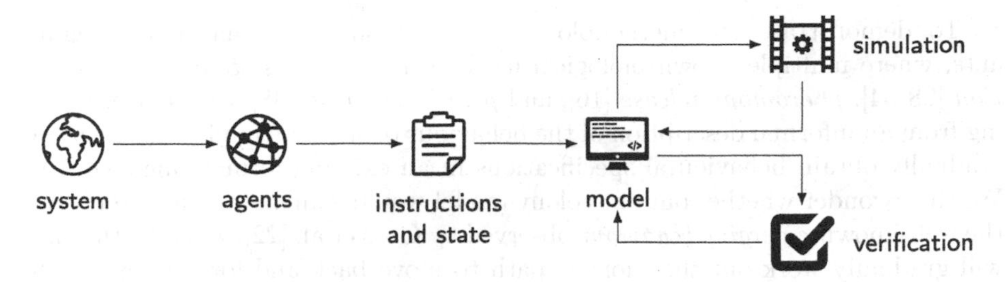

Fig. 1. A visualization of our proposed methodology.

through mathematical formalisms, e.g., interconnected continuous- or discrete-time dynamical systems. This does have its benefits, as it allows leveraging powerful analytical tools developed in well-established areas of mathematical research, and in particular allow simulation through efficient numerical methods. However, it requires a considerable learning curve, and is hardly suited to incremental or compositional modelling.

We argue that adopting a tailored formal language may provide better affordances, enabling an intuitive modelling methodology where informal ideas and concepts related to the behaviour of a system gradually evolve into machine-tractable models expressed through formal specifications. Such models can be maintained and refined with limited effort, for instance to introduce small variations in the attempt to prove or disprove specific hypotheses on a system of interest. Besides, adopting a language with formal semantics opens up to a variety of possibilities in terms of formal analysis techniques, and can thus offer considerably more rigorous forms of reasoning than possible with massive simulation [14].

Our methodology (sketched in Fig. 1) aims at supporting intuitive agent-based modelling of the system of interest, while reducing the overall technical effort for automated analysis. In practice, one starts by identifying the relevant characteristics of the environment and of the agents populating it. Then, one sketches an informal description of the behaviour of the agents in natural language. Such a description identifies the main actions carried out by the agents, and the effect of such actions on their own characteristics and on the environment. Then, taking advantage of a domain-specific language, one progressively expands such main actions into more detailed sequences of operations, eventually working out full formal specifications. Thanks to a mechanised procedure, at this point the formal specifications can either be simulated or formally verified. Simulation is less computationally expensive than verification, and allows for quick feedback initially. By inspecting simulated system evolutions, one can quickly identify and fix possible problems, such as situations that are not realistic, or trivial violations of some simple property of interest. After one or more rounds of refinements, one can start to analyse the emerging behaviour of interest, again via more extensive simulations, or formal verification, and so on, possibly until the verification verdict is satisfactory.

To demonstrate our methodology, we focus on the scenario of foraging ants, where multiple known biological mechanisms, such as *stigmergic interaction* [28,34], *pheromone release* [16], and *path integration* [38], take place. Starting from an informal description of the behaviour of the ants in plain English, we gradually obtain behavioural specifications in an existing formal language [13]. We then wonder whether our ant colony model would somehow be able replicate the well-known *emerging behaviour* observed by Goss et al. [22], whereby the ants will gradually work out the shortest path to move back and forth between the nest and the food repository. To that end, we discuss how state-of-the-art general-purpose techniques for automated analysis [17] can fit within our methodology and be automatically re-used to either *simulate* or *formally verify* the specifications. The possibility of integrating simulation and formal analysis within the same workflow is a particularly relevant element of novelty in this context: on the one hand simulation allows to quickly evaluate possible hypotheses of interest on the system under analysis; on the other hand, formal analysis provides a considerably more rigorous formalism that can effectively pinpoint so-called *rare events* which usually are difficult to pinpoint, or prove their absence for good.

The paper is structured as follows. In Sect. 2 we introduce the scenario of a colony of foraging ants and develop our formal specifications for it. In Sect. 3 we show how to analyse the specifications in order to assess the emergence of the mentioned collective behaviour of interest in the colony. We discuss related work in Sect. 4 and provide our concluding remarks in Sect. 5.

2 Modelling

In this section, we develop a model to replicate the collective behaviour of a colony of foraging ants, where different mechanisms extensively studied in biology take place. Starting from an informal description of the behaviour of interest in English, we gradually obtain intuitive specifications in a formal language known as LAbS [13].

Let us consider a colony of ants initially moving chaotically while searching for food. Over time, the ants are progressively influenced by the presence of pheromone in the environment: an ant may smell traces of pheromone nearby, and move towards it, following the mechanism of *stigmergic interaction* observed in several species [28,34]. Having picked up food, the ant will return to its nest using the most efficient route thanks to a well-known ability known as *path integration* [38]; at the same time, the ant releases pheromone on its way back to the nest, following the pattern of *pheromone release* observed in [16].

Let us now focus on the *environment* in which the colony operates. We assume that the space is a two-dimensional *arena* of regular shape, initially without traces of pheromone; we also assume for simplicity that there exists a single source of food, and that there are no obstacles. In Listing 1, we model the arena as a square grid of size *size* and the coordinates of the food repository as (*foodx*, *foody*). We represent the amount of pheromone at each position of the arena as an array initially containing only zeros (line 3). We eventually specify that our colony consists of n ants (line 4). Note that *foodx*, *foody*, *size*, and n are parameters of the specifications whose value is to be set by the user at the beginning of the analysis; the same holds for m (that constrains the range of movement for every ant), and for k (discussed later).

Listing 1. Definition of the environment and external parameters.

```
1 system {
2   extern = size, n, foodx, foody, m, k
3   environment = ph[size, size]: 0
4   spawn = Ant: n
5 }
```

Listing 2. Attributes of an ant.

```
1 agent Ant {
2   interface = x: 0..size; y: 0..size;
3              nextX: 0; nextY: 0
4   ...
5 }
```

To model the position and movement of an ant, we introduce separate features (or *attributes*) to keep track of its current position and the position it wishes to move to. We assume that initially every ant starts from any position within the arena. Intuitively, here we are assuming that the ants have already been exploring the arena for a while, and thus might have ended up anywhere in the arena. Listing 2 shows how this is formalized in LAbS. At line 2, x and y indicate the coordinates of the ant within the grid; the notation 0..*size* indicates a non-deterministic value in the interval ranging from 0, inclusive, to *size*, exclusive. Lines 3 initialize the coordinates of the position where the ant wishes to move.

We can now describe the behaviour of an ant. Following the informal description at the beginning of the section, an ant repeatedly explores the arena until it finds some food, heads back to its nest, and finally starts over. This is formalized in Listing 3, where an ant's behaviour (Behavior) is described as a sequence of steps, or, more formally, *processes*: one for exploring (Explore), one for returning to the nest (GoHome), and one (Behavior) which is actually a recursive call to repeat the same sequence indefinitely. The rest of the section is dedicated to formalizing such three processes; from now on our listings will omit the other parts of the ants' specification and only show the behavioural definitions.

The Explore process is composed of a sensing phase during which the ant looks for nearby pheromone, and a moving phase when the ant either goes to a pheromone-marked position or, if none is sensed, moves chaotically. When following the GoHome behaviour, the ant marks its current position with pheromone and then moves one step closer to the nest (as mentioned earlier, we can assume that an ant knows the right way home thanks to path integration).

Listing 4 expands the specifications of the process Explore and GoHome. Lines 6–10 shows the details for Explore. The exploration consists in first searching for pheromone and then moving, but only as long as the ant is not at the same position as the food source. We enforce this requirement at line 7: if it does not hold, the rest of the process is skipped. The definition of SmellPheromone in Listing 4 involves two additional processes, namely SmellPheromone and Move

Listing 3. High-level behaviour of an ant.

```
1 agent Ant {
2   interface = ...
3
4   Behavior =
5     Explore;
6     GoHome;
7     Behavior
8 }
```

(lines 8–9), followed by a self-loop, to make sure the process loops until the ant finds itself at the food repository, at which point it will break out of the process.

Let us now expand on SmellPheromone. Initially, an ant searches for nearby traces of pheromone considering some position at maximum sensing distance m and then checking whether it contains a pheromone marking. Listing 4 (lines 24–40) encodes such behaviour. Observe that the process description is enclosed in curly braces: this specifies that all actions therein are executed by the ant in a single step. Lines 25–26 pick two a non-deterministic value between 1 and $m+1$, which are used to select two positions within the sensing distance of the ant (textx1, texty1) and (textx2, texty2) (lines 27–30). To ensure that they are within the grid, their values are compared and only those between 0 and $size - 1$ are accepted. Lines 32–39 define the position (nextX, nextY) that the ant will move to if it detects pheromone. Note that, if the ant does not detect pheromone at either of the two observed positions, the desired position is the current position of the ant.

Listing 4 (lines 12–21) defines the actual movement of the ant. If the ant has detected pheromone at one of the two previous positions, the content of the block is ignored and the position of the ant updated with the desired position (line 22). In case no pheromone was detected, the ant chooses a reachable position nearby (lines 14–16) within the grid (lines 17–20) and updates its position with the chosen one (line 22). It is worth to note that the way we implement the pheromone detection allows for some uncertainty: even when very close to a pheromone-marked location, an ant may still fail to detect it and stray from the trail. Once the ant reaches the food, it starts heading back to the nest. For simplicity, we position the nest at one of the edges of the arena, at $(0, foody)$, so that the shortest path between the nest and food is a straight horizontal segment. Lines 42–46 implement the homing behaviour: the ant releases a quantity of pheromone equal to 1 (line 44) and moves towards the nest (line 45), and repeats these two steps until the nest is reached.

Listing 4. Full behavioural description of an ant.

```
1  Behavior =
2    Explore;
3    GoHome;
4    Behavior
5
6  Explore =
7    x ≠ foodx or y ≠ foody ⇒ (
8      SmellPheromone;
9      Move;
10     Explore)
11
12 Move =
13   (nextX = x and nextY = y ⇒ {
14     dX, dY := [−m..m+1], [−m..m+1];
15     nextX ← x+dX;
16     nextY ← y+dY;
17     nextX ← max(nextX, 0);
18     nextY ← max(nextY, 0);
19     nextX ← min(nextX, size−1);
20     nextY ← min(nextY, size−1)
21   });
22   x, y ← nextX, nextY
23
24 SmellPheromone = {
25   dX := [1..m+1];
26   dY := [1..m+1];
27   testx1, testy1 := min(x+dX, size−1),
28                     min(y+dY, size−1);
29   testx2, testy2 := max(x−dX, 0),
30                     max(y−dY, 0);
31
32   nextX ←
33     if ph[testx1, testy1] then testx1 else
34     if ph[testx2, testy2] then testx2
35     else x;
36   nextY ←
37     if ph[testx1, testy1] then testy1 else
38     if ph[testx2, testy2] then testy2
39     else y
40 }
41
42 GoHome =
43   x ≠ 0 or y ≠ foody ⇒ ({
44   ph[x,y] ⇐1;
45   x ← max(0, x−1)
46   }; GoHome)
```

3 Analysis

We would now like to assess whether, following the behaviour formalised in Sect. 2, our ants will keep wandering chaotically through the arena, or instead will gradually prefer the shortest path between their nest and the food source, thus replicating an emerging phenomenon similar to that observed by Goss et al. [22].

To enable automated reasoning, we decorate the specifications with appropriate properties that allow to evaluate whether such behaviour of interest emerges. Knowing that the shortest path is the segment from $(0, \mathit{foodx})$ to $(\mathit{foodx}, \mathit{foody})$, we can look for evidence of emergence by evaluating how far our ants are from this segment as the system evolves. If the system is able to self-organize, this distance should gradually decrease with time; otherwise, it should fluctuate chaotically. For an ant located at position (x, y), we define its distance from the path as:

$$d(\mathsf{x}, \mathsf{y}) = |\mathsf{y} - \mathit{foody}| + \max(0, \mathsf{x} - \mathit{foodx}),$$

meaning that the distance is just $|\mathsf{y} - \mathit{foody}|$ if the ant's x-coordinate is not greater than foodx; otherwise, it is the Manhattan distance [7] between the ant and the food source.

We can now proceed to analysing the colony of ants in different ways. We first generate arbitrary simulation traces, to gather some *empirical* evidence about our hypothesis; then, we use formal verification to understand whether the collective behaviour in consideration is *guaranteed* to happen.

Let us define some concepts used in the rest of the section. A *trace* of a system is a sequence of *steps* performed by its agents. A step corresponds to the execution of a single statement, or a compound atomic block enclosed in braces. We focus on finite-length traces, where ants perform their steps one at a time in a fixed order. Thus, a trace of length $q \cdot n$ for a system composed of n ants will consist of q steps per ant, and we say that this execution consists of q *epochs*.

Listing 5. Assumptions on the initial state of the ants.

```
1 assume {
2   FoodAnt = exists Ant a,
3     (x of a = foodx) and (y of a = foody)
4   FarFromThePath = forall Ant a,
5     ((x of a = foodx) and (y of a = foody)) or
6     (x of a > foodx + k) or
7     (y of a > foody + k) or
8     (y of a < foody - k)
9 }
```

Listing 6. Checking that ants aggregate close to the shortest path after B steps have elapsed.

```
1 check {
2   ShortestPath =
3     after B forall Ant a,
4     (x of a <= foodx + k) and
5     (y of a >= foody - k) and
6     (y of a <= foody + k)
7 }
```

Verification Technology. To perform our experiments, we use a tool called SLiVER[1] that enables *simulation* as well as *formal verification* of LAbS models using a variety of techniques. The core idea behind SLiVER is to turn the LAbS specification given as input into a symbolic intermediate representation that can be translated into different target languages, so as to re-use off-the-shelf tools developed for those languages as back ends for the analysis [17]. For the scope of this work, we rely on symbolic bounded model checking of [10], which in turn reduces to Boolean satisfiability (SAT). We would like to emphasize that we use the same encoding and rely on the same back end technique for both simulation and verification: for the former, we override some of the heuristics of the SAT solver in order to introduce extra randomness and thus produce different simulations at each run.

[1] SLiVER is available at https://github.com/labs-lang/sliver.

Listing 7. Ensuring that a single ant is initially located at the food source.

```
1  assume {
2    UniqueFoodAnt = exists-unique Ant a,
3      onFood of a = 1
4
5    OneOnFood = forall Ant a,
6      (onFood of a = 0) or
7      ((x of a = foodx) and (y of a = foody))
8
9    OthersFarFromThePath = forall Ant a,
10     (onFood of a = 1) or
11     (x of a > foodx + k) or
12     (y of a > foody + k) or
13     (y of a < foody - k)
14 }
```

Listing 8. The negation of the property shown in Listing 6.

```
1 check {
2   NegShortestPath = after B
3     exists Ant a,
4     (x of a > foodx + k) or
5     (y of a > foody + k) or
6     (y of a < foody - k)
7 }
```

Assumptions on the Initial State. In our specifications, ants are initially scattered through the arena (see line 2 of Listing 2). At the same time, we want to start our simulation at a moment where *at least* one ant has found the food source: after all, we already know that until some pheromone appears in the arena, ants can only carry on with their chaotic exploration. Finally, we want to ignore those simulations where the ants that have not found food are too close to the shortest path between the food repository and the nest. This last requirement is simply there so that we can focus on more interesting traces.

We formalize these requirements in a separate section of the specifications, as shown in Listing 5. This separation enables us to quickly evaluate different scenarios without having to alter the behaviour of the ants. Here, we say that at least one ant starts at the same position of the food source (AntOnFood), and we define a region around the shortest food-nest path that ants cannot occupy in the initial state (FarFromThePath). This region is depicted in Fig. 2.

Simulation. Given a file `ants.labs` containing our specifications, we can ask SLiVER to generate s traces of length B by invoking it as follows:

```
./sliver.py ants.labs <parameters> --fair --simulate+ s --steps+ B
        --concretization=sat
```

Here, `<parameters>` is a list of values to assign to each external parameter, and `--fair` enforces that agents execute their actions in ordered epochs. The `--concretization=sat` option enables a *concretization* step that alleviates the load on the SAT solver underlying our workflow. This step determines a feasible, random initial state of the system, and instruments the solver to start from that specific state, rather than considering all feasible initial states indifferently. The

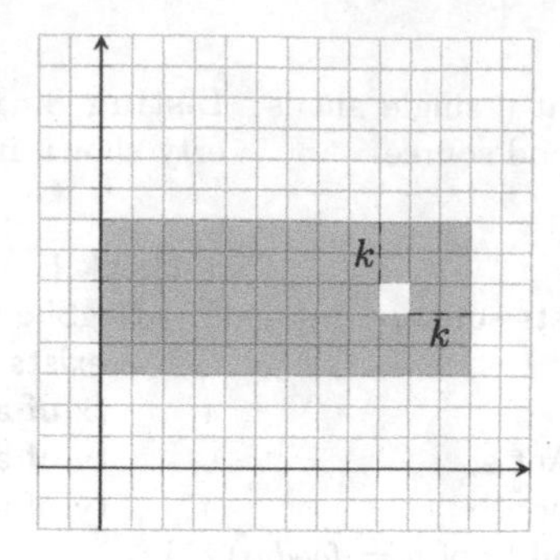

Fig. 2. The FarFromThePath assumption of Listing 5 enforces that no ant will occupy the shaded area in the initial state. The unshaded square within the shaded area is the location of the food source.

same instrumentation also involves other parts of the model, e.g., the operations at lines 25–26 and 14 of Listing 4, which pick values nondeterministically.[2]

Table 1. Values for our model's parameters during the experimental evaluation.

Name	Description	Value
foodx, *foody*	Position of the food source	(10, 10)
k	Initial distance from the shortest path	2
n	Number of ants	10
m	Maximum range of an ant's movement	1

We ran 200 simulations in a virtualized environment on a dedicated machine, running 64-bit GNU/Linux with kernel 5.4.0 and equipped with four 2-GHz Xeon E7-4830v4 10-core processors and 512 GB of physical memory. Table 1 shows the parameters used to instantiate the experiments. Each generated trace had a length of 800 steps, i.e., 80 epochs. By running 16 instances of SLiVER in parallel, we were able to simulate our system at an approximate rate of one trace every 13 min. This will seem quite slow for simulation. However, we would like to remark that the focus here is on integrating simulation and formal analysis within the same workflow; our workflow was initially conceived for exhaustive formal analysis and is not optimised for simulation yet.

The results of our simulations are summarized in Fig. 3. Each line depicts the average ant-path distance for one trace. Circle markers denote the mean value across all traces. The plot does seem to indicate that the average distance between the ants and the shortest path decreases with time. Apart from some outliers, this distance appears to drop below 2 rather quickly: the mean across all traces becomes less than 2 after 21 epochs, and becomes 0.6 after 80 epochs.

[2] Notice that this is not the same as solving the formula under assumptions: if the instrumentation makes the formula unsatisfiable, we leave the solver free to drop part of it and resume its search.

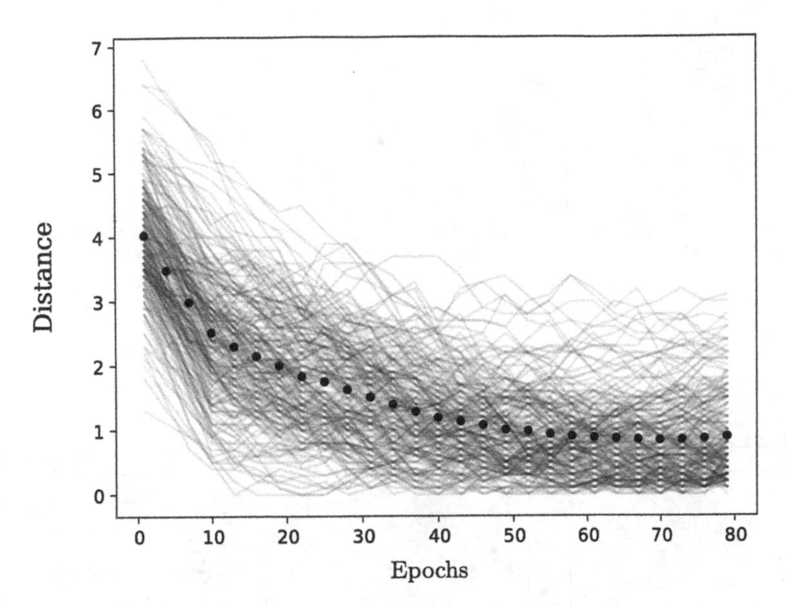

Fig. 3. Average distance of the ants from the shortest nest-food path.

The initial average distance may vary across traces because we allow one, two or more ants to start at the food source location: thus, their initial distance from the shortest path is nought.

Formal Verification. We now start to wonder whether the behaviour observed in the simulations is *inevitably* going to happen. In other words, will *every* trace of this system show the emergence of this feature?

To answer this question, we decorate our specifications with an additional section, containing the property that we wish to verify. In this case, we want all ants to be close enough to the shortest path after they have performed a certain number of actions B, which in our case corresponds to the verification bound. By "close enough" we mean that their value of x should be at most $foodx + k$ and that their distance from the path should not exceed k: (Listing 6). In other words, they should all lie within the area they were forbidden to occupy in the initial state, which we depicted in Fig. 2.

To verify this property, we invoke SLiVER with the following command line:

```
./sliver.py ants.labs <parameters> --fair --steps B.
```

We ran this task using the parameters in Table 1 and the same bound of $B =$ 800 steps, i.e., 80 epochs. Our verification task fails and SLiVER produces a counterexample as proof that the system can indeed violate the property under investigation, in other words, the ants will not necessarily adopt an increasingly optimised food-nest path.

Figure 4 shows a graphical visualization of the counterexample (we only report one state every 10 epochs). Each circle represents an ant; the triangle

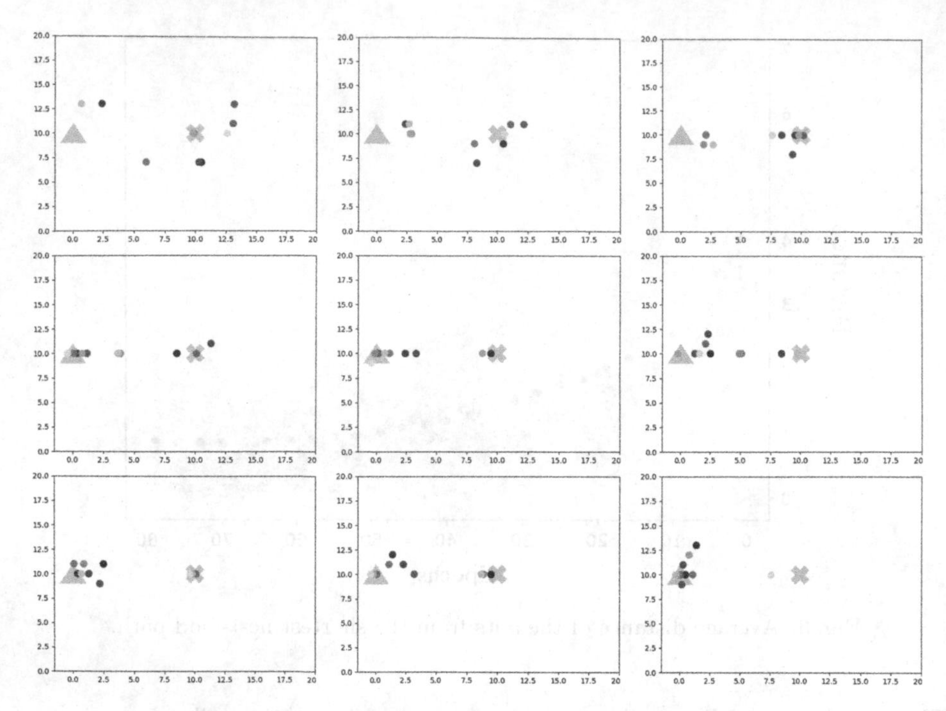

Fig. 4. A counterexample to the property of Listing 6.

and cross mark the position of the nest and food source, respectively. As the trace shows, the ants still appear to aggregate around the shortest path, but at some point one of them strays away and ends up in position (1, 13), i.e., at a distance of 3 from the path. This is enough to violate the property. This counterexample is a result of the uncertainty in the ants' pheromone detecting behaviour. Probabilistically, it is rather likely that an ant close to the pheromone trail will detect it at least once in a while, and thus keep close to it: this explains why our previous simulations seem to converge rather effortlessly. However, formal verification treats all potential traces alike, regardless from their likelihood in a probabilistic sense. This allows SLiVER to detect traces that could be hard to spot through simulation alone.

Looking for Interesting Traces via Verification. We close this section by highlighting another potential use of verification: that is, to assess the *existence* of a trace satisfying some specific requirements. For instance, let us consider again our specifications and ask ourselves: assuming that *a single* ant starts at the food source, is there *any* trace where all ants manage to aggregate along the shortest path? We can get an answer by asking SLiVER to verify that *all* traces satisfy the *negation* of this property: if this task fails, we will be presented with a trace that satisfies our requirements. Otherwise, it means that no such trace exists.

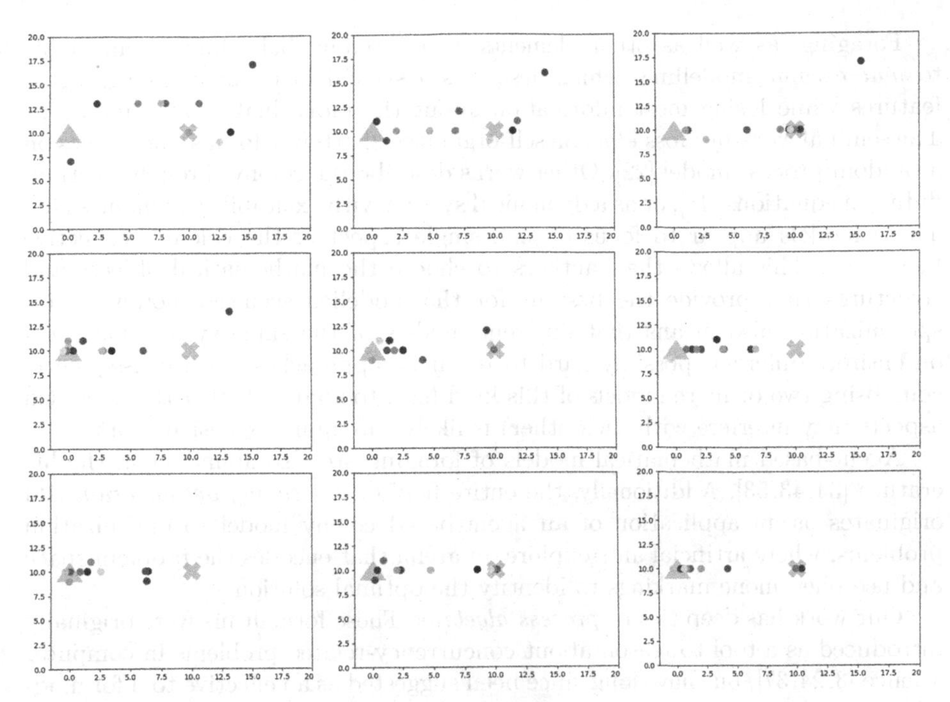

Fig. 5. A counterexample to the property of Listing 8.

To encode the new assumptions, we equip all ants with an attribute onFood that may be initialized to either 0 or 1. Then, we replace our previous assumptions with new ones (Listing 7) stating the following: only one ant starts with onFood set to 1 (UniqueFoodAnt); this ant starts at the food location (OneOnFood); and all other ants start somewhere far from the shortest path (OthersFarFromThePath).[3] Lastly, we have to negate the property that we have previously verified: that is, we want to check that after B steps, there is at least one ant that is still far from the shortest path. The resulting property is shown in Listing 8.

We verified this property on the same machine used for the previous verification task, with the same parameters and verification bound. Again, the task failed and SLiVER returned the counterexample depicted in Fig. 5.

4 Related Work

Extensive overviews of the foraging strategies followed by different species of ants can be found in [16,52]. For additional discussion about current research on ant foraging, we refer the reader to [28].

[3] Adding the attribute onFood is not strictly necessary, but makes for more readable specifications.

Foraging, as well as other elements of ant colony behaviour, is amenable to *macroscopic* modelling, which describes a system in terms of its aggregate features while losing most information about the individuals that compose it. The seminal work by Goss *et al.* on self-organized shortcuts, for instance, relies on a random process model [22]. Other works describe the colony through reaction-diffusion equations [48], or as a dynamical system with excitability dynamics [40]. These models appear to focus on one single aspect of the colony's collective behaviour. This allows their authors to choose the mathematical objects and structures that provide the best fit for the modelled scenario; however, this specialization also means that different models of the same system may rely on entirely different, possibly hard to reconcile approaches. As a consequence, composing two or more models of this kind (say, to study whether the modelled aspects may interfere with each other) is likely to require extensive effort.

Agent-based mathematical models of foraging are also abundant in the literature [34,43,53]. Additionally, the entire field of *ant colony optimization* [18] originates as an application of an agent-based colony model to optimization problems, where artificial ants explore an arena that encodes the problem space and use pheromone markings to identify the optimal solution.

Our work has deep ties to *process algebras*. These formalisms were originally introduced as a tool to reason about concurrency-related problems in computer science [3,24,37], but have long since been suggested as an effective tool for modelling complex natural systems. Among other things, it is argued that process algebras are inherently compositional (so they lend themselves well to describing collections of interacting agents), and that the specifications can be formally validated [51].

Our formal specification language LAbS aims at exploiting these perks, while offering a more high-level and intuitive set of primitives and constructs for the specification of collective systems. Other forms of natural behaviour that have been modelled with LAbS include bird flocking [14] and a simplified reproduction of the experiment from [22] on self-organization in colonies of the Argentinian ant *Iridomyrmex humilis* [15]. However, these models were somewhat limited since they used an earlier version of the language without conditional processes and multidimensional arrays.

The semantics of a process algebra is commonly defined as some form of transition system (LAbS also falls in this category). This allows for verification of qualitative properties by exhaustively exploring the state space of the system. It is also possible to equip a process algebra with probabilistic or quantitative semantics, thus enabling other forms of automated reasoning: relevant examples include WSCCS [49], which has a Markov-chain semantics and has been used to model and analyse behaviours commonly observed in ant colonies, from synchronization to activity recruiting [47,50] and Bio-PEPA [9], with an ODE-based semantics that has been exploited for fluid flow analysis of ant foraging behaviour [36]. For an exhaustive list of Process Algebra-based methods, please refer to Sect. 2 of [2].

The programming and modelling language literature provides a wide choice of languages tailored towards collective, agent-based systems [29]. Examples include ASCAPE [26], SCAMP [42], and MASON [35]; see [41] for a MASON model of foraging ants that bears some similarity to our work.

Arguably, the most notable specimens in this category are LOGO [20] and its derivatives, most notably StarLogo [44] and NetLogo [54], which have found a rather wide adoption as a tool to create and simulate agent-based models across multiple research areas, including systems biology [8,43]. StarLogo and NetLogo are engineered to handle thousands of agents, and NetLogo also integrates powerful visualization features. Additionally, StarLogo and NetLogo allow specifying complex, dynamic environments by defining *patches*, a concept introduced in the artificial life community [25]. The environment is assumed to be an n-dimensional grid partitioned into non-overlapping patches or cells, where each patch may interact with neighbouring patches and with agents that are inhabiting it at a given moment. Thus, the concept of patches reconciles agent-based models with cellular automata [31]. In contrast, the world inhabited by LAbS agents has no predefined structure and is essentially just a store of shared variables (e.g., the pheromone matrix in our case study): the structure, if any, emerges from the behavioural rules of the individual agents.

5 Conclusions

In this work we have advocated for the application of language-based, bottom-up methodologies to support research in computational biology, and highlighted the benefits of integrating simulation and formal analysis within the same workflow. We have shown that high-level languages with adequate constructs and primitives allow for an intuitive process where an informal description of the individual behaviour of a complex biological system (in our example, an ant colony) is gradually turned into a formal specification. Compared to lower-level mathematical models, these specifications are easier to refine and extend; they support reuse and composition; furthermore, they are amenable to formal verification, which may provide further insight into the system's emergent properties.

To illustrate this, we have first produced a collection of simulation traces apparently confirming the emergence of self-organized pathfinding in the colony. Then, we have shown how formal verification can disprove that this form of behaviour emerges inevitably. Intuitively, this is due to the fact that ants in our model are able to take a sequence of bad choices that lead them farther, not closer, to the path. Lastly, we have used verification to show that a single ant finding a food source may be enough for the self-organizing behaviour to emerge. Crucially, these tasks were all carried out starting from the same model, with only minimal and intuitive changes being required for the last task.

We plan on improving our methodology along several directions. The modelling language should be further extended with primitives promoting composability and code reuse, such as parameterized process invocations and modules. We would like to foster cooperation with the computational biology community, to gather realistic case studies and to get valuable feedback to motivate further refinements of our language and methodologies.

On the analysis side, we intend to address the main drawback of our simulation workflow, namely its limited efficiency with respect to traditional approaches based either on executing the model (when the development platform includes a compiler or runtime environment) or on numerical methods. In this respect, SLiVER's reliance on existing verification technologies allows it to evolve with little effort as the state of the art in general-purpose automated reasoning advances. For instance, we could improve the performance by integrating recent optimised SAT decision procedures [21], or by exploiting techniques for deep counterexample detection, e.g., transition power abstractions [4]. This might be especially beneficial when simulating long traces.

Probabilistic reasoning may be enabled by leveraging existing probabilistic model checkers [30]. To support this use case, we might extend the language so that the choices of agents follow a probabilistic distribution. Lightweight simulation-based formal techniques, including statistical model checking [45] and runtime verification [32], may provide some statistical assurance on the behaviour of systems that are currently out of the reach of formal analysis.

In this work, we have focused on bounded executions of systems with a fixed and predefined number of agents. We may support unbounded verification either by computing a *completeness threshold*, i.e., a bound large enough to encompass the whole state space of the system under analysis [11], or by resorting to inductive reasoning techniques for program analysis, such as k-induction [46] or property-directed reachability (PDR, also known as IC3) [6]. We could move past systems of predefined size by considering the number of agents as a fixed, but of arbitrary value. In principle, we could even add language primitives allowing an agent to *spawn* others, or to *leave* the system, making the size of our models dynamic. At the analysis level, these changes would likely require describing our agents through dynamically-allocated data structures; while this is feasible in principle, we suspect that it would result in further challenges to tractability.

References

1. Attanasi, A., et al.: Information transfer and behavioural inertia in starling flocks. Nat. Phys. **10** (2014). https://doi.org/10.1038/nphys3035
2. Bartocci, E., Lió, P.: Computational modeling, formal analysis, and tools for systems biology. PLoS Comput. Biol. **12**(1), e1004591 (2016)
3. Bergstra, J.A., Klop, J.W., Tucker, J.V.: Algebraic tools for system construction. In: Clarke, E., Kozen, D. (eds.) Logic of Programs 1983. LNCS, vol. 164, pp. 34–44. Springer, Heidelberg (1984). https://doi.org/10.1007/3-540-12896-4_353
4. Blicha, M., Fedyukovich, G., Hyvärinen, A.E.J., Sharygina, N.: Transition power abstractions for deep counterexample detection. In: TACAS 2022. LNCS, vol. 13243, pp. 524–542. Springer, Cham (2022). https://doi.org/10.1007/978-3-030-99524-9_29
5. Bonabeau, E.: Agent-based modeling: Methods and techniques for simulating human systems. PNAS **99** (2002). https://doi.org/10.1073/pnas.082080899

6. Bradley, A.R.: SAT-based model checking without unrolling. In: Jhala, R., Schmidt, D. (eds.) VMCAI 2011. LNCS, vol. 6538, pp. 70–87. Springer, Heidelberg (2011). https://doi.org/10.1007/978-3-642-18275-4_7
7. Brezis, H., Brézis, H.: Functional Analysis, Sobolev Spaces and Partial Differential Equations, vol. 2. Springer, New York (2011). https://doi.org/10.1007/978-0-387-70914-7
8. Chiacchio, F., Pennisi, M., Russo, G., Motta, S., Pappalardo, F.: Agent-based modeling of the immune system: NetLogo, a promising framework. BioMed Res. Int. (2014). https://doi.org/10.1155/2014/907171
9. Ciocchetta, F., Hillston, J.: Bio-PEPA: an extension of the process algebra PEPA for biochemical networks. Electr. Notes Theor. Comput. Sci. **194** (2008). https://doi.org/10.1016/j.entcs.2007.12.008
10. Clarke, E., Kroening, D., Lerda, F.: A tool for checking ANSI-C programs. In: Jensen, K., Podelski, A. (eds.) TACAS 2004. LNCS, vol. 2988, pp. 168–176. Springer, Heidelberg (2004). https://doi.org/10.1007/978-3-540-24730-2_15
11. Clarke, E., Kroening, D., Ouaknine, J., Strichman, O.: Completeness and complexity of bounded model checking. In: Steffen, B., Levi, G. (eds.) VMCAI 2004. LNCS, vol. 2937, pp. 85–96. Springer, Heidelberg (2004). https://doi.org/10.1007/978-3-540-24622-0_9
12. Cristiani, E., Menci, M., Papi, M., Brafman, L.: An all-leader agent-based model for turning and flocking birds. J. Math. Biol. **83** (2021). https://doi.org/10.1007/s00285-021-01675-2
13. De Nicola, R., Di Stefano, L., Inverso, O.: Multi-agent systems with virtual stigmergy. Sci. Comput. Program. **187** (2020). https://doi.org/10.1016/j.scico.2019.102345
14. De Nicola, R., Di Stefano, L., Inverso, O., Valiani, S.: Modelling flocks of birds from the bottom up. In: Margaria, T., Steffen, B. (eds.) ISoLA 2022. LNCS, vol. 13703, pp. 82–96. Springer, Cham (2022). https://doi.org/10.1007/978-3-031-19759-8_6
15. De Nicola, R., Di Stefano, L., Inverso, O., Valiani, S.: Modelling flocks of birds and colonies of ants from the bottom up. Int. J. Softw. Tools Technol. Transf. (2023, to appear)
16. Deneubourg, J.L., Aron, S., Goss, S., Pasteels, J.M., Duerinck, G.: Random behaviour, amplification processes and number of participants: how they contribute to the foraging properties of ants. Physica D **22**(1), 176–186 (1986). https://doi.org/10.1016/0167-2789(86)90239-3
17. Di Stefano, L., De Nicola, R., Inverso, O.: Verification of distributed systems via sequential emulation. ACM Trans. Softw. Eng. Methodol. **31** (2022). https://doi.org/10.1145/3490387
18. Dorigo, M., Birattari, M., Stützle, T.: Ant colony optimization. IEEE Comput. Intell. Mag. **1**, 28–39 (2006)
19. Farmer, J.D., Foley, D.: The economy needs agent-based modelling. Nature **460** (2009). https://doi.org/10.1038/460685a
20. Feurzeig, W., Papert, S.: Programming-languages as a conceptual framework for teaching mathematics. In: NATO Conference on Computers and Learning, pp. 37–42 (1968)
21. Froleyks, N., Heule, M., Iser, M., Järvisalo, M., Suda, M.: SAT competition 2020. Artif. Intell. **301**, 103572 (2021). https://doi.org/10.1016/j.artint.2021.103572
22. Goss, S., Aron, S., Deneubourg, J.L., Pasteels, J.M.: Self-organized shortcuts in the Argentine ant. Naturwissenschaften **76**(12), 579–581 (1989). https://doi.org/10.1007/BF00462870

23. Grauwin, S., Bertin, E., Lemoy, R., Jensen, P.: Competition between collective and individual dynamics. PNAS **106** (2009). https://doi.org/10.1073/pnas.0906263106
24. Hoare, C.A.R.: Communicating Sequential Processes. Prentice-Hall, London (1985)
25. Hogeweg, P.: Mirror beyond mirror: puddles of life. In: ALIFE. Santa Fe Institute Studies in the Sciences of Complexity, vol. 6, pp. 297–316. Addison-Wesley (1987)
26. Inchiosa, M.E., Parker, M.T.: Overcoming design and development challenges in agent-based modeling using ASCAPE. PNAS **99** (2002). https://doi.org/10.1073/pnas.082081199
27. Kaul, H., Ventikos, Y.: Investigating biocomplexity through the agent-based paradigm. Br. Bioinf. **16** (2015). https://doi.org/10.1093/bib/bbt077
28. Kolay, S., Boulay, R., d'Ettorre, P.: Regulation of ant foraging: a review of the role of information use and personality. Front. Psychol. **11**, 734 (2020). https://doi.org/10.3389/fpsyg.2020.00734
29. Kravari, K., Bassiliades, N.: A survey of agent platforms. J. Artif. Soc. Soc. Simul. **18** (2015). https://doi.org/10.18564/jasss.2661
30. Kwiatkowska, M., Norman, G., Parker, D.: PRISM 4.0: verification of probabilistic real-time systems. In: Gopalakrishnan, G., Qadeer, S. (eds.) CAV 2011. LNCS, vol. 6806, pp. 585–591. Springer, Heidelberg (2011). https://doi.org/10.1007/978-3-642-22110-1_47
31. Langton, C.G.: Studying artificial life with cellular automata. Physica D **22**(1), 120–149 (1986). https://doi.org/10.1016/0167-2789(86)90237-X
32. Leucker, M., Schallhart, C.: A brief account of runtime verification. J. Log. Algebr. Program. **78** (2009). https://doi.org/10.1016/j.jlap.2008.08.004
33. Levin, S.: Complex adaptive systems: exploring the known, the unknown and the unknowable. Bull. Amer. Math. Soc. **40** (2003). https://doi.org/10.1090/S0273-0979-02-00965-5
34. Li, L., Peng, H., Kurths, J., Yang, Y., Schellnhuber, H.J.: Chaos–order transition in foraging behavior of ants. PNAS **111** (2014). https://doi.org/10.1073/pnas.1407083111
35. Luke, S., Cioffi-Revilla, C., Panait, L., Sullivan, K., Balan, G.C.: MASON: a multiagent simulation environment. Simulation **81** (2005). https://doi.org/10.1177/0037549705058073
36. Massink, M., Latella, D.: Fluid analysis of foraging ants. In: Sirjani, M. (ed.) COORDINATION 2012. LNCS, vol. 7274, pp. 152–165. Springer, Heidelberg (2012). https://doi.org/10.1007/978-3-642-30829-1_11
37. Milner, R.: A Calculus of Communicating Systems. LNCS, vol. 92. Springer, Heidelberg (1980). https://doi.org/10.1007/3-540-10235-3
38. Müller, M., Wehner, R.: Path integration in desert ants, cataglyphis fortis. PNAS **85**(14), 5287–5290 (1988). https://doi.org/10.1073/pnas.85.14.5287
39. Olfati-Saber, R.: Flocking for multi-agent dynamic systems: Algorithms and theory. IEEE Trans. Automat. Contr. **51** (2006). https://doi.org/10.1109/TAC.2005.864190
40. Pagliara, R., Gordon, D.M., Leonard, N.E.: Regulation of harvester ant foraging as a closed-loop excitable system. PLoS Comput. Biol. **14**(12) (2018). https://doi.org/10.1371/journal.pcbi.1006200
41. Panait, L.A., Luke, S.: Ant foraging revisited. In: ALIFE, pp. 569–574. MIT Press (2004). https://doi.org/10.7551/mitpress/1429.003.0096
42. Parunak, H.V.D.: Social simulation for non-hackers. In: Van Dam, K.H., Verstaevel, N. (eds.) MABS 2021. LNCS (LNAI), vol. 13128, pp. 1–14. Springer, Cham (2022). https://doi.org/10.1007/978-3-030-94548-0_1

43. Perna, A., et al.: Individual rules for trail pattern formation in argentine ants (Linepithema Humile). PLoS Comput. Biol. **8** (2012). https://doi.org/10.1371/journal.pcbi.1002592
44. Resnick, M.: Turtles, Termites, and Traffic Jams - Explorations in Massively Parallel Microworlds. MIT Press (1998)
45. Sen, K., Viswanathan, M., Agha, G.: Statistical model checking of black-box probabilistic systems. In: Alur, R., Peled, D.A. (eds.) CAV 2004. LNCS, vol. 3114, pp. 202–215. Springer, Heidelberg (2004). https://doi.org/10.1007/978-3-540-27813-9_16
46. Sheeran, M., Singh, S., Stålmarck, G.: Checking safety properties using induction and a SAT-solver. In: Hunt, W.A., Johnson, S.D. (eds.) FMCAD 2000. LNCS, vol. 1954, pp. 127–144. Springer, Heidelberg (2000). https://doi.org/10.1007/3-540-40922-X_8
47. Sumpter, D.J., Blanchard, G.B., Broomhead, D.S.: Ants and agents: a process algebra approach to modelling ant colony behaviour. Bull. Math. Biol. **63** (2001). https://doi.org/10.1006/bulm.2001.0252
48. Theraulaz, G., et al.: Spatial patterns in ant colonies. PNAS **99** (2002). https://doi.org/10.1073/pnas.152302199
49. Tofts, C.: A synchronous calculus of relative frequency. In: Baeten, J.C.M., Klop, J.W. (eds.) CONCUR 1990. LNCS, vol. 458, pp. 467–480. Springer, Heidelberg (1990). https://doi.org/10.1007/BFb0039078
50. Tofts, C.M.N.: Describing social insect behaviour using process algebra. Trans. Soc. Comput. Simul. **9**, 227 (1992)
51. Tofts, C.M.N.: Process algebra as modelling. Electr. Notes Theor. Comput. Sci. **162** (2006). https://doi.org/10.1016/j.entcs.2005.12.114
52. Traniello, J.F.A.: Foraging strategies of ants. Annu. Rev. Entomol. **34**(1), 191–210 (1989). https://doi.org/10.1146/annurev.en.34.010189.001203
53. Vittori, K., Talbot, G., Gautrais, J., Fourcassié, V., Araújo, A.F.R., Theraulaz, G.: Path efficiency of ant foraging trails in an artificial network. J. theor. Biol. **239** (2006). https://doi.org/10.1016/j.jtbi.2005.08.017
54. Wilensky, U.: Modeling nature's emergent patterns with multi-agent languages. In: EuroLogo (2001)

Cell-Level Pathway Scoring Comparison with a Biologically Constrained Variational Autoencoder

Pelin Gundogdu[1,2], Miriam Payá-Milans[1,2], Inmaculada Alamo-Alvarez[1,3], Isabel A. Nepomuceno-Chamorro[4(✉)], Joaquin Dopazo[1,2,5,6(✉)], and Carlos Loucera[1,2(✉)]

[1] Andalusian Platform for Computational Medicine, Andalusian Public Foundation Progress and Health-FPS, Sevilla, Spain
cloucera@juntadeandalucia.es

[2] Computational Systems Medicine, Institute of Biomedicine of Seville (IBIS), Hospital Virgen del Rocio, Sevilla, Spain

[3] Department of Immunology, Institute of Biomedicine of Seville (IBIS), Hospital Virgen del Rocio, Sevilla, Spain

[4] Dpto. de Lenguajes y Sistemas Informaticos, University of Seville, Sevilla, Spain
inepomuceno@us.es

[5] Centro de Investigación Biomédica en Red de Enfermedades Raras (CIBERER), FPS, Hospital Virgen del Rocío, Sevilla, Spain

[6] FPS/ELIXIR-es, Hospital Virgen del Rocío, Sevilla, Spain
joaquin.dopazo@juntadeandalucia.es

Abstract. Unsupervised techniques are ubiquitous to study and understand the complex patterns that arise when analyzing genomic data at single-cell resolution. Particularly, unsupervised deep learning models provide state-of-the-art solutions for the most common tasks that arise when dealing with scRNA-seq data. However, the biological usefulness of these complex models is burdened by their black-box nature. To address such limitations several lines of research have emerged, from post hoc approximations to ante hoc modeling. In this work, we study the behavior of two biologically-constrained variational autoencoders (ante hoc modeling). On the one hand, we use a one-layer architecture where the constraints come from the signaling pathways, and, on the other hand, we propose a two-layer architecture following the recent trends in mechanistic models of signal transduction. We use the representations learned by the model as proxies of the signaling activity at the single-cell level. We check the performance of the scoring model using a known scRNA-seq public dataset with a clearly established ground truth. Although both models capture the relevant signals, the most pronounced differences are better captured by the one-layer architecture, while the two-layer design is able to learn more fine-grained features that can expose less prominent aspects of the data.

Keywords: variational autoencoder (VAE) · single-cell RNA sequencing (scRNA-seq) · latent space representation · interpretable neural network (xAI)

J. Pang and J. Niehren (Eds.): CMSB 2023, LNBI 14137, pp. 62–77, 2023.
https://doi.org/10.1007/978-3-031-42697-1_5

1 Introduction

The emergence of single-cell RNA sequencing (scRNA-seq) technologies has enabled the study of the complexity and heterogeneity at the transcriptomic level with an unprecedented resolution [25]. However, as the technologies advance so does the computational needs, in order to produce tools, protocols, and models that can cope with the increasing data dimensionality [17,33]: at the sample (more cells), variable (e.g. more genes), and modality axes (different kinds of measurements).

From a data science point of view, there are innumerable challenges that arise when trying to decipher the complex patterns across the myriad of single-cell datasets released on a yearly basis: from cell-type clustering, annotation, visualization, quality control, dataset integration, to name a few [17]. Machine Learning [24], and in particular Deep Learning-based solutions [30], are especially well suited for scRNA-seq data-driven tasks.

However, the latent representations learned by deep-learning-based methods are not useful for interpreting the underlying biology [31], which is a major drawback from a systems-biology point of view. To address these limitations several ante hoc, which make the model more interpretable a priori, and post hoc, which explain the model a posteriori using surrogate interpretable models, methods have been proposed. In this work, we are interested in domain-constrained models [6], where the wiring of the neural networks (NNs) is conditioned by one or multiple sources of a priori knowledge of the domain (biology in our case).

Among the plethora of explainable modeling solutions, domain-constrained deep learning models have drawn the attention of the research community. For instance, Visible Neural Networks (VNNs), NNs where the layers are coupled with representations of the biological components of human cells, have been used to better understand eukaryotic cells [21] or to model the human cell structure to predict anti-cancer drug responses [16]. In [20] they propose ExiMap, an interpretable network for data integration and gene program discovery. Although a review of the different explainability and interpretability methods and terminology is beyond the scope of this work, we refer the reader to consult [8] for a general review and [34] for a biology-focused one.

In this work, we aim to analyze the biological utility of the representations learned by a variational autoencoder (VAE) constrained by cell signaling entities. More precisely, we are interested in studying how the source of a priori knowledge conditions the biological results offered by the model. To this aim, we train the informed VAE using two different sources of domain knowledge: i) the Reactome pathway database [7] and ii), the Kyoto Encyclopedia of Genes and Genomes database (KEGG) [23]. While on Reactome we use the gene sets as is, to make it easily comparable with the literature [12], when using the KEGG database we follow the recent trends in mechanistic modeling to decompose the pathways into smaller functional units (the so-called circuits) [5,13]. The models are fit to a dataset of human peripheral blood mononuclear cells (PBMCs) of lupus patients where the ground truth of an interferon-β perturbation experiment is known [14].

We empirically demonstrate how the biological signal learned by a simple known VAE architecture can be refined by using more fine-grained constraints.

2 Materials and Methods

2.1 Dataset

In this work, we evaluate the proposed model's biological usefulness by scoring the pathway and circuit activities in a publicly available dataset that contains untreated and IFN-β stimulated human peripheral blood mononuclear cells (PBMC) from eight patients with Lupus [14], see Table 1 for a brief summary of the cell type distribution. The data was obtained using the `scanpy` library [32], as described in the book "Single-cell best practices" [12].

The dataset allows us to easily measure the performance of unsupervised methods since we have the ground truth for the perturbations, thus we can use an unsupervised method for inferring the signaling activity and compare the results across the two groups: *control* and *stimulated*. Due to the nature of the perturbation, we expect that interferon-based pathways are scored higher in the *stimulated* population than in the *control* (*non-stimulated*) cells. This is a known result [12,20] when studying the dataset with the Reactome pathway database [7].

Table 1. Cell type distribution of the untreated and IFN-stimulated human PBMC cells dataset by Kang et al. [14]

Condition	B cells	CD14+ Monocytes	CD4 T cells	CD8 T cells	Dendritic cells	FCGR3A+ Monocytes	Megakar-yocytes	NK
control	1316	2932	5560	811	258	520	63	855
stimulated	1335	2765	5678	810	271	569	69	861

2.2 Sources of Biological Information

In order to elucidate if the VAE activity scoring is stable under changes of the *a priori* knowledge used to inform the network, we trained two Neural Networks using different sources of biological knowledge. On the one hand, we have trained a VAE informed by the Reactome pathway knowledgebase [7] and, on the other hand, we have used the Kyoto Encyclopedia of Genes and Genomes database (KEGG) [23] to inform the neural network architecture. See Sect. 2.3 for a concise explanation of how the information is used to inform the variational autoencoder design.

2.3 Model Design

In this work, we proposed a biologically constrained artificial neural network, i.e. a NN whose weights and kernels (the informed layers) are conditioned by a set of *a priori* knowledge. We aim to improve the explainability of the model by learning biologically interpretable scoring functions from each informed layer.

Variational Autoencoders. To learn a latent representation of the data we adopt a variational autoencoder (VAE) architecture (see Fig. 1), which is a type of deep generative model used for unsupervised learning that learns a probability distribution over the latent representation. It aims to reconstruct data points ($\mathbf{x}$), let's call ($\hat{\mathbf{x}}$) the reconstructed approximation, with a minimum error as possible by means of a composite network conformed by an encoder (e) and a decoder (d) block joined in a non-trivial way (i.e. there is a bottleneck layer where the information is necessarily compressed).

The model assumes that there is an underlying data distribution, where the encoder gives us the distribution (Gaussians μ_x, σ_x) of this latent representation of the data (Eq. 1a), while the decoder block samples from the distribution generating new data points (Eq. 1b), the approximation ($\hat{\mathbf{x}}$). These latent variables are used to sample a vector ($\mathbf{z}$) (see Eq. 2) which is used to feed to the decoder block (see Eq. 1b) to reconstruct the input data.

To optimize the model, two terms are needed to compose the final loss function (Eq. 3a): the *reconstruction loss* (Eq. 3b) and the Kullback-Leibler divergence or *similarity loss*. These terms are derived from the probabilistic model, in our case, given the Gaussian assumptions, the reconstruction loss takes the form of the mean squared error between the input ($\mathbf{x}$) and the approximation ($\hat{\mathbf{x}}$) data points, while the KL loss guides the model to *become* a unit normal distribution (Eq. 3c). To compute the expectation value of the *reconstruction* term we use the sampling approximation.

$$\textit{encoded data}(e) = \mathcal{N}(\mu_x, \sigma_x) \tag{1a}$$

$$\textit{decoded data} = d(z) \tag{1b}$$

$$\textit{sampling} = z \sim \mathcal{N}(\mu_x, \sigma_x) \tag{2}$$

$$\textit{loss} = \textit{reconstruction loss} + \textit{similarity loss} \tag{3a}$$

$$\textit{reconstruction loss} = \frac{1}{N}\sum_{i=1}^{N}(\mathbf{x}_i - \hat{\mathbf{x}}_i)^2 \tag{3b}$$

$$\textit{similarity loss} = KL[\ \mathcal{N}(\mu_x, \sigma_x), \mathcal{N}(0,1)\ ] \tag{3c}$$

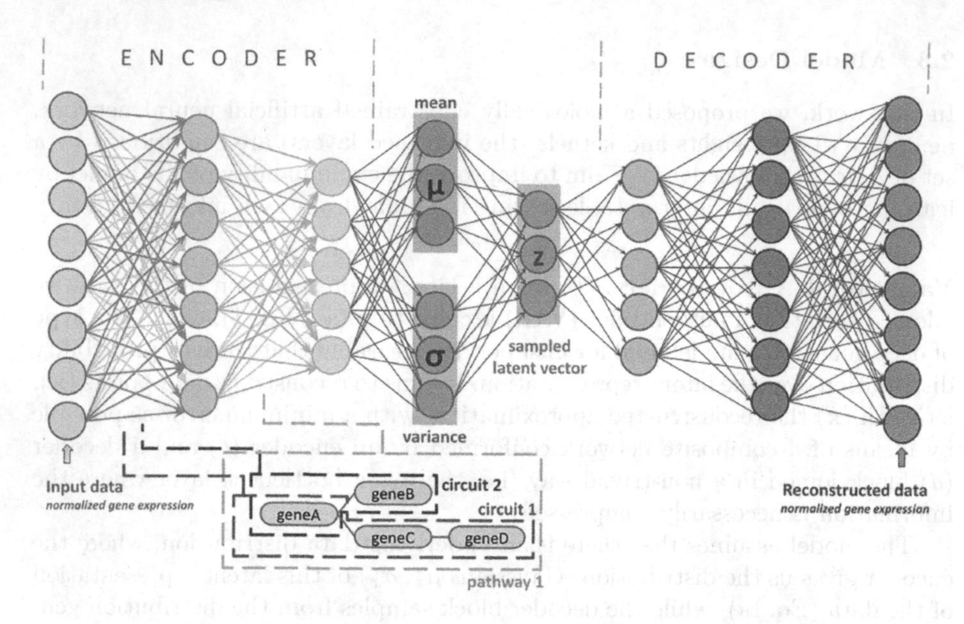

Fig. 1. Proposed Variational autoencoder (VAE) architecture

Signaling-Primed Layers. In this work, we constrain the VAE architecture by using *a priori* biological knowledge that informs the construction and operations of the layers. An informed layer behaves as a linear layer except that the kernel is informed by an indicator matrix ($\mathbf{I}_S$) that informs, using a collection of biological entities $S = \{s_j\}_{j=1}^{ns}$, which inputs of the previous layer should be used. Thus, if the previous layer has the biological entities $\mathcal{A} = \{a_1, \ldots, a_k\}$, $\mathbf{I}_S(i,j) = 1$ if the entity a_i belongs to the entity s_j, otherwise $\mathbf{I}_S(i,j) = 0$.

An S-informed layer $\mathbf{H}_S^\star$ is updated using the following formula:

$$\mathbf{H}_S^\star = \texttt{activation}\left((\mathbf{W} \odot \mathbf{I}_S)\,\mathbf{H} + \mathbf{b}\right) \tag{4}$$

where $\mathbf{W}$ and $\mathbf{b}$ denote the corresponding weight and bias tensors, respectively. $\mathbf{I}_S$ is the indicator matrix of the signalization set S, $\mathbf{H}$ is the previous layer outputs, `activation` refers to the layer activation function and, $\odot$ represents the element wise (Hadamard) product.

As has been mentioned before, in this work we define two different architecture designs. On the one hand, we use the Reactome pathway database to inform the first hidden layer of the VAE, thus we only have one informed layer, where $\mathcal{A}$ are the genes, S is the collection of the pathway gene sets and $\mathbf{I}_S$ encodes the information of which gene belongs to every gene set. On the other hand, the architecture for the KEGG database is constructed using the circuit decomposition proposed in [13], where each pathway is decomposed into effector sub-pathways (so-called circuits) that represent the minimal functional units. We inform the first hidden layer of the architecture using the gene sets defined by the circuits in a similar fashion as with Reactome pathways, where $\mathcal{A}, S$ represent the genes circuits, respectively. However, given that each pathway is decomposed into as many circuits as effector genes, the resulting neuronal wiring could be too sparse, since each circuit has potentially new genes compared to whole pathways. To overcome such limitations, we create a second hidden layer based on the pathways, a node of the first hidden layer is connected to a node of the second layer if the circuit that it represents belongs to the pathway represented by the node of the second hidden layer. To avoid connecting the layers in blocks, we also connect a pathway ℓ with a circuit j if they share a gene. This KEGG-based design improves upon other works by the authors when conditioning supervised neural networks to identify cell types [9,10], which showcased competitive results when compared with state-of-the-art methods such as Scibet [19].

The circuit decomposition and gene set extraction for the KEGG database has been done using the HiPathia R package (v 2.11.4) [13] for the *Homo sapiens* organism.

Architectures Tested. As previously stated, we conducted tests on two informed VAE models that utilize either the KEGG or Reactome pathway databases as a knowledge source. In order to establish a baseline for comparison, we created two models that mirror the informed architecture, using the same number of layers and nodes but with a dense wiring. Table 2 displays the nodes and trainable parameters for all models that were tested.

Table 2. Complexity decomposition for the different architectures.

		Integrated knowledge information			
		KEGG database	Reactome database	Dense (KEGG)	Dense (Reactome)
VAE network	No. of hidden layer	6	4	6	4
	No. of nodes in input layer	1,820	7,573	1,820	7,573
	No. of parameters	2,438,070	16,219,527	4,688,920	28,383,122
Encoder Block	No. of hidden layer	2	1	2	1
	No. of nodes in layers	1,820 / 1,221 / 93	7,573 / 1,615	1,820 / 1,221 / 93	7,573 / 1,615
	No. of parameters	86,237	68,415	2,337,087	12,232,010
Latent Block	No. of hidden layer	2	2	2	2
	No. of nodes in layers	46 / 46	807 / 807	46 / 46	807 / 807
	No. of parameters	8,648	2,608,224	8,648	2,608,224
Decoder Block	No. of hidden layer	2	1	2	1
	No. of nodes in layers	93 / 1,221 / 1,820	1,615 / 7,573	93 / 1,221 / 1,820	1,615 / 7,573
	No. of parameters	2,343,185	13,542,888	2,343,185	13,542,888

Hyperparameter Selection. We trained the networks for 100 epochs in batches of size 32, used Hyperbolic tangent activation functions for all layers except the output layer for the decoder which uses a linear activation function, and the informed layers were regularized using ℓ_2-based activity regularizers. We used the ADAM optimizer [15] for the optimization with a learning rate of 1e-5.

It was implemented in Python 3.10 using numpy (v 1.23.5) [11], scipy (v 1.10.1) [29] and TensorFlow 2.10 [1].

2.4 Visualization and Comparison of the Signaling Activity

Since the signaling entities, either from circuits or pathways, are first-class entities of our model, we can use standard statistical tools to analyze the inferred signaling activity at the single-cell resolution.

The `Scanpy` library [32] has been used to visualize the signaling activities inferred by the VAE at single-cell resolution following the standard procedure to visualize gene expression: i) we create an `AnnData` object [27,28] where the bottleneck layer substitutes the slot of gene expression data, ii) compute the neighbor graph, iii) cluster the cells using the `leiden` algorithm [26] as proposed in [18], iv) use the UMAP dimensionality reduction technique [22] to produce a two-dimensional space that is easy to visualize, v) add new layers with the inferred activities, and vi) color the cells by the inferred activity of any given circuit or pathway.

In this work, we have used the tools provided by the `Scanpy` library to perform Wilcoxon rank-sum tests to compare the inferred signaling activity across different groups of cells: IFN-β *stimulated* versus *control* (i.e. non-stimulated) cells. In all cases, p values have been corrected for multiple testing with False Discovery Rate (FDR) [4]. We report circuit/pathway names, the test scores (scores), FDR-adjusted p values (pvals_adj), and the log fold change (logFC) between the conditions.

2.5 Code Availability

The code required to train the networks and execute all of the analyses reported in this work can be found at https://github.com/babelomics/ivae_scorer.

3 Results

In this section, we present the results of our signaling scoring model trained with KEGG or Reactome as the source of *a priori* knowledge (see Sect. 2.3). All the models have been fitted using the same hyperparameter schema (see Sect. 2.3) to the dataset of PBMCs cells described in Sect. 2.1. The ground truth is labeled as *stimulated* or *control* to indicate if the cells have been treated with interferon-β (IFN-β) or not, respectively.

3.1 Pathway Activity at Single-Cell Resolution Using KEGG as Prior Knowledge

Here we present the results on the IFN-β perturbed dataset [14] (see Sect. 2.1) for the proposed VAE scoring model using the KEGG pathway database as the source of a priori information (see Fig. 2). Once the model has been fitted, we compute the circuit activities using the first hidden layer of the encoder (see Sect. 2.3) and use them to carry out a Wilcoxon rank-sums test between the *control* and *stimulated* cell groups. The results for the top-10 ranked circuits are summarized in Table 3. The nomenclature for the circuits is "pathway: effector genes".

As expected, the top-ranked circuits detected by our method capture parts of interferon-related signaling pathways. The list is dominated by the "Rig-I-like receptor" pathway since most of its circuits are represented, which could be driven by the fact that RIG-I-like receptors (RLRs) recruit particular intracellular adaptor proteins to activate signaling pathways that result in the creation of type I interferon, among other inflammatory cytokines. Moreover, the circuits that have IFN-α or IFN-β as effectors, namely the "Toll-like receptor signaling pathway: IFN-α" and "Toll-like receptor signaling pathway: IFN-β" circuits, are among the top-ranked circuits, which should be expected given the interferon-based treatment.

Table 3. Top 10 differentially activated KEGG circuits with respect to the *stimulated* versus *control* comparison

circuits	scores	pvals_adj	logFC
RIG-I-like receptor: MAVS TMEM173	104.57	0.00	2.63
RIG-I-like receptor: MAPK14	103.98	0.00	2.63
RIG-I-like receptor: MAPK8	103.82	0.00	2.62
RIG-I-like receptor: CHUK IKBKB IKBKG	100.95	0.00	2.34
RIG-I-like receptor: IRF7	100.26	0.00	2.46
RIG-I-like receptor: NFKB1	97.65	0.00	2.06
Natural killer cell mediated cytotoxicity: TNFRSF10D	62.89	0.00	0.55
RIG-I-like receptor: IRF3 PIN1	62.15	0.00	1.36
Cytosolic DNA-sensing pathway: TBK1	61.26	0.00	1.36
Toll-like receptor: IFNA1	58.35	0.00	1.37

One of the advantages of cell-level gene-set scoring mechanisms is that we can identify cells where specific gene sets are active [2], in our case signaling circuits/pathways. Although a complete analysis of the scoring performance across the different cell groups is beyond the scope of the present work, Fig. 2 shows the activity at the single-cell resolution of the major sources of variation across the *stimulated* and *control* groups. As outlined in Sect. 2.4 each axis represents a component of the UMAP reduction of the latent space learned by the VAE. The top sub-figures represent the cells grouped by either the condition (*control* or *stimulated*) or the cell type. Instead, the rest of the sub-figures show the change in the level of activity of each signaling circuit across the cells.

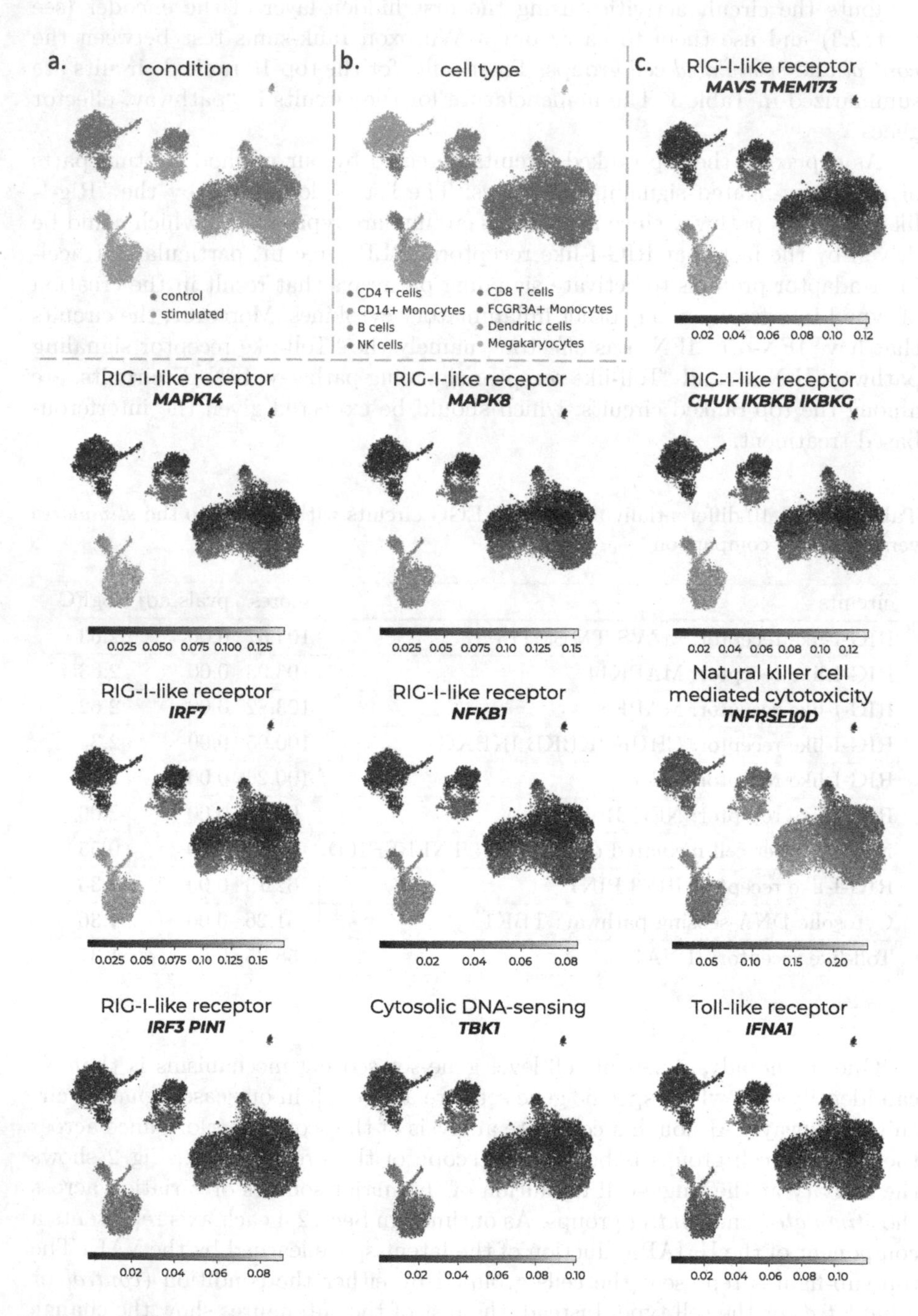

Fig. 2. Circuit activity scoring using the KEGG-informed VAE.

3.2 Pathway Activity at Single-Cell Resolution Using Reactome as Prior Knowledge

Following an analogous procedure to the one presented in the previous section for the KEGG database, we fitted a Reactome-informed VAE to the same IFN-β perturbed dataset [14].

Table 4 shows the top 10 deferentially activated Reactome pathways with respect to the *stimulated* versus *control* comparison following the Wilcoxon procedure (see Sect. 2.4). Given that a group of cells has been stimulated with IFN-β, it is expected that interferon-related pathways should be ranked higher in the *stimulated* cells with respect to the *control* population [12,14], as it is the case: the top expressed pathways in *stimulated* cells according to the Reactome-informed VAE include the "Interferon α, β Signaling", "Interferon Signaling", "Interferon Gamma Signaling", and "Antiviral Mechanism By IFN Stimulated Genes" pathways, among others.

Table 4. Top 10 differentially activated Reactome pathways with respect to the *stimulated* versus *control* comparison

pathways	scores	pvals_adj	logFC
Interferon α, β Signaling	109.32	0.00	3.75
Interferon Signaling	105.27	0.00	2.42
Cytokine Signaling In Immune System	104.96	0.00	2.43
Negative Regulators Of DDX58 IFIH1 Signaling	102.65	0.00	2.39
Antiviral Mechanism By IFN Stimulated Genes	101.54	0.00	2.38
NS1 Mediated Effects On Host Pathways	95.99	0.00	2.15
Interferon Gamma Signaling	88.25	0.00	0.97
OAS Antiviral Response	87.39	0.00	2.44
Post Translational Modification...	87.00	0.00	2.19
DNA Damage Bypass	84.06	0.00	1.83

Figure 3 shows the pathway activity scores at the single-cell level for the top-scored interferon-related pathways. The figure is composed as in the KEGG-informed case, the axes represent the UMAP reduction of the cell representation learned by the Reactome-informed VAE, we show the cell type and condition groups, as well as the activity of each top-ranked Reactome pathway across the cells.

The results of the Reactome-informed VAE are on par with those presented in the "Single-cell best practices" manual [12] where they use the same IFN-β perturbed dataset to assert the performance of the `decoupleR` tool [3] when inferring the activity of the pathways at the single-cell resolution by means of the `AUCell` method [2].

Furthermore, the difference at the signaling level between the *stimulated* and *control* condition is clearer and easier to interpret when using Reactome to inform the VAE since it includes a specific entity for the interferon signaling pathway.

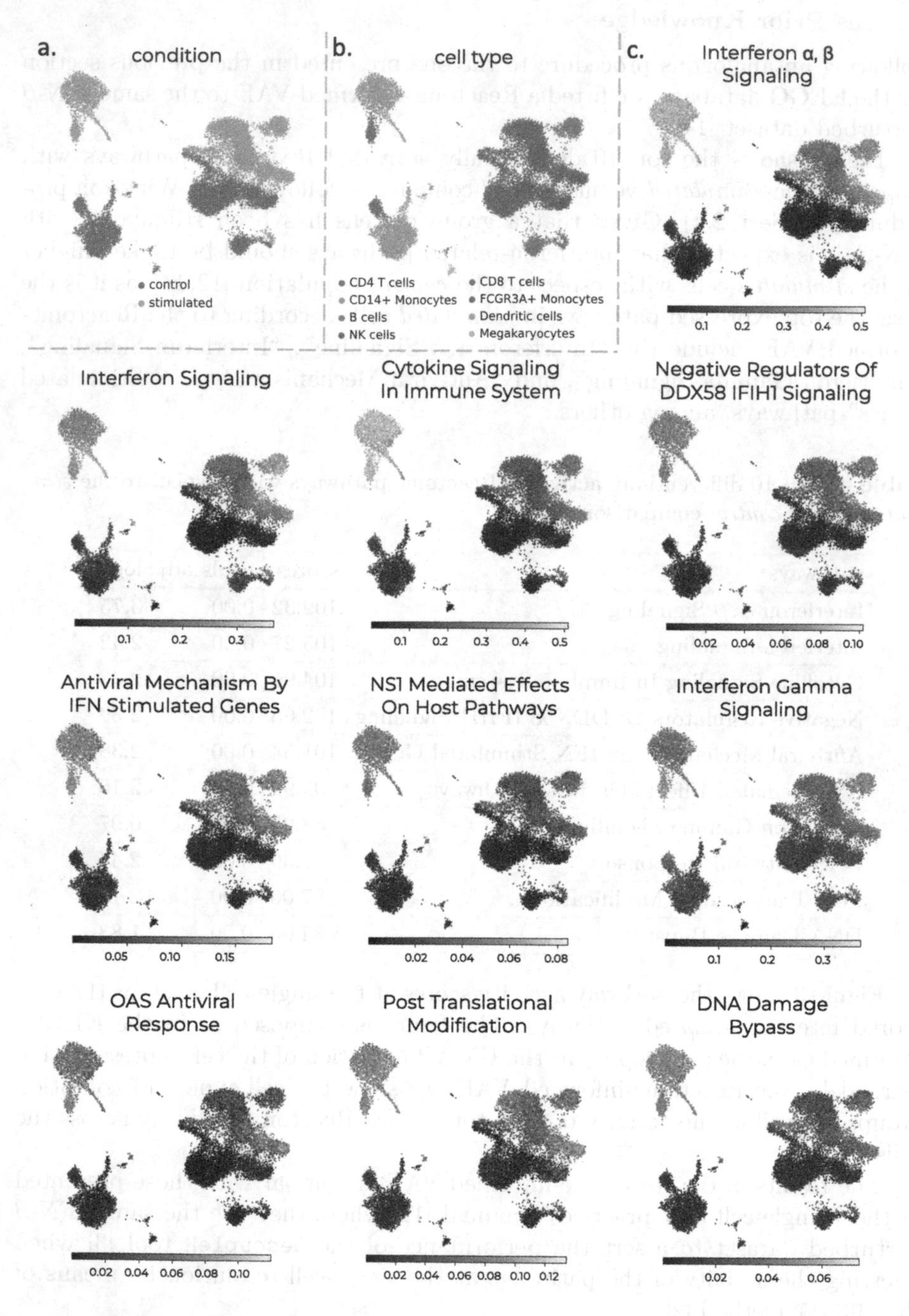

Fig. 3. Pathway activity scoring using the Reactome-informed VAE.

3.3 Gene-Set Precision

In this section, we use the previously fitted models (KEGG and Reactome informed VAE) to study the Influenza pathway, whose enrichment has been already described for the dataset under study in [14].

When using the KEGG-informed VAE we observe a significant alteration of the "Influenza A: IRF7" circuit (part of the "Influenza A" KEGG pathway) between the *control* and *stimulated* condition (FDR-adjusted p-value < 0.05, $\log FC > 1$). Interestingly, the "Influenza A: IRF7" circuit leads to the interferon-based T cell activation and antibody response. Figure 4 shows the "Influenza A: IRF7" circuit scores across the cells using the same representation schema of the previous section.

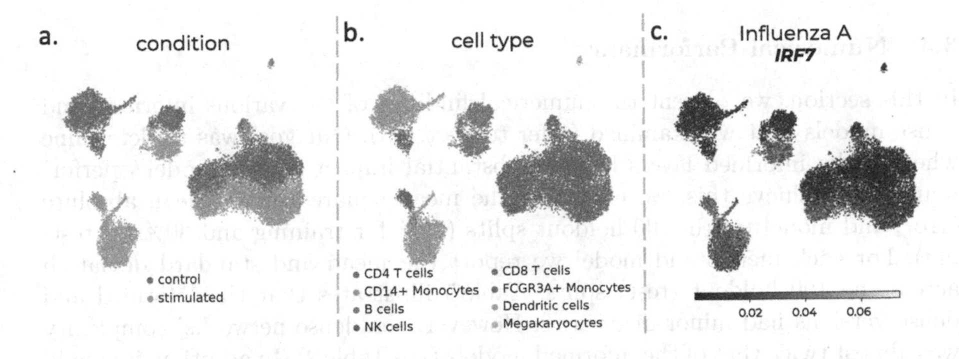

Fig. 4. Influenza circuit scoring using the KEGG-informed VAE.

However, when using the broader Reactome pathways to inform the VAE, the model losses the ability to capture the Influenza-based differences that require more precise gene sets. Instead, the model captures differences at the cell-type level, as can be seen in Fig. 5.

Note that, these differences in performance could not be related to differences in the intrinsic quality of the databases. Further research is needed since it is likely that the Influenza signal could be captured with the Reactome database by finding a way to decompose each pathway into functional subunits akin to what we have done with the KEGG database.

In our analysis, we discovered that out of the 7573 genes in the Reactome gene space after normalization, only 268 were involved in the most significant pathways that differentiate between the *stimulated* and *control* conditions (as shown in Table 4). Similarly, for the KEGG database, 60 genes out of 1820 were involved in the most significant circuits (as shown in Table 3). Interestingly, we found that only 24 genes were shared between the Reactome and KEGG most representative sets.

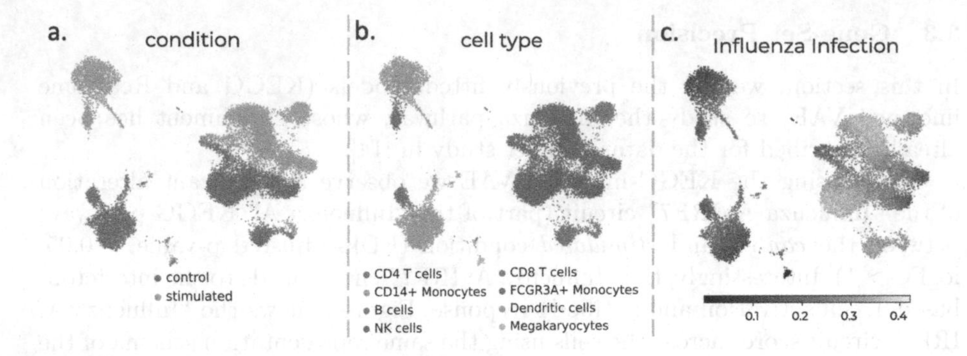

Fig. 5. Influenza pathway activity scoring using the Reactome-informed VAE.

3.4 Numerical Performance

In this section, we present the numerical findings of the various informed and dense models that we examined (refer to Sect. 2.3). Our goal was to determine whether the informed layers have a substantial impact on the model's performance. To achieve this, we measured the mean squared error, mean absolute error, and model loss in 100 holdout splits (70% for training and 30% for testing). For each metric and model we report the mean and standard deviation across the 100 holdout (test) splits. Table 5 illustrates that the informed and dense versions had minor disparities. However, the dense networks' complexity was almost twice that of the informed models (see Table 2). In addition, it should be noted that all the models are very stable with respect to each of the metrics, as evidenced by the low deviations.

Table 5. Comparison of informed VAE and its dense counterparts. For each model and metric we report the mean and the standard deviation.

		Integrated knowledge information			
		KEGG database	Reactome database	Dense (KEGG)	Dense (Reactome)
metrics	Model Loss	6.97675 ± 2.69E-02	31.49675 ± 3.10E-02	6.74558 ± 2.15E-02	30.72749 ± 2.94E-02
	Mean squared error	0.00357 ± 1.30E-05	0.00380 ± 8.00E-06	0.00337 ± 1.00E-05	0.00348 ± 0.00E+00
	Mean absolute error	0.01997 ± 3.80E-05	0.02150 ± 2.80E-05	0.01950 ± 3.30E-05	0.02209 ± 2.80E-05

4 Conclusions

This work provides an empirical evaluation of how unsupervised neural networks can be conditioned with a priori biological knowledge to infer cell-level pathway activities. Our proposed signaling-informed variational autoencoder provides results comparable to those found in the literature while being easily extensible to include other biological entities. However, as expected, our model works better if the knowledge base used to inform the wiring of the neural network includes entities that are better aligned with the underlying biological conditions.

Acknowledgements. This work has been partially supported by grants PID2020-117979RB-I00 and PID2020-117954RB-C22 from the Spanish Ministry of Science and Innovation, IMP/00019 from the Instituto de Salud Carlos III (ISCIII), PIP-0087-2021 from Junta de Andalucía, co-funded with European Regional Development Funds (ERDF); grant H2020 Programme of the European Union grants Marie Curie Innovative Training Network "Machine Learning Frontiers in Precision Medicine" (MLFPM) (GA 813533). The authors also acknowledge Junta de Andalucía for the postdoctoral contract of Carlos Loucera (PAIDI2020-DOC_00350) co-funded by the European Social Fund (FSE) 2014-2020.

References

1. Abadi, M., et al.: TensorFlow: large-scale machine learning on heterogeneous distributed systems, March 2016. https://doi.org/10.48550/arXiv.1603.04467
2. Aibar, S., et al.: SCENIC: single-cell regulatory network inference and clustering. Nat. Methods **14**(11), 1083–1086 (2017). https://doi.org/10.1038/nmeth.4463
3. Badia-i-Mompel, P., et al.: decoupleR: ensemble of computational methods to infer biological activities from omics data. Bioinf. Adv. **2**(1), vbac016 (2022). https://doi.org/10.1093/bioadv/vbac016
4. Benjamini, Y., Hochberg, Y.: Controlling the false discovery rate: a practical and powerful approach to multiple testing. J. Roy. Stat. Soc. Ser. B (Methodological) **57**(1), 289–300 (1995). https://doi.org/10.1111/j.2517-6161.1995.tb02031.x
5. Çubuk, C., Loucera, C., Peña-Chilet, M., Dopazo, J.: Crosstalk between metabolite production and signaling activity in breast cancer. Int. J. Mol. Sci. **24**(8), 7450 (2023). https://doi.org/10.3390/ijms24087450
6. Dash, T., Chitlangia, S., Ahuja, A., Srinivasan, A.: A review of some techniques for inclusion of domain-knowledge into deep neural networks. Sci. Rep. **12**(1), 1040 (2022). https://doi.org/10.1038/s41598-021-04590-0
7. Gillespie, M., et al.: The Reactome pathway knowledgebase 2022. Nucleic Acids Res. **50**(D1), D687–D692 (2022). https://doi.org/10.1093/nar/gkab1028
8. Graziani, M., et al.: A global taxonomy of interpretable AI: unifying the terminology for the technical and social sciences. Artif. Intell. Rev. **56**(4), 3473–3504 (2023). https://doi.org/10.1007/s10462-022-10256-8
9. Gundogdu, P., Alamo, I., Nepomuceno-Chamorro, I.A., Dopazo, J., Loucera, C.: SigPrimedNet: a signaling-informed neural network for scRNA-seq annotation of known and unknown cell types. Biology **12**(4), 579 (2023). https://doi.org/10.3390/biology12040579
10. Gundogdu, P., Loucera, C., Alamo-Alvarez, I., Dopazo, J., Nepomuceno, I.: Integrating pathway knowledge with deep neural networks to reduce the dimensionality in single-cell RNA-seq data. BioData Mining **15**(1), 1 (2022). https://doi.org/10.1186/s13040-021-00285-4
11. Harris, C.R., et al.: Array programming with NumPy. Nature **585**(7825), 357–362 (2020). https://doi.org/10.1038/s41586-020-2649-2
12. Heumos, L., et al.: Best practices for single-cell analysis across modalities. Nat. Rev. Genet. (2023). https://doi.org/10.1038/s41576-023-00586-w

13. Hidalgo, M.R., Cubuk, C., Amadoz, A., Salavert, F., Carbonell-Caballero, J., Dopazo, J.: High throughput estimation of functional cell activities reveals disease mechanisms and predicts relevant clinical outcomes. Oncotarget **8**(3), 5160–5178 (2016). https://doi.org/10.18632/oncotarget.14107
14. Kang, H.M., et al.: Multiplexed droplet single-cell RNA-sequencing using natural genetic variation. Nat. Biotechnol. **36**(1), 89–94 (2018). https://doi.org/10.1038/nbt.4042
15. Kingma, D.P., Ba, J.: Adam: a method for stochastic optimization, January 2017. https://doi.org/10.48550/arXiv.1412.6980
16. Kuenzi, B.M., et al.: Predicting drug response and synergy using a deep learning model of human cancer cells. Cancer Cell **38**(5), 672-684.e6 (2020). https://doi.org/10.1016/j.ccell.2020.09.014
17. Lähnemann, D., et al.: Eleven grand challenges in single-cell data science. Genome Biol. **21**(1), 31 (2020). https://doi.org/10.1186/s13059-020-1926-6
18. Levine, J.H., et al.: Data-driven phenotypic dissection of AML reveals progenitor-like cells that correlate with prognosis. Cell **162**(1), 184–197 (2015). https://doi.org/10.1016/j.cell.2015.05.047
19. Li, C., et al.: SciBet as a portable and fast single cell type identifier. Nat. Commun. **11**(1), 1818 (2020). https://doi.org/10.1038/s41467-020-15523-2. https://www.nature.com/articles/s41467-020-15523-2, bandiera_abtest: a Cc_license_type: cc_by Cg_type: Nature Research Journals Number: 1 Primary_atype: Research Publisher: Nature Publishing Group Subject_term: Machine learning;Transcriptomics Subject_term_id: machine-learning;transcriptomics
20. Lotfollahi, M., et al.: Biologically informed deep learning to query gene programs in single-cell atlases. Nat. Cell Biol. **25**(2), 337–350 (2023). https://doi.org/10.1038/s41556-022-01072-x
21. Ma, J., et al.: Using deep learning to model the hierarchical structure and function of a cell. Nat. Methods **15**(4), 290–298 (2018). https://doi.org/10.1038/nmeth.4627
22. McInnes, L., Healy, J., Melville, J.: UMAP: uniform manifold approximation and projection for dimension reduction, September 2020. https://doi.org/10.48550/arXiv.1802.03426
23. Ogata, H., Goto, S., Sato, K., Fujibuchi, W., Bono, H., Kanehisa, M.: KEGG: Kyoto encyclopedia of genes and genomes. Nucleic Acids Res. **27**(1), 29–34 (1999). https://doi.org/10.1093/nar/27.1.29
24. Petegrosso, R., Li, Z., Kuang, R.: Machine learning and statistical methods for clustering single-cell RNA-sequencing data. Brief. Bioinform. **21**(4), 1209–1223 (2020). https://doi.org/10.1093/bib/bbz063
25. Regev, A., et al.: Human cell atlas meeting participants: the human cell atlas. eLife **6**, e27041 (2017). https://doi.org/10.7554/eLife.27041
26. Traag, V., Waltman, L., van Eck, N.J.: From Louvain to Leiden: guaranteeing well-connected communities. Sci. Rep. **9**(1), 5233 (2019). https://doi.org/10.1038/s41598-019-41695-z
27. Virshup, I., et al.: The scverse project provides a computational ecosystem for single-cell omics data analysis. Nat. Biotechnol., 1–3 (2023). https://doi.org/10.1038/s41587-023-01733-8
28. Virshup, I., Rybakov, S., Theis, F.J., Angerer, P., Wolf, F.A.: Anndata: annotated data, December 2021. https://doi.org/10.1101/2021.12.16.473007
29. Virtanen, P., et al.: SciPy 1.0: fundamental algorithms for scientific computing in Python. Nat. Methods **17**(3), 261–272 (2020). https://doi.org/10.1038/s41592-019-0686-2

30. Wang, J., Zou, Q., Lin, C.: A comparison of deep learning-based pre-processing and clustering approaches for single-cell RNA sequencing data. Briefings Bioinf. **23**(1), bbab345 (2022). https://doi.org/10.1093/bib/bbab345
31. Way, G.P., Greene, C.S.: Discovering pathway and cell type signatures in transcriptomic compendia with machine learning. Ann. Rev. Biomed. Data Sci. **2**(1), 1–17 (2019). https://doi.org/10.1146/annurev-biodatasci-072018-021348
32. Wolf, F.A., Angerer, P., Theis, F.J.: SCANPY: large-scale single-cell gene expression data analysis. Genome Biol. **19**(1), 15 (2018). https://doi.org/10.1186/s13059-017-1382-0
33. Zappia, L., Theis, F.J.: Over 1000 tools reveal trends in the single-cell RNA-seq analysis landscape. Genome Biol. **22**(1), 301 (2021). https://doi.org/10.1186/s13059-021-02519-4
34. Zhao, Y., Shao, J., Asmann, Y.W.: Assessment and optimization of explainable machine learning models applied to transcriptomic data. Genomics Proteomics Bioinf. **20**(5), 899–911 (2022). https://doi.org/10.1016/j.gpb.2022.07.003

On Estimating Derivatives of Input Signals in Biochemistry

Mathieu Hemery and François Fages(✉)

Lifeware Project-Team, Inria Saclay, Palaiseau, France
{mathieu.hemery,Francois.Fages}@inria.fr

Abstract. The online estimation of the derivative of an input signal is widespread in control theory and engineering. In the realm of chemical reaction networks (CRN), this raises however a number of specific issues on the different ways to achieve it. A CRN pattern for implementing a derivative block has already been proposed for the PID control of biochemical processes, and proved correct using Tikhonov's limit theorem. In this paper, we give a detailed mathematical analysis of that CRN, thus clarifying the computed quantity and quantifying the error done as a function of the reaction kinetic parameters. In a synthetic biology perspective, we show how this can be used to compute online functions with CRNs augmented with an error correcting delay for derivatives. In the systems biology perspective, we give the list of models in BioModels containing (in the sense of subgraph epimorphisms) the core derivative CRN, most of which being models of oscillators and control systems in the cell, and discuss in detail two such examples: one model of the circadian clock and one model of a bistable switch.

1 Introduction

Sensing the presence of molecular compounds in a cell compartment is a necessary task of living cells to maintain themselves in their environment, and achieve high-level functions as the result of low-level processes of basic biomolecular interactions. The formalism of chemical reaction networks (CRN) [11] is both a useful abstraction to describe such complex systems in the perspective of systems biology [18], and a possible molecular programming language in the perspective of synthetic biology [8,23].

Sensing the concentration levels of molecular compounds has been well-studied in the domain of signal transduction networks. For instance, the ubiquitous CRN structure of MAPK signaling networks has been shown to provide a way to implement analog-digital converters in our cells, by transforming a continuous input signal, such as the concentration of an external hormone activating membrane receptors, into an almost all-or-nothing output signal according to some threshold value of the input, i.e. using a stiff sigmoid as dose-response input-output function [17].

The analysis of input/output functions fits well with the computational theory of CRNs. In particular, the Turing-completeness result shown in [8] for

J. Pang and J. Niehren (Eds.): CMSB 2023, LNBI 14137, pp. 78–96, 2023.
https://doi.org/10.1007/978-3-031-42697-1_6

the interpretation by Ordinary Differential Equations (ODE) of CRNs, possibly restricted to elementary CRNs using mass-action law kinetics and at most bimolecular reactions, demonstrates the generality of this approach to biomolecular programming. Furthermore, it comes with an algorithm to automatically generate a finite CRN for implementing any computable real function. Such a compiler is implemented in our CRN modeling software BIOCHAM [4] in several forms, including a theoretically more limited but practically more interesting framework for robust *online computation* [14].

Sensing the derivative of an input molecular concentration is nevertheless beyond the scope of this computational paradigm since it assumes that the input molecular concentrations are stabilized at some fixed values which makes no sense for computing the derivative. Furthermore, it is well-known that the derivative of a computable real function is not necessarily computable [20]. We must thus content ourselves with *estimating* the derivative of an input with some error, instead of *computing* it with arbitrary precision as computability theory requires.

In control theory and engineering, online estimations of input signal derivatives are used in many places. Proportional Integral Derivative (PID) controllers adjust a target variable to some desired value by monitoring three components: the error, that is the difference between the current value and the target, its integral over a past time slice, and its current derivative. The derivative term can improve the performance of the controller by avoiding overshoots and solving some problematic cases of instability.

Following early work on the General Purpose Analog Computer (GPAC) [22], the integral terms can be implemented with CRNs using simple catalytic synthesis reactions such as $A \rightarrow A + B$ for integrating A over time, indeed $B(T) = \int_O^T A(t)dt$. Difference terms can be implemented using the annihilation reaction $A_+ + A_- \rightarrow \emptyset$ which is also used in [7,8,21] to encode negative values by the difference of two molecular concentrations, i.e. dual-rail encoding. This is at the basis of the CRN implementations of, for instance, antithetic PI controllers presented in [3].

For the CRN implementation of PID controllers, to the best of our knowledge three different CRN templates have been proposed to estimate derivative terms. The first one by Chevalier et al. [5] is inspired by bacteria's chemotaxis, but relies on strong restrictions upon the parameters and the structure of the input function making it apparently limited in scope. A second one proposed by Alexis et al. [1] uses tools from signal theory to design a derivative circuit with offset coding of negative values and to provide analytic expressions for its response. The third one developed by Whitby et al. [24] is practically similar in its functioning to the one we study here, differing only on minor implementation details, and proven correct through Tikhonov's limit theorem. This result ensures that when the appropriate kinetic rates tend to infinity, the output is precisely the derivative of the input.

In this paper, we give a detailed mathematical analysis of that third derivative CRN and quantify the error done as a function of the reaction kinetic parameters, by providing a first-order correction term. We illustrate the precision of this analysis on several examples, and show how this estimation of the derivative can be actively used to compute elementary mathematical functions online with CRNs augmented with an error-correcting delay. Furthermore, we compare our core derivative CRN to the CRN models in the curated part of BioModels.net model repository. For this, we use the theory of subgraph epimorphisms (SEPI) [12,13] and its implementation in BIOCHAM [4], to identify the models in BioModels which contain the derivative CRN structure. We discuss with some details the SEPIs found on two such models: `biomodels 170`, one of the smallest eukaryotes circadian clock model [2], and `biomodels 318`, a model of the bistable switch at the restriction point of the cell cycle [25].

The rest of the article is organized as follow. In Sect. 2, we provide some preliminaries on CRNs and their interpretation by ODEs. We present the core differentiation CRN in Sect. 3, in terms of both of some of its different possible biological interpretations, and of its mathematical properties. Section 4 develops the mathematical analysis to bound the error done by that core CRN, and in Sect. 5 we give some examples to test the validity of our estimation and its use to compute functions online with CRNs augmented with an error-correcting delay. Section 6 is then devoted to the search of that derivative CRN pattern in BioModels repository and the analysis of those matching in two cases. Finally, we conclude on the perspectives of our approach to both CRN design at an abstract mathematical level, and comparison to natural CRNs to help understanding their functions.

2 Preliminaries on CRNs

2.1 Reactions and Equations

The CRN formalism allows us to represent the molecular interactions that occur on a finite set of molecular compounds or species, $\{X_i\}_{i\in 1\ldots n}$, through a finite set of formal (bio)chemical reactions, without prejudging their interpretation in the differential, stochastic, Petri Net and Boolean semantics hierarchy [10]. Each reaction is a triplet (R, P, f), also written $R \xrightarrow{f} P$, where R and P are multisets of respectively reactant and product species in $\{X_i\}$, and $f : \mathbb{R}^n_+ \mapsto \mathbb{R}_+$ is a kinetic rate function of the reactant species. A CRN is thus entirely described by the two sets of n species and m reactions: $\{X_i\}, \{R_s \xrightarrow{f_s} P_s\}$.

The differential semantics of a CRN associates positive real valued molecular concentrations, also noted X_i by abuse of notation, and the following ODEs which define the time evolution of those concentrations:

$$\frac{dX_i}{dt} = \sum_{s\in S}(P_s(X_i) - R_s(X_i))f_s(X), \tag{1}$$

where $P_s(X_i)$ (resp. $R_s(X_i)$) denotes the multiplicity (stoichiometry) of X_i in the multiset of products (resp. reactants) of reaction s.

In the case of a mass action law kinetics, the rate function is a monomial, $f_s = k_s \prod_{x \in R_s} x$, composed of the product of the concentrations of the reactants by some positive constant k_s. If all reactions have mass action law kinetics, we write the rate constant in place of the rate function $R \xrightarrow{k} P$, and the differential semantics of the CRN is defined by a Polynomial Ordinary Differential Equation (PODE).

From the point of view of the computational theory of CRNs, there is no loss of generality to restrict ourselves to elementary CRNs composed of at most bimolecular reactions with mass action law kinetics. Indeed, [8] shows that any computable real functions (in the sense of computable analysis, i.e. with arbitrary finite precision by a Turing machine), can be computed by such a CRN, using the dual-rail encoding of real values by the difference of molecular concentrations, $x = X_+ - X_-$. While our compiler ensures that the quantity $X_+ - X_-$ behaves properly, it is also important to degrade both of them with an annihilation reaction, $X_+ + X_- \xrightarrow{fast} \emptyset$, to avoid a spurious increase of their concentration. Those annihilation reactions are supposed to be faster than the other reactions of the CRN.

Example 1. The first example given in [8] showed the compilation of the cosine function of time, $y = cos(t)$ in the following CRN:

$$\begin{array}{lll} A_p \rightarrow A_p + y_p & A_m \rightarrow A_m + y_m & A_m(0) = 0,\ A_p(0) = 0 \\ y_m \rightarrow A_p + y_m & y_p \rightarrow A_m + y_p & y_m(0) = 0,\ y_p(0) = 1 \\ y_m + y_p \xrightarrow{fast} \emptyset & A_m + A_p \xrightarrow{fast} \emptyset & \end{array} \quad (2)$$

The last two reactions are necessary to avoid an exponential increase of the species concentration. The associated PODE is:

$$\begin{array}{ll} d(A_m)/dt = y_p - fast * A_m * A_p & A_m(0) = 0 \\ d(A_p)/dt = y_m - fast * A_m * A_p & A_p(0) = 0 \\ d(y_m)/dt = A_m - fast * y_m * y_p & y_m(0) = 0 \\ d(y_p)/dt = A_p - fast * y_m * y_p & y_p(0) = 1 \end{array} \quad (3)$$

2.2 CRN Computational Frameworks

The notions of CRN computation proposed in [8] and [14] for computing input/output functions, do not provide however a suitable framework for computing derivative functions. Both rely on a computation at the limit, meaning that the output converges to the result of the computation whenever the CRN is either properly initialized [8], or the inputs are stable for a sufficient period of time [14]. To compute a derivative, we cannot ask that the input stay fixed for any period of time as this would imply a null derivative. We want the output to follow "at run time" the derivative of the input.

Our question is thus as follows. Given an input species X following a time course imposed by the environment $X(t)$, is it possible to perform an online computation such that we can approximate the derivative $\frac{dX}{dt}$ on the concentration of 2 output species using a dual-rail encoding?

The idea is to approximate the left derivative by getting back to its very mathematical definition:

$$\frac{dX}{dt}(t) = \lim_{\epsilon \to 0+} \frac{X(t) - X(t-\epsilon)}{\epsilon}, \tag{4}$$

but how can we measure $X(t-\epsilon)$?

3 Differentiation CRN

3.1 Biological Intuition Using a Membrane

One biological intuition we may have to measure a value in a previous time is to use a membrane with a fast diffusive constant. Indeed, if we suppose that the input is the outside species, the inside species equilibrates to follow the concentration of the outside one (the input) but also suffers a lag due to the diffusion. Building upon this simple trick leads to the CRN presented in Fig. 1. As the derivative may be positive or negative, a dual-rail encoding is used for the derivative. This CRN is mainly equivalent to the derivative block proposed in [24] apart from the fact that we suppose (for the sake of clarity) that the input stay positive and no dual-rail encoding is used for it. In the case of a dual-rail encoded input, the two species need to have the same permeability through the membrane, otherwise the delay is not the same for the positive and negative parts.

The delay is thus introduced through a membrane under the assumption that the outside concentration is imposed by the environment. This conveniently explains why the kinetic rates are the same for the two monomials in the derivative of X_{in}, but this is not mandatory. Indeed two other settings can be used to construct such a CRN without relying on a membrane. We could use a phosphorylation and a dephosphorylation reactions where X_{in} would be the phosphorylated species. Or we could, as in [24], rely on a catalytic production of X_{in} by X_{ext} and a degradation reaction of X_{in}. A drawback of these two other implementations is that they need to be tuned to minimize the difference between the rates of the two monomials in the derivative of X_{in}. Otherwise a proportional constant is introduced between X_{ext} and X_{in}, and needs to be corrected by adjusting the production rates of D_+ and D_-.

However, the membrane implementation also has its own drawback as it requires the reaction $X_{ext} \rightarrow X_{ext} + D_+$ to occur through the membrane. We may think of a membrane protein M that mediates this reaction ($X_{ext} + M \rightarrow X_{ext} + M + D_+$). Then, since its concentration is constant, it can simply be wrap up in the kinetic constant of the reaction. Which of this three implementations should be chosen may depend on the exact details of the system to be build.

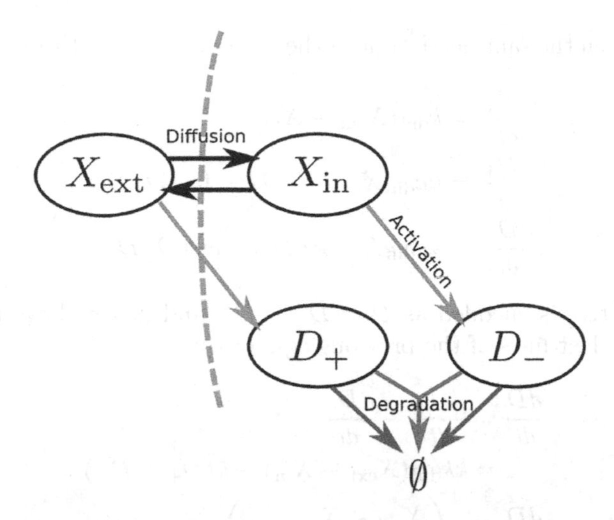

Fig. 1. Hypergraph representation of the core differentiation CRN composed of one input, two outputs and one intermediate species. The input X_{ext} is outside of the membrane and thus present in so large quantity that its concentration is not modified by the dynamics, once it crossed the membrane it is labelled as X_{in}. Each species X_{ext} (resp. X_{in}) activates the synthesis of its part of the input: D_+ (resp. D_-). Finally a fast annihilation reaction eliminates both D_+ and D_- so that only the highest of the two remains present, depending on the sign of the derivative of X_{ext}.

3.2 Core Differentiation CRN

Our core differentiation CRN schematized in Fig. 1 is more precisely composed of the following 7 reactions:

$$\begin{aligned}
&X_{ext} \xrightarrow{k_{diff}} X_{in} && X_{in} \xrightarrow{k_{diff}} X_{ext} \\
&X_{ext} \xrightarrow{k.k_{diff}} X_{ext} + D_+ && X_{in} \xrightarrow{k.k_{diff}} X_{in} + D_- \\
&D_+ \xrightarrow{k} \emptyset && D_- \xrightarrow{k} \emptyset \\
&D_+ + D_- \xrightarrow{fast} \emptyset &&
\end{aligned} \tag{5}$$

The diffusion through the membrane is symmetrical with a constant k_{diff} and both activations should have the same constant product $k.k_{diff}$ while the degradation of the outputs should have a rate k. We make the assumption that the outside species X_{ext} is present in large quantity so that its concentration is not affected by the dynamics of the CRN. Under this assumption, the differential semantics is then the same as the one of the differentiation CRN proposed in [24]:

$$\begin{aligned} \frac{dX_{\text{in}}}{dt} &= k_{\text{diff}}(X_{\text{ext}} - X_{\text{in}}) \\ \frac{dD_+}{dt} &= kk_{\text{diff}}X_{\text{ext}} - kD_+ - \text{fast}D_+D_- \\ \frac{dD_-}{dt} &= kk_{\text{diff}}X_{\text{in}} - kD_- - \text{fast}D_+D_- \end{aligned} \tag{6}$$

The derivative is encoded as $D = D_+ - D_-$ and hence obeys the equation (using the two last lines of the previous equation):

$$\begin{aligned} \frac{dD}{dt} &= \frac{dD_+}{dt} - \frac{dD_-}{dt} \\ &= kk_{\text{diff}}(X_{\text{ext}} - X_{\text{in}}) - k(D_+ - D_-) \\ \frac{dD}{dt} &= k\left(\frac{X_{\text{ext}} - X_{\text{in}}}{\frac{1}{k_{\text{diff}}}} - D\right) \end{aligned} \tag{7}$$

In the next section, we prove that X_{in} is equal to X_{ext} with a delay ϵ, hence giving us our second time point $X(t-\epsilon)$, up to the first order in $\epsilon = \frac{1}{k_{\text{diff}}}$. The fractional part of the last equation is thus precisely an estimate of the derivative of X_{ext} as defined in Eq. 4, with a finite value for ϵ.

It is also worth remarking that such derivative circuits can in principle be connected to compute higher-order derivatives, with a dual-rail encoded input. It is well known that such estimations of higher-order derivatives can be very sensitive to noise and error, and are thus not reliable for precise computation but may be good enough for biological purposes. We will see a biological example of this kind in Sect. 6.2 on a simple model of the circadian clock.

4 Mathematical Analysis of the Quality of the Estimation

Our first goal is to determine precisely the relation between X_{in} and X_{ext} when the later is enforced by the environment. Using the first line of Eq. 6, we obtain by symbolic integration:

$$X_{\text{in}}(t) = k_{\text{diff}} \int_0^{\infty} \exp(-k_{\text{diff}}s) X_{\text{ext}}(t-s)ds, \tag{8}$$

where we can see that X_{in} is the convolution of X_{ext} with a decreasing exponential. This convolution is not without reminding the notion of *evaluation* in the theory of distribution and has important properties of regularisation of the input function. In particular, whatever the input function is, this ensures that the internal representation is continuous and differentiable.

The interesting limit for us is when $k_{\text{diff}} \to \infty$, that is when $\epsilon = \frac{1}{k_{\text{diff}}} \to 0$. In this case, the exponential is neglectable except in a neighbourhood of the current time and supposing that X_{ext} is infinitely differentiable[1], we obtain by Taylor expansion:

$$\begin{aligned} X_{\text{in}}(t) &= \int_0^\infty k_{\text{diff}} \exp(-k_{\text{diff}} s) \sum_{n=0}^{\infty} \frac{(-s)^n}{n!} X_{\text{ext}}^{(n)}(t) ds \\ &= \sum_{n=0}^{\infty} \frac{k_{\text{diff}}}{n!} X_{\text{ext}}^{(n)}(t) \int_0^\infty (-s)^n \exp(-k_{\text{diff}} s) ds \end{aligned} \tag{9}$$

The integral may be evaluated separately using integration by parts and recursion:

$$\begin{aligned} I_n &= \int_0^\infty (-s)^n \exp(-k_{\text{diff}} s) ds = -n\epsilon I_{n-1} \\ &= (-1)^n (\epsilon)^{n+1} n! \end{aligned} \tag{10}$$

We thus have:

$$\begin{aligned} X_{\text{in}}(t) &= \sum_{n=0}^{\infty} \frac{k_{\text{diff}}}{n!} X_{\text{ext}}^{(n)}(t) (-1)^n n! \epsilon^{n+1} \\ &= \sum_n (-\epsilon)^n X_{\text{ext}}^{(n)}(t) \\ &= X_{\text{ext}}(t) - \epsilon X'_{\text{ext}}(t) + \epsilon^2 X''_{\text{ext}}(t) + \ldots \\ X_{\text{in}}(t) &= X_{\text{ext}}(t - \epsilon) + o(\epsilon^2). \end{aligned} \tag{11}$$

Using Taylor expansion once again in the last equation somehow formalizes our intuition: the concentration of the internal species X_{in} follows the time course of the external one with a delay equal to the inverse of the diffusive constant k_{diff}. This validates our formulation of the derivative.

Now, it is sufficient to remark that Eq. 7 has exactly the same form as the first line of Eq. 6 that we just study in length. Just replace X_{ext} by the estimation of the left derivative, X_{ext} by the output D and the rate constant k instead of k_{diff}. The delay approximation is thus also possible in this step and, introducing the delay $\tau = \frac{1}{k}$, we immediately obtain a precise expression for D:

$$D(t) = \frac{X_{\text{ext}}(t-\tau) - X_{\text{ext}}(t-\epsilon-\tau)}{\epsilon} + o(\epsilon) + o(\tau^2). \tag{12}$$

We can see this as the secant approximation of the derivative of X_{ext} with a step size ϵ and a delay τ. Moreover we also know that the residual error on this expression are of first order in ϵ and second order in τ.

It is well known in the field of numerical computation that the secant method provides a rather poor approximation, but it has the benefit to be the simplest one, and thus gives here a small size derivative circuit. In the hope of improving

[1] We also explore in Figs. 2D and 3C what a non analyticity of X_{ext} imply for our model.

the precision, one could implement higher-order methods using several "membranes" to access the value of the function on several time points before performing the adapted computation. Increasing the complexity would however also increase the delay between the input and output function.

5 Validation on Simple Examples

5.1 Verification of the Delay-Approximation

In this first subsection, we want to validate the approximation expressed by Eq. 11. For this, we focus on the diffusion part of our CRN: $X_{ext} \leftrightarrow X_{in}$. We make numerical simulation for 2 different values of ϵ and 2 different input functions: a sine wave and an absolute value signals. The second allowing us to see how well the delay approximation works in presence of non analyticity.

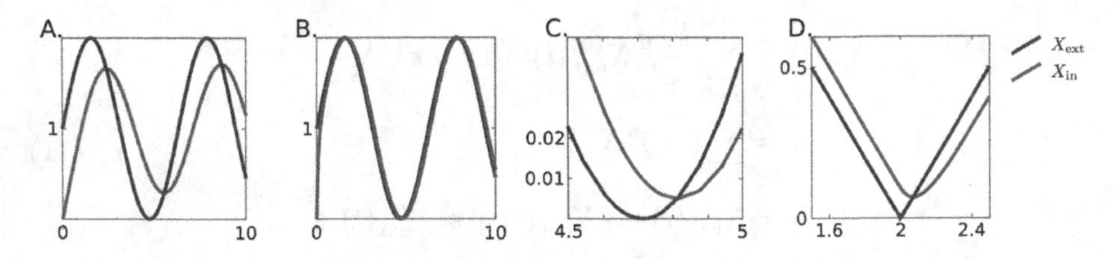

Fig. 2. Behaviour of the simple diffusion model when fed with different input functions. We have $k_{diff} = 1$ in panel **A** and $k_{diff} = 10$ in the three others. The input function of the three first panels is an offset sine wave: $X_{ext}(t) = 1 + \sin(t)$. Panel **C** is a focus on part of the panel **B** and Panel **D** present as input a shifted absolute value: $X_{ext}(t) = |t - 2|$ as a study case of non-differentiable function. See the main text for the discussion.

Figure 2 shows the response of X_{in} in that different condition. In panel **A**, the kinetic constant is very low so we expect our approximation to fail. Indeed, one can see that in addition to having an important delay, the output is strongly smoothed, this tends to average the variation of the input, bringing back X_{in} to the average value of the input. In panel **B** the diffusion constant is increased by a factor 10. The delay approximation is now very good and we only expect an error of order $\epsilon^2 = 10^{-2}$ which can be checked with good accuracy on panel **C**. Panel **D** shows a case of a non-differentiable function in which an error of order $\epsilon = 0.1$ is visible shortly after the discontinuity and vanishes in a similar timescale.

5.2 Approximation of the Derivative

Let us now check the behaviour of the derivative circuit. On Fig. 3, we can see the response of our derivative circuit for a sine wave and an absolute value input functions. In panels **A** and **B** we see that when the first and second order derivatives of the input are smaller than the kinetic reaction rates, the delay approximation gives a very good picture of the response. On a complementary point of view, the panel **C** shows that in front of singularity, the system adapts after an exponential transient phase with a characteristic time $\tau = \frac{1}{k}$.

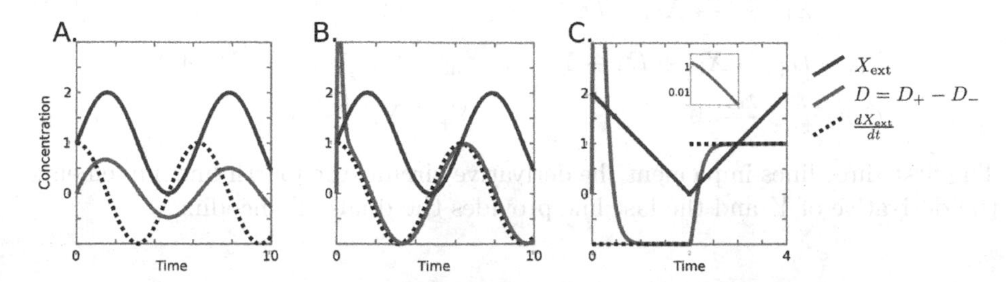

Fig. 3. Simulation of the derivative circuit for $k_{\text{diff}} = k = 1$ (panel **A**) and $k_{\text{diff}} = k = 10$ (panels **B** and **C**) when fed with two different input signal: a sine wave (panels **A** and **B**) and an offset absolute value (panel **C**). In **A** the characteristic time of the diffusion is too large ($\epsilon = 1$) so that the delay approximation fail. In **B** $\epsilon = 10^{-1}$, the approximation is sounded and the output $D = D_+ - D_-$ correctly follows a cosine (in dark) with a delay of $\tau = 10^{-1}$. In **C** the singularity makes the derivative harder to compute. As for the diffusion, we see a typical decreasing exponential toward its correct value as shown in the inset that depict the absolute difference between D and its correct value (1 in this case) in logarithmic scale for the time between 2 and 3. (That is precisely the same time as the main figure where the inset is placed.)

5.3 Using Signal Derivatives for Online Computations

Our main motivation for analyzing the differentiation CRN is to compute a function f of some unknown input signal, $X_{\text{ext}}(t)$, online. That is, given a function f, to compute the function $f(X_{\text{ext}}(t))$. Yet the differentiation CRN only allows us to approximate the derivative of the input signal. The idea is thus to implement the PIVP:

$$\frac{dY}{dt} = f'(X_{\text{ext}}(t))\frac{dX_{\text{ext}}}{dt}, \qquad Y(0) = f(X(0)) \tag{13}$$

and provides the result online on a set of internal species $Y(t) = Y_+ - Y_-$. This necessitates to compute the function f' and estimate the derivative of the input. Using the formalism developed in [15,16] we know that there exist an elementary CRN (i.e. quadratic PODE) computing $f'(X_{\text{ext}})$ for any elementary function f and we just have shown that $\frac{dX_{\text{ext}}}{dt}$ can be approximated by the differentiation CRN. Therefore, in principle, any elementary function of input signals can be approximated online by a CRN.

As a toy example, let us consider the square function, $\frac{dY}{dt} = 2X_{\text{ext}}(D_+ - D_-)$, and as input, a sine wave offset to stay positive : $X_{\text{ext}}(t) = 1 + \sin(t)$.

The CRN generated by BIOCHAM according to these principles, to compute the square of the input online is:

$$\begin{array}{ll}
X_{\text{ext}} \xrightarrow{k_{\text{diff}}} X_{\text{in}}, & X_{\text{in}} \xrightarrow{k_{\text{diff}}} X_{\text{ext}}, \\
X_{\text{ext}} \xrightarrow{k_{\text{diff}}k} X_{\text{ext}} + D_+ & D_+ \xrightarrow{k} \emptyset \\
X_{\text{in}} \xrightarrow{k_{\text{diff}}k} X_{\text{in}} + D_- & D_- \xrightarrow{k} \emptyset \\
X_{\text{in}} + D_+ \xrightarrow{2} X_{\text{in}} + D_+ + Y_+ & X_{\text{in}} + D_- \xrightarrow{2} X_{\text{in}} + D_- + Y_- \\
D_+ + D_- \xrightarrow{\text{fast}} \emptyset & Y_+ + Y_- \xrightarrow{\text{fast}} \emptyset
\end{array} \tag{14}$$

The first three lines implement the derivative circuit, the fourth line implements the derivative of Y and the last line provides the dual-rail encoding.

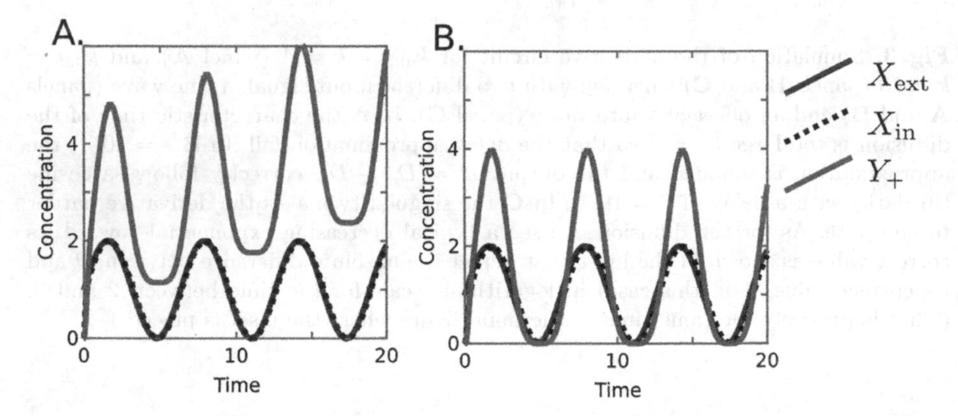

Fig. 4. Square computation CRN when fed with an offset sine wave $X_{\text{ext}}(t) = 1+\sin(t)$. The parameters are: $k_{\text{diff}} = k = 10, \text{fast} = 10^6$. Panel **A** shows a simulation of the naive CRN while panel **B** shows the corrected one where the derivative is integrated using a delayed input which eliminates the drift presented in the first panel.

The numerical simulation of this CRN is depicted in Fig. 4**A** One can see that while it effectively computes the square of the input, it also suffers from a strong drift. To verify if this drift comes from the delay between the input and the output, we can compute analytically the output of our network with our approximation of derivative with a delay (see the full computation in Appendix).

$$\begin{aligned}
y(t) &= \int 2x(s)x'(s-\tau)ds \\
&\simeq (1+\sin(t))^2 + \tau t.
\end{aligned} \tag{15}$$

This is precisely the behaviour that can be seen on the time course of Fig. 4 **A**. After the integration of 20 time units, the offset is of order 2 which is exactly what is predicted for a delay $\tau = \frac{1}{k} = 0.1$. Therefore, while it is always possible to get rid of such errors by increasing k_{diff}, the identification of the cause of the drift, gives us a potentially simpler path to eliminate it: using a representation of the input that is itself delayed: $X_{\text{in}} \leftrightarrow X_{\text{delay}}$, and use this delayed signal as the catalyst for the production of Y_+ and Y_- in the place of X_{in}. This leads to the CRN given in Appendix (Eq. 20) for which the numerical integration (Fig. 4**B**) shows that we indeed have get rid of the drift, or said otherwise, the correct implementation for online computation is given by:

$$\frac{dY}{dt} = f'(X_{\text{ext}}(t-\tau))\frac{dX_{\text{ext}}}{dt}(t-\tau), \tag{16}$$

where the delays has to be equal for the two pieces of the derivative.

6 Biological Examples

6.1 BioModels Repository

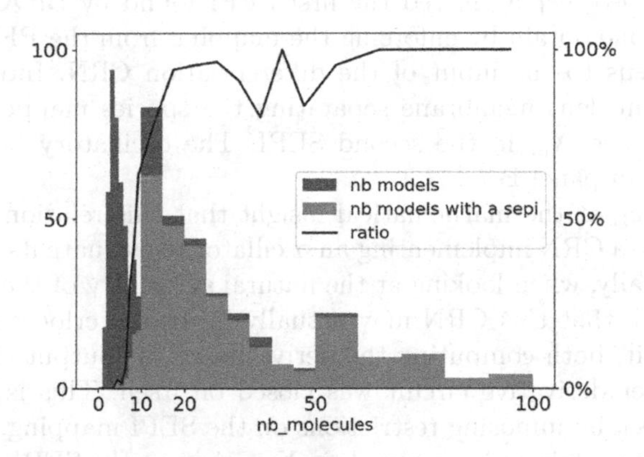

Fig. 5. Number of models in the curated part of BioModels per number of species, given with the number of models having a SEPI reduction to the differentiation CRN and ratio (black curve) between these two quantities.

To explore the possibility that natural biochemical systems already implement a form or another of the core differentiation CRN, one can try to scan the CRN models of the BioModels repository [19]. This can be automated with the general graph matching notion of Subgraph EPImorphism (SEPI) introduced in [12,13] to compare CRN models and identify model reduction relationships based on their graph structures. SEPI generalizes the classical notion of subgraph isomorphism by introducing an operation of node merging in addition to node deletion. Considering two bipartite graphs of species and reactions, there exists a SEPI from G_A to G_B if there exists a sequence of mergings[2] and deletions of nodes in G_A such that the resulting graph is isomorphic to G_B.

More precisely, we used the SEPI detection algorithm of BIOCHAM to scan the curated models in Biomodels (after automatic rewriting with well-formed reactions [9]) and check the existence of a SEPI from each model graph to the differentiation CRN graph. Figure 5 shows that our small differentiation CRN with 4 species is frequently found in large models. It is thus reasonable to restrict to models with no more than 10 species. Table 1 lists the models with no more than 10 species in the 700 first models of BioModels that contain our differentiation CRN. The predominance of models exhibiting oscillatory dynamics, and in particular circadian clock models is striking.

6.2 Circadian Clock

Model `biomodels 170` of the eukaryotes circadian clock proposed by Becker-Weimann et al. [2] is among the smallest models of the circadian clock displaying a SEPI reduction toward our differentiation CRN. Its influence graph is depicted in Fig. 6**A**, we also display in red the first SEPI found by BIOCHAM, and in green a second one obtain by enforcing the mapping from the PER/Cry species inside the nucleus to the input of the differentiation CRN. Interestingly, this model has the nucleus membrane separating the species mapped to X_{ext} and the one mapped to X_{in} in the second SEPI. The oscillatory behavior of this model is shown in panel **B**.

Now, thinking at the mathematical insight that this relation provides, it is quite natural for a CRN implementing an oscillator to evaluate its own derivative on the fly. Actually, when looking at the natural symmetry of the model, we are inclined to think that this CRN may actually be two interlocked CRNs of the derivative circuit, both computing the derivative of the output of the other, as if a second order derivative circuit was closed on itself. This is something we could easily check by imposing restrictions on the SEPI mapping. Enforcing the nucleus PerCRY protein to be mapped on X_{ext} gives us the SEPI shown in green in Fig. 6**A**. To validate the preservation of the function of the derivative CRN given by this SEPI, we can verify that the quantities defined by summing the species that are mapped together are effectively linked by the desired derivative relation. As can be seen in Fig. 6**B**, the agreement is striking. One can even note that the delay of the chemical derivative is the one predicted by our theory.

The case of Fig. 6**C** is more complex as this part of the model seems to compute the opposite of the derivative. It is however worth noting that there is absolutely no degree of freedom in our choice of the species used in Fig. 6**B** and **C** that are entirely constrained by the SEPI given by BIOCHAM. Taking both SEPI together we see that $\text{Bmal1}_{\text{protein}}^{\text{nucleus}}$ and $\text{Bmal1}_{\text{mRNA}}^{\text{cytoplasm}}$ play symmetrical roles, being the input and derivative of the two displayed SEPI. Given that the

[2] A species (resp. reaction) node can only be merged with another species (resp. reaction) node and the resulting node inherits of all the incoming and outcomming edges of the two nodes.

Table 1. List of models having a SEPI reduction to the differentiation CRN, given with model ID, number of species, number of reactions, and process modeled, among the first 700 models of the curated part of Biomodels with no more than 10 species.

Model ID	# Species	# reactions	Topic
0021	10	30	Circadian clock
0022	10	34	Circadian clock
0034	9	22	Circadian clock
0035	9	15	Circadian clock
0041	10	17	Creatine kinase
0065	8	16	Operon lactose
0067	7	16	Circadian clock
0075	10	13	Phosphoinositide turnover
0084	8	16	ERK Cascade
0107	9	23	Cell cycle
0108	9	18	Superoxide dismutase overexpression
0170	7	17	Circadian clock
0171	10	27	Circadian clock
0179	7	17	Cellular memory
0185	8	20	Circadian clock
0206	9	22	Circadian clock
0206	8	15	Glycolytic oscillations
0216	5	17	Circadian clock
0228	9	22	Cell cycle
0229	7	28	Circadian clock
0240	6	14	DegU transcriptional regulator
0257	8	19	Self-maintaining Metabolism
0262	9	14	AkT Signalling
0263	9	14	AkT Signalling
0269	9	22	Hormonal crosstalk in plant
0318	7	17	Bistable switch
0355	9	17	Calcium signalling
0359	9	15	Tissue factor pathway inhibitor
0360	9	15	Tissue factor pathway inhibitor
0495	8	18	Phospholipid synthetic pathways
0530	10	17	Cooperative gene regulation
0539	6	11	Mixed feedback loop
0563	10	17	Plant-microbe interaction
0586	10	23	Genetic oscillatory network
0587	10	23	Genetic oscillatory network
0590	9	40	Biosynthesis of pyrimidines
0615	4	34	Aggregation kinetics in Parkinson's Disease
0616	4	20	Resolution of inflammation
0619	10	13	Basic model of Acetaminophen
0622	10	19	Ubiquitination oscillatory dynamics
0632	8	14	Cell fate decision
0665	7	13	Interleukin-2 dynamics
0696	9	23	Incoherent type 1 feed-forward loop

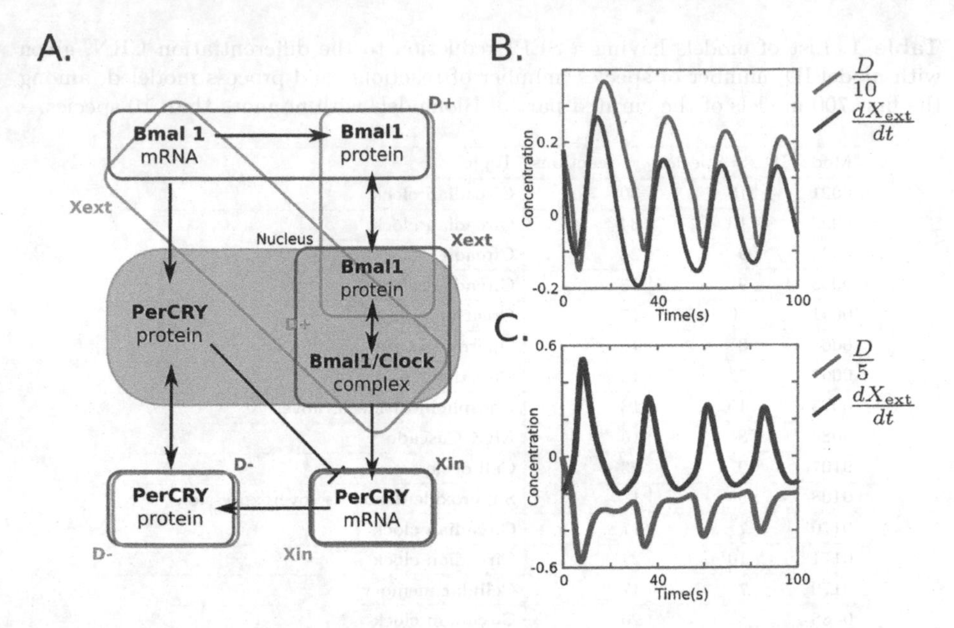

Fig. 6. Circadian clock model proposed by Becker-Weimann et al. **A** Graph of the model, the grey block indicates the nucleus while the other species are present in the cytoplasm. The first SEPI found by BIOCHAM is displayed in red. A second SEPI is displayed in green where the Per/CRY complex in the nucleus is mapped to the input of our differentiation CRN. **B** Validation of the first (red) SEPI: we display the derivative of the input species here considered as the sum of its part: $X_{\text{ext}} = \text{Bmal1}^{\text{nucleus}}_{\text{protein}} + \text{Bmal1/Clock}^{\text{nucleus}}_{\text{protein}}$ and similarly for the output: $D = D_+ - D_- = \text{Bmal1}^{\text{cytoplasm}}_{\text{mRNA}} + \text{Bmal1}^{\text{cytoplasm}}_{\text{protein}} - \text{Per/CRY}^{\text{cytoplasm}}_{\text{protein}}$. **C** Validation of the second (green) SEPI. As for the previous panel, species that are mapped together are simply summed for this validation. The qualitative matching of the two quantities in these two graphs are a good indication that these SEPI are meaningful. (Color figure online)

second SEPI introduces a negative sign, we may see this as:

$$\begin{aligned} \text{Bmal1}^{\text{cytoplasm}}_{\text{mRNA}} &= \frac{d}{dt}\text{Bmal1}^{\text{nucleus}}_{\text{protein}} \\ \text{Bmal1}^{\text{nucleus}}_{\text{protein}} &= -\frac{d}{dt}\text{Bmal1}^{\text{cytoplasm}}_{\text{mRNA}} \end{aligned} \tag{17}$$

The solution of this well known equation are the sine and cosine functions, and this perfectly fits the oscillatory behaviour of this CRN. To confirm this hypothesis, we check for the presence of a SEPI from the clock model to the compiled cosine CRN presented in Eq. 2 which is effectively the case. On the other hand, there is no SEPI relation between the compiled cosine and the derivative circuit.

6.3 Bistable Switch

The model `biomodels 318` of a bistable switch in the context of the restriction point [25] displays a SEPI toward our derivative circuit. This model, presented in Fig. 7**A**, study the Rb-E2F pathway as an example of bistable switch where the presence of a (not modeled) growth factor activates the MyC protein, starting the pathway until it reach the E2F factor that constitute the output of the model. Yao et al. show that once E2F reaches a threshold, its activation becomes self sustained hence the notion of switch.

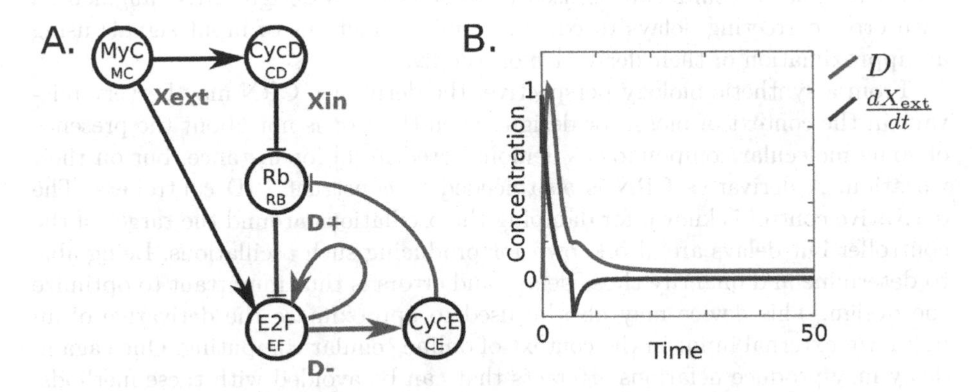

Fig. 7. Analysis of the bistable switch. **A** Schematic representation of the model and its SEPI, the smaller fonts for the species corresponds to the names used in the model provided online. Two other species are also present in the model (the phosphorylated form of RB and the Rb-E2F complex) but they are not presented in the figure of the original article and are deleted during the SEPI, we thus choose to not display them either. **B** True derivative of MyC (in blue) along with the ones computed by the CRN (in red), see main text for the exact definition of D. (Color figure online)

The SEPI given by Biocham is worth of interest as it does not merge any species and only three reactions into one leaving all the other either untouched or deleted, thus indicating that the pattern of the derivative is already well present. Moreover, MyC is mapped to the input and E2F to one part of the output, reinforcing our intuition that the discovered SEPI is closed from the natural functioning of the CRN.

To conform this, we run the simulation as provided by the models and display the derivative of the MyC protein against a scaled difference of the D_+ and D_- species: $D = aRB - bE2F$ where a and b are positive constant adjusted so that D goes to 0 at final time and are of the same magnitude as $\frac{dMyC}{dt}$. (This gives $a = 6.3, b = 0.063$.) Clearly, D is a delayed and smoothed version of the input derivative exactly as our derivative device would provide.

7 Conclusion and Perspectives

We have presented a mathematical analysis of the core differentiation CRN introduced by Whitby et al. [24]. In particular, we have shown that what is computed is an approximation of the left derivative given a small time in the past with a time constant determined by the diffusion constant between the input and its internal representation: $\epsilon = \frac{1}{k_{\text{diff}}}$. Moreover, there is a delay τ due to the computation time that can also be precisely estimated given the rate of activation and degradation of the species encoding the derivative: $\tau = \frac{1}{k}$. We have shown that such results can be used in some cases to design CRNs augmented with error-correcting delays to compute online functions of input signals using an approximation of their derivative on the fly.

From a synthetic biology perspective, the derivative CRN may be very relevant in the context of biosensor design, when the test is not about the presence of some molecular compounds as demonstrated in [6] for instance, but on their variation. A derivative CRN is also needed to construct PID controllers. The derivative control is known for damping the oscillations around the target of the controller but delays are also known for producing such oscillations. Being able to determine and quantify those delays and errors is thus important to optimize the design. This device may also be used to approximate the derivative of an unknown external input in the context of online cellular computing. Once again, delay may produce nefarious artefacts that can be avoided with these methods.

Furthermore, using the notion of SEPI to scan the BioModels repository, we were able to highlight a certain number of CRN models containing the core differentiation CRN structure. A high number of these models occur in models presenting oscillations. We have shown on one such example, a circadian clock model, why it makes sense for an oscillator to sense its own derivative, and to reproduce what a mathematician would produce in a more direct way for the most basic oscillatory functions: sine and cosine.

Acknowledgment. This work benefited from ANR-20-CE48-0002 δifference project grant.

Appendix: Computation of Integration with a Delay

To prove that the drift of the output is a direct consequence of the delay, we first compute the input and the approximate derivative for our choice of input:

$$
\begin{aligned}
x(t) &= 1 + \sin(t) \\
x'(t - \tau) &= \cos(t - \tau) \\
&= \cos(t) + \tau \sin(t) + o(\tau^2)
\end{aligned}
\tag{18}
$$

Then we can compute the output up to the first order:

$$\begin{aligned} y(t) &= \int 2x(s)x'(s-\tau)ds \\ &= \int 2\left(1+\sin(s)\right)\cos(s)ds + \int 2\tau(\sin(s)+\sin^2(s))ds \\ &= (1+\sin(t))^2 + 2\tau \int \sin(s)+\sin^2(s)ds \\ y(t) &\simeq (1+\sin(t))^2 + \tau t \end{aligned} \tag{19}$$

Then, to correct the observed drift, we propose to introduce a delay signal and use it in the computation to produce the output species Y_+ and Y_-, with the following CRN:

$$\begin{aligned} X_{ext} &\xrightarrow{k_{diff}} X_{in}, & X_{in} &\xrightarrow{k_{diff}} X_{ext}, \\ X_{in} &\xrightarrow{k} X_{in} + X_{delay}, & X_{delay} &\xrightarrow{k} \emptyset, \\ X_{ext} &\xrightarrow{k_{diff}k} X_{ext} + D_+ & D_+ &\xrightarrow{k} \emptyset \\ X_{in} &\xrightarrow{k_{diff}k} X_{in} + D_- & D_- &\xrightarrow{k} \emptyset \\ X_{delay} + D_+ &\xrightarrow{2} X_{delay} + D_+ + Y_+ & X_{delay} + D_- &\xrightarrow{2} X_{delay} + D_- + Y_- \\ D_+ + D_- &\xrightarrow{fast} \emptyset & Y_+ + Y_- &\xrightarrow{fast} \emptyset \end{aligned} \tag{20}$$

References

1. Alexis, E., Schulte, C.C.M., Cardelli, L., Papachristodoulou, A.: Biomolecular mechanisms for signal differentiation. Iscience **24**(12), 103462 (2021)
2. Becker-Weimann, S., Wolf, J., Herzel, H., Kramer, A.: Modeling feedback loops of the mammalian circadian oscillator. Biophys. J . **87**(5), 3023–3034 (2004)
3. Briat, C., Gupta, A., Khammash, M.: Antithetic integral feedback ensures robust perfect adaptation in noisy biomolecular networks. Cell Syst. **2**(1), 15–26 (2016)
4. Calzone, L., Fages, F., Soliman, S.: BIOCHAM: an environment for modeling biological systems and formalizing experimental knowledge. Bioinformatics **22**(14), 1805–1807 (2006)
5. Chevalier, M., Gómez-Schiavon, M., Ng, A.H., El-Samad, H.: Design and analysis of a proportional-integral-derivative controller with biological molecules. Cell Syst. **9**(4), 338–353 (2019)
6. Courbet, A., Amar, P., Fages, F., Renard, E., Molina, F.: Computer-aided biochemical programming of synthetic microreactors as diagnostic devices. Mol. Syst. Biol. **14**(4) (2018)
7. Érdi, P., Tóth, J.: Mathematical Models of Chemical Reactions: Theory and Applications of Deterministic and Stochastic Models. Nonlinear Science: Theory and Applications. Manchester University Press (1989)

8. Fages, F., Le Guludec, G., Bournez, O., Pouly, A.: Strong Turing completeness of continuous chemical reaction networks and compilation of mixed analog-digital programs. In: Feret, J., Koeppl, H. (eds.) CMSB 2017. LNCS, vol. 10545, pp. 108–127. Springer, Cham (2017). https://doi.org/10.1007/978-3-319-67471-1_7
9. Fages, F., Gay, S., Soliman, S.: Inferring reaction systems from ordinary differential equations. Theor. Comput. Sci. **599**, 64–78 (2015)
10. Fages, F., Soliman, S.: Abstract interpretation and types for systems biology. Theor. Comput. Sci. **403**(1), 52–70 (2008)
11. Feinberg, M.: Mathematical aspects of mass action kinetics. In: Lapidus, L., Amundson, N.R. (eds.) Chemical Reactor Theory: A Review, chapter 1, pp. 1–78. Prentice-Hall, Upper Saddle River (1977)
12. Gay, S., Fages, F., Martinez, T., Soliman, S., Solnon, C.: On the subgraph epimorphism problem. Discret. Appl. Math. **162**, 214–228 (2014)
13. Gay, S., Soliman, S., Fages, F.: A graphical method for reducing and relating models in systems biology. Bioinformatics **26**(18), i575–i581 (2010). special issue ECCB'10
14. Hemery, M., Fages, F.: Algebraic biochemistry: a framework for analog online computation in cells. In: Petre, I., Păun, A. (eds.) CMSB 2022. LNCS, vol. 13447, pp. 3–20. Springer, Cham (2022). https://doi.org/10.1007/978-3-031-15034-0_1
15. Hemery, M., Fages, F., Soliman, S.: On the complexity of quadratization for polynomial differential equations. In: Abate, A., Petrov, T., Wolf, V. (eds.) CMSB 2020. LNCS, vol. 12314, pp. 120–140. Springer, Cham (2020). https://doi.org/10.1007/978-3-030-60327-4_7
16. Hemery, M., Fages, F., Soliman, S.: Compiling elementary mathematical functions into finite chemical reaction networks via a polynomialization algorithm for ODEs. In: Cinquemani, E., Paulevé, L. (eds.) CMSB 2021. LNCS, vol. 12881, pp. 74–90. Springer, Cham (2021). https://doi.org/10.1007/978-3-030-85633-5_5
17. Huang, C.-Y., Ferrell, J.E.: Ultrasensitivity in the mitogen-activated protein kinase cascade. PNAS **93**(19), 10078–10083 (1996)
18. Kitano, H.: Systems biology: a brief overview. Science **295**(5560), 1662–1664 (2002)
19. le Novère, N., et al.: BioModels database: a free, centralized database of curated, published, quantitative kinetic models of biochemical and cellular systems. Nucleic Acid Res. **1**(34), D689–D691 (2006)
20. Myhill, J.: A recursive function defined on a compact interval and having a continuous derivative that is not recursive. Mich. Math. J. **18**(2), 97–98 (1971)
21. Oishi, K., Klavins, E.: Biomolecular implementation of linear i/o systems. IET Syst. Biol. **5**(4), 252–260 (2011)
22. Shannon, C.E.: Mathematical theory of the differential analyser. J. Math. Phys. **20**, 337–354 (1941)
23. Vasic, M., Soloveichik, D., Khurshid, S.: CRN++: molecular programming language. In: Doty, D., Dietz, H. (eds.) DNA 2018. LNCS, vol. 11145, pp. 1–18. Springer, Cham (2018). https://doi.org/10.1007/978-3-030-00030-1_1
24. Whitby, M., Cardelli, L., Kwiatkowska, M., Laurenti, L., Tribastone, M., Tschaikowski, M.: Pid control of biochemical reaction networks. IEEE Trans. Autom. Control **67**(2), 1023–1030 (2021)
25. Yao, G., Lee, T.J., Mori, S., Nevins, J.R., You, L.: A bistable rb-e2f switch underlies the restriction point. Nat. Cell Biol. **10**(4), 476–482 (2008)

Harissa: Stochastic Simulation and Inference of Gene Regulatory Networks Based on Transcriptional Bursting

Ulysse Herbach(✉)

Université de Lorraine, CNRS, Inria, IECL, 54000 Nancy, France
ulysse.herbach@inria.fr

Abstract. Gene regulatory networks, as a powerful abstraction for describing complex biological interactions between genes through their expression products within a cell, are often regarded as virtually deterministic dynamical systems. However, this view is now being challenged by the fundamentally stochastic, 'bursty' nature of gene expression revealed at the single cell level. We present a Python package called Harissa which is dedicated to simulation and inference of such networks, based upon an underlying stochastic dynamical model driven by the transcriptional bursting phenomenon. As part of this tool, network inference can be interpreted as a calibration procedure for a mechanistic model: once calibrated, the model is able to capture the typical variability of single-cell data without requiring ad hoc external noise, unlike ordinary or even stochastic differential equations frequently used in this context. Therefore, Harissa can be used both as an inference tool, to reconstruct biologically relevant networks from time-course scRNA-seq data, and as a simulation tool, to generate quantitative gene expression profiles in a non-trivial way through gene interactions.

Keywords: Gene expression · Regulatory networks · Single-cell data

1 Introduction

Inferring graphs of interactions between genes has become a standard task for high-dimensional statistics, while mechanistic models describing gene expression at the molecular level have come into their own with the advent of single-cell data. Linking these two approaches seems crucial today, but the dialogue is far from obvious: statistical models often suffer from a lack of biological interpretability, and mechanistic models are known to be difficult to calibrate from real data.

J. Pang and J. Niehren (Eds.): CMSB 2023, LNBI 14137, pp. 97–105, 2023.
https://doi.org/10.1007/978-3-031-42697-1_7

Here, we present a Python package for both network simulation and inference from single-cell gene expression data (typically scRNA-seq), called Harissa ('HARtree approximation for Inference along with a Stochastic Simulation Algorithm'). It was implemented in the context of a mechanistic approach to gene regulatory network inference from single-cell data [5] and is based upon an underlying stochastic dynamical model driven by the transcriptional bursting phenomenon. In this tool paper, we introduce briefly the main concepts behind the package, and detail its usage through some application examples.

2 Theory

Consider a network of n genes. Our starting point is the well-known 'two-state model' of gene expression [11], which corresponds to the following set of elementary chemical reactions [5] for each gene $i \in \{1, \dots, n\}$:

$$\begin{aligned} G_i \underset{k_{\mathrm{off},i}}{\overset{k_{\mathrm{on},i}}{\rightleftharpoons}} G_i{}^*, \qquad G_i{}^* &\xrightarrow{s_{0,i}} G_i{}^* + X_i, \qquad X_i \xrightarrow{d_{0,i}} \varnothing, \\ X_i &\xrightarrow{s_{1,i}} X_i + Z_i, \qquad Z_i \xrightarrow{d_{1,i}} \varnothing, \end{aligned} \tag{1}$$

where G_i, $G_i{}^*$, X_i and Z_i respectively denote 'inactive promoter', 'active promoter', mRNA and protein copy numbers for gene i. These reactions describe the main two stages of gene expression, namely *transcription* (rate $s_{0,i}$) and *translation* (rate $s_{1,i}$), along with degradation of mRNA (rate $d_{0,i}$) and protein (rate $d_{1,i}$) molecules. Note that $[G_i] + [G_i{}^*] = 1$ is a conserved quantity: the particularity of the two-state model is that transcription of gene i can only occur when its promoter is active, corresponding to $[G_i{}^*] = 1$.

More specifically, experimental data consistently suggest a particular regime for this model [13]: $k_{\mathrm{off},i} \gg k_{\mathrm{on},i}$ and $s_{0,i} \gg d_{0,i}$ with the ratio $k_{\mathrm{on},i}s_{0,i}/(k_{\mathrm{off},i}d_{0,i})$ remaining fixed, corresponding to short active periods during which many mRNA molecules are produced. In this regime, mRNA is transcribed by 'bursts' of tens to hundreds of molecules. Moreover, mRNA and protein copy numbers are not conserved quantities in the model and can be reasonably described in a continuous way using standard 'mass action' kinetics. This leads to a hybrid dynamical model, consisting of ordinary differential equations subject to random jumps (Fig. 1) where the related quantities are denoted by $X(t) = (X_1(t), \dots, X_n(t)) \in \mathbb{R}_+^n$ and $Z(t) = (Z_1(t), \dots, Z_n(t)) \in \mathbb{R}_+^n$.

So far the genes are not interacting: the main point of our approach is to consider the burst frequency $k_{\mathrm{on},i}$ as a function of proteins levels $Z_1, \dots, Z_n$ [5]. In the current version of Harissa, this function takes the following form:

$$k_{\mathrm{on},j}(z) = \frac{k_{0,j} + k_{1,j}\exp(\beta_j + \sum_{i=1}^n \theta_{ij} z_i)}{1 + \exp(\beta_j + \sum_{i=1}^n \theta_{ij} z_i)} \qquad \forall j \in \{1, \dots, n\} \tag{2}$$

with $k_{0,j} \ll k_{1,j}$, so that $(\theta_{ij})_{1 \le i,j \le n}$ can be easily interpreted as the *network interaction matrix* while β_j encodes basal activity of gene j (see also Table 1). The rate $k_{\mathrm{off},i}$ (inverse of mean burst duration) is kept constant.

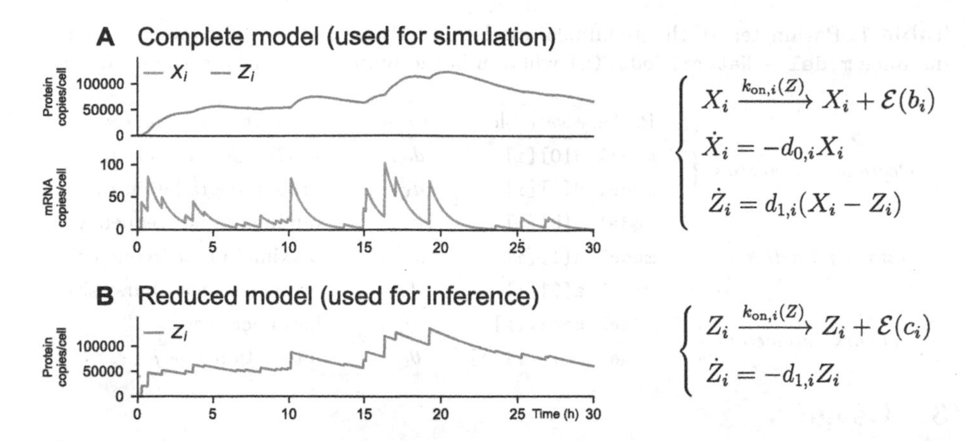

Fig. 1. The stochastic dynamical model underlying Harissa. (A) Complete model with mRNA levels $X(t)$ and protein levels $Z(t)$, used to perform simulations. Transcription of gene i occurs in bursts at random times with rate $k_{\text{on},i}(Z(t))$ and the burst size follows an exponential distribution $\mathcal{E}(b_i)$ where b_i corresponds to a scaling factor. (B) Reduced model involving only protein levels $Z(t)$. Without loss of generality, the parameter c_i is set to an arbitrary value as protein levels are not measured. This model serves as a basis for deriving a pseudo-likelihood whose maximization is at the core of inference.

Mathematically speaking, the complete set of variables $(X(t), Z(t))_{t\geq 0} \in \mathbb{R}_+^n \times \mathbb{R}_+^n$ forms a *piecewise-deterministic Markov process* (PDMP) that is fully characterized by its (continuous) master equation:

$$\begin{aligned}\frac{\partial}{\partial t}p(x,z,t) = \sum_{i=1}^{n}\Bigg[& d_{0,i}\frac{\partial}{\partial x_i}\{x_i p(x,z,t)\} + d_{1,i}\frac{\partial}{\partial z_i}\{(z_i - x_i)p(x,z,t)\} \\ & + k_{\text{on},i}(z)\left(\int_0^{x_i} p(x - he_i, z, t) b_i e^{-b_i h}\mathrm{d}h - p(x,z,t)\right)\Bigg], \end{aligned} \tag{3}$$

where $p(x, z, t)$ is the probability distribution of $(X(t), Z(t))$. The master equation describes the time evolution of this probability and is related to the trajectory dynamics (see Fig. 1A). Notably, each Z_i is rescaled so that parameters $s_{1,i}$ do not appear anymore, and $s_{0,i}$ and $k_{\text{off},i}$ aggregate in the bursty regime into the 'burst size' parameter $1/b_i = s_{0,i}/k_{\text{off},i}$ (Table 1).

This 'complete' model is simulated in Harissa using an efficient acceptance-rejection method, similarly to the explicit construction given in [1]. A great advantage of this method is that it is guaranteed to be exact without requiring any numerical integration, contrary to the basic algorithm [6].

We also consider a 'protein-only' model (Fig. 1B), which can be seen as a first-order approximation (Appendix A.1): this reduced model turns out to provide a tractable inference procedure based on analytical results (Appendix A.2).

Table 1. Parameters of the dynamical model underlying Harissa. Here we consider an instance `model = NetworkModel(n)` where `n` is the number of genes in the network.

	Package variable	Notation	Interpretation (gene i)
Degradation kinetics	`model.d[0][i]`	$d_{0,i}$	mRNA degradation rate
	`model.d[1][i]`	$d_{1,i}$	protein degradation rate
Bursting kinetics	`model.a[0][i]`	$k_{0,i}$	minimal burst frequency
	`model.a[1][i]`	$k_{1,i}$	maximal burst frequency
	`model.a[2][i]`	b_i	inverse of mean burst size
Network parameters	`model.basal[i]`	β_i	basal activity
	`model.inter[i,j]`	θ_{ij}	interaction $i \rightarrow j$

3 Usage

The Harissa package has two main functionalities: network inference interpreted as calibration of the stochastic dynamical model defined by (3), and data simulation from the same model (e.g., scRNA-seq counts [10] but also RT-qPCR levels [7,12]). Besides, the package also allows for basic network visualization (directed graphs with positive or negative edge weights) as well as data binarization (using gene-specific thresholds derived from the data-calibrated dynamical model).

The first step is to create an instance of the model:

```
model = NetworkModel(n)
```

where `n` is the number of genes. After setting the parameter values (see Table 1), the model can be simulated using `model.simulate(time)`, where `time` is either a single time or a list of time points.

Importantly, `simulate()` does not depend on time discretization and always returns exact stochastic simulations: the resulting continuous-time trajectories are simply extracted at user-specified time points. It is also possible to consider a stimulus, represented by an additional protein Z_0 that receives no feedback and verifies $Z_0(t) = 0$ for $t \leq 0$ and $Z_0(t) = 1$ for $t > 0$. In order to reach a pre-stimulus steady state before perturbation, an optional `burnin` parameter sets the time during which the model is simulated with $Z_0(t) = 0$.

Example: Repressilator Network. As an example, we consider a 'repressilator' network made of 3 genes forming a directed cycle of negative interactions (Fig. 2).

Some critical parameters of the dynamical model are degradation rates $d_{0,i}$ and $d_{1,i}$, which characterize the 'responsiveness' of mRNA and protein levels. Here we set $d_{0,i}/d_{1,i} = 10$, meaning that proteins are ten times more stable than mRNAs. This is biologically realistic, but note that this ratio is known to span a very wide range [9] so there is no single choice.

An example of simulated single-cell trajectory for this network is shown in Fig. 3. It is worth noticing that despite the strong level of stochasticity, a robust periodic pattern is already emerging.

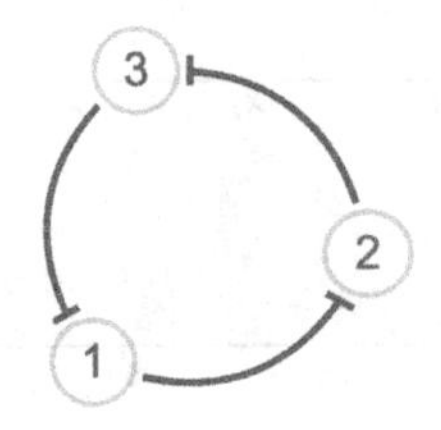

Fig. 2. The repressilator network used in Fig. 3. This plot was made with `plot_network` from the `harissa.utils` module.

More stable proteins—with respect to mRNA—will lead to less 'intrinsic noise' in the system. We increase mRNA degradation rates and burst frequencies instead, which is equivalent to a zoom-out regarding the time scale. Since mRNA and protein scales are normalized, the overall levels do not depend on degradation rates (but the *dynamics* does).

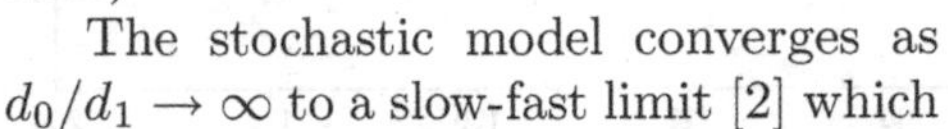

The stochastic model converges as $d_0/d_1 \to \infty$ to a slow-fast limit [2] which turns out to be a nonlinear ODE system involving only proteins (bottom plot of Fig. 3). It appears here that due to the deterministic dynamics, the initial protein levels need to be perturbed so as not to stay in a trivial manifold. This limit model can be simulated with `model.simulate_ode(time)` and is generally useful to gain insight into the system attractors (note however that it is a rough approximation of the stochastic model, see Remark 1).

In this slow-fast limit, which is part of the rationale behind the reduced model (Appendix A.1), mRNA levels become *independent conditionally on protein levels* such that $X_i(t) \sim \text{Gamma}(k_{\text{on},i}(Z(t))/d_{0,i}, b_i)$ for $i = 1, 2, 3$. This quasi-steady-state (QSS) behavior can be understood intuitively from the top plot of Fig. 3. Regarding mRNA levels, `simulate_ode` only returns the mean of the QSS distribution conditionally on protein levels (the true limit model would consist in sampling from this distribution independently for every $t > 0$).

Network Inference. Here the main function is `model.fit()`, which takes as input a time-course single-cell dataset (Fig. 4). Inference can be performed by creating a new instance `model = NetworkModel()` without any size parameter, then loading a dataset `x` and using `model.fit(x)`. This will update all parameters of the model (Table 1) except $d_{0,i}$ and $d_{1,i}$, which need to be provided by the user from external data—or left to their default values, see remark 1—as they cannot be inferred without seriously compromising identifiability of other parameters. As an important feature, the updated `model` instance is ready for simulation, to assess reproducibility of the original data or to predict the outcome of network modifications. From a network inference viewpoint, the only important parameter is `model.inter` which is a non-symmetric signed weight matrix.

Remark 1. Degradation rates $d_{0,i}$ and $d_{1,i}$ are in fact not required for the *inference* part, but they are important for the *simulation* part. They are currently set by default to $d_{0,i} = \ln(2)/9 \approx 0.077\ \text{h}^{-1}$ and $d_{1,i} = \ln(2)/46 \approx 0.015\ \text{h}^{-1}$ as median values from thousands of genes [9]. This leads in particular to $d_{0,i}/d_{1,i} \approx 5.11$.

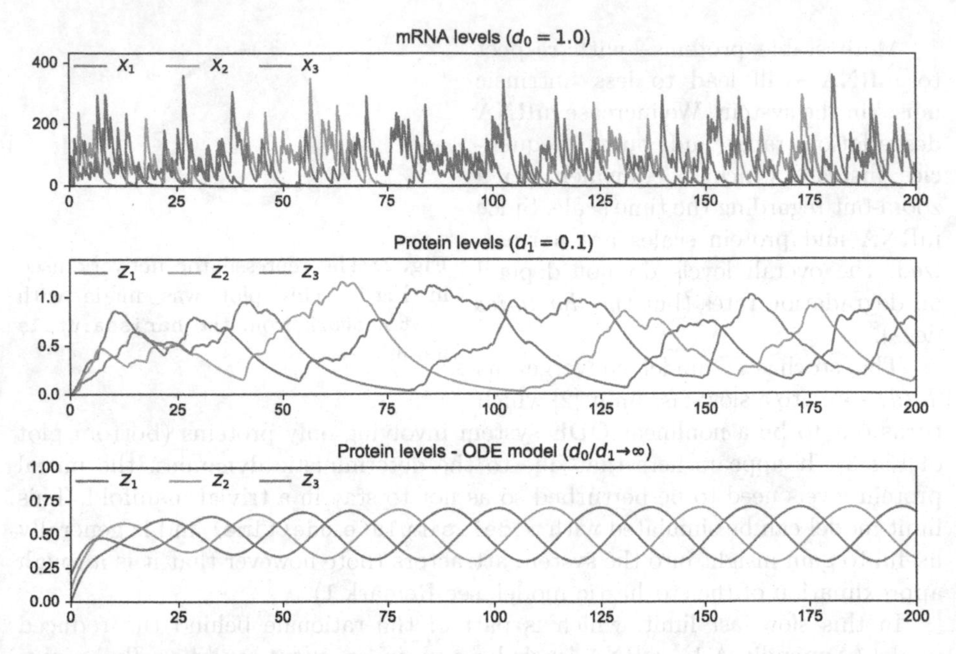

Fig. 3. Simulation of a repressilator network using Harissa. The first two plots show an example of single-cell trajectory of mRNA levels $X(t)$ and protein levels $Z(t)$ from the full stochastic model (3). The bottom plot shows a trajectory of $Z(t)$ from the related deterministic model corresponding to the slow-fast limit (very stable proteins compared to mRNA, i.e. $d_0/d_1 \to \infty$). Importantly, the periodic pattern is emerging and robust well before the limit is reached (here $d_0/d_1 = 10$). The burst parameters (Table 1) are $k_{0,i} = 0$, $k_{1,i} = 2$ and $b_i = 0.02$ for all $i \in \{1, 2, 3\}$, which are current default values. The network parameters are given in Appendix A.3.

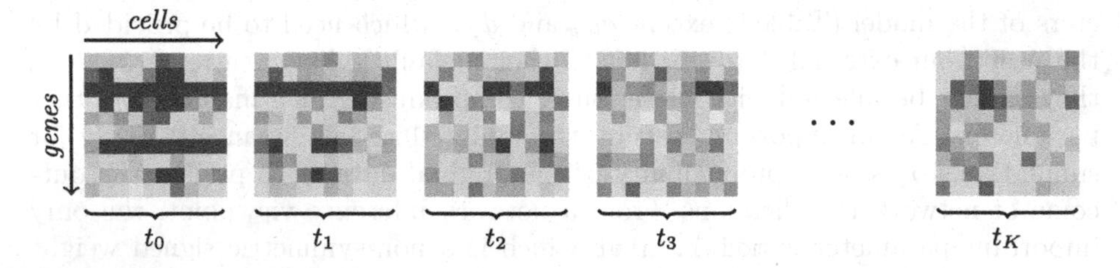

Fig. 4. Typical structure of time-course scRNA-seq data, as required by the inference module (here each row is a gene and each column is a cell). Each group of cells collected at the same experimental time t_k forms a *snapshot* of the biological heterogeneity at time t_k. Due to the destructive nature of the measurement process, snapshots are made of different cells. This data is therefore different from so-called 'pseudotime' trajectories, which attempt to reorder cells according to some smoothness hypotheses.

4 Conclusions

While the development of Harissa started a few years ago, the tool has recently become more mature and easy to use. Following early versions [3,5], an alternative method for the inference part called Cardamom was developed in [14], which in turn influenced the current version of Harissa. The two inference methods remain complementary at this stage and may be merged into the same package in future development. They were evaluated in a recent benchmark [16], giving encouraging results and showing, most importantly, that the stochastic dynamical model (3) can indeed reproduce real time-course scRNA-seq data through gene interactions. While the simulation part is quite well optimized and can be considered as stable, the inference part is much more challenging. Future directions include a better use of the dynamical information contained in time-course transcriptional profiles, which has great potential for performance improvement.

Acknowledgements. The author is very grateful to Elias Ventre and Olivier Gandrillon for fruitful discussions which led to improve the Harissa package.

Code Availability. The code of the package, a tutorial and some basic usage scripts are available at https://github.com/ulysseherbach/harissa. In addition, Harissa is indexed in the Python Package Index and can be installed via `pip`.

A Appendices

A.1 Reduced Model

The inference procedure is based on analytical results which are not available for the two-stage 'mRNA-protein' model (3). On the other hand, such results exist for a one-stage 'protein-only' model that is a valid approximation of the former when proteins are more stable than mRNA (i.e. $d_{0,i}/d_{1,i} \gg 1$). The resulting process $(Z(t))_{t\geq 0} \in \mathbb{R}_+^n$ is also a PDMP, whose master equation can be interpreted in terms of simplified trajectories (Fig. 1B):

$$\begin{aligned}\frac{\partial}{\partial t}p(z,t) = \sum_{i=1}^{n} \Bigg[& d_{1,i}\frac{\partial}{\partial z_i}\{z_i p(z,t)\} \\ & + \int_0^{z_i} k_{\mathrm{on},i}(z-he_i)p(z-he_i,t)c_i e^{-c_i h}\mathrm{d}h - k_{\mathrm{on},i}(z)p(z,t)\Bigg].\end{aligned} \tag{4}$$

Given $Z(t)$, mRNA levels $X(t)$ are obtained by sampling independently for every $i \in \{1,\dots,n\}$ and $t > 0$ from $X_i(t) \sim \mathrm{Gamma}(k_{\mathrm{on},i}(Z(t))/d_{0,i}, b_i)$, which is the *quasi-steady-state* (QSS) distribution of the complete model [5,6].

A.2 Inference Algorithm

Now consider mRNA counts measured in m cells, assumed independent, along a time-course experiment following a stimulus. Each cell $k = 1,\dots,m$ is associated with an experimental time point t_k. We introduce the following notation:

$\mathbf{x}_k = (x_{ki}) \in \{0, 1, 2, \dots\}^n$: mRNA counts (cell k, gene i);
$\mathbf{z}_k = (z_{ki}) \in (0, +\infty)^n$: latent protein levels (cell k, gene i);
$\alpha = (\alpha_{ij}(t_k)) \in \mathbb{R}^{n \times n}$: effective interaction $i \to j$ at time t_k.

A stimulus is represented as gene $i = 0$ and we therefore add parameters $\alpha_{0j}(t_k)$ for $j = 1, \dots, n$ and $k = 1, \dots, m$. We further set $z_{k0} = 0$ if $t_k \leq 0$ (before stimulus) and $z_{k0} = 1$ if $t_k > 0$ (after stimulus). Then, writing $a_i = k_{1,i}/d_{0,i}$, the underlying statistical model of Harissa is defined by

$$p(\mathbf{z}_k) = \prod_{i=1}^{n} {z_{ki}}^{c_i \sigma_{ki} - 1} e^{-c_i z_{ki}} \frac{{c_i}^{c_i \sigma_{ki}}}{\Gamma(c_i \sigma_{ki})}, \tag{5}$$

$$p(\mathbf{x}_k | \mathbf{z}_k) = \prod_{i=1}^{n} \frac{1}{x_{ki}!} \frac{\Gamma(a_i z_{ki} + x_{ki})}{\Gamma(a_i z_{ki})} \frac{{b_i}^{a_i z_{ki}}}{(b_i + 1)^{a_i z_{ki} + x_{ki}}}, \tag{6}$$

with

$$\sigma_{ki} = \Big[1 + \exp(-\{\beta_i + \alpha_{0i}(t_k) z_{k0} + \textstyle\sum_{j=1}^{n} \alpha_{ji}(t_k) z_{kj}\})\Big]^{-1}. \tag{7}$$

Details of this derivation can be found in [4,5,8,14]. Roughly, (5) comes from a 'Hartree' approximation of (4), while (6) corresponds to a Poisson distribution with random parameter sampled from the QSS distribution of $X(t)$ given $Z(t)$. Note that $p(\mathbf{z}_k)$ is in general only a pseudo-likelihood as σ_{ki} depends on $\mathbf{z}_k$.

Since the preliminary version of Harissa [5], the global inference procedure has been heavily improved using important identifiability results from [14,15]. The final algorithm consists of three steps:

1. *Model calibration:* estimate a_i and b_i for each gene individually from (6);
2. *Bursting mode inference:* estimate the frequency mode ($k_{0,i}$ or $k_{1,i}$) for each gene in each cell (can be seen as a binarization step with specific thresholds);
3. *Network inference:* consider $\mathbf{z}_k$ as observed from step 2 and maximize (5) with respect to α after adding an appropriate penalization term [14]. Each parameter θ_{ij} is then set to $\alpha_{ij}(t_k)$ with t_k that maximizes $|\alpha_{ij}(t_k)|$.

A.3 Repressilator Network

Considering an instance `model = NetworkModel(3)`, the repressilator network simulated in Fig. 3 is defined as follows:

```
model.d[0] = 1 # mRNA degradation rates
model.d[1] = 0.1 # Protein degradation rates
model.basal[1] = 5 # Basal activity of gene 1
model.basal[2] = 5 # Basal activity of gene 2
model.basal[3] = 5 # Basal activity of gene 3
model.inter[1,2] = -10 # Interaction 1 -> 2
model.inter[2,3] = -10 # Interaction 2 -> 3
model.inter[3,1] = -10 # Interaction 3 -> 1
```

References

1. Benaïm, M., Le Borgne, S., Malrieu, F., Zitt, P.A.: Qualitative properties of certain piecewise deterministic Markov processes. Annales de l'Institut Henri Poincaré - Probabilités et Statistiques **51**(3), 1040–1075 (2015). https://doi.org/10.1214/14-AIHP619
2. Faggionato, A., Gabrielli, D., Crivellari, M.: Averaging and large deviation principles for fully-coupled piecewise deterministic Markov processes and applications to molecular motors. Markov Process. Rel. Fields **16**(3), 497–548 (2010). https://doi.org/10.48550/arXiv.0808.1910
3. Herbach, U.: Modélisation stochastique de l'expression des gènes et inférence de réseaux de régulation. Ph.D. thesis, Université de Lyon (2018)
4. Herbach, U.: Stochastic gene expression with a multistate promoter: breaking down exact distributions. SIAM J. Appl. Math. **79**(3), 1007–1029 (2019). https://doi.org/10.1137/18M1181006
5. Herbach, U., Bonnaffoux, A., Espinasse, T., Gandrillon, O.: Inferring gene regulatory networks from single-cell data: a mechanistic approach. BMC Syst. Biol. **11**(1), 105 (2017). https://doi.org/10.1186/s12918-017-0487-0
6. Malrieu, F.: Some simple but challenging Markov processes. Annales de la Faculté de Sciences de Toulouse **24**(4), 857–883 (2015). https://doi.org/10.5802/afst.1468
7. Richard, A., et al.: Single-cell-based analysis highlights a surge in cell-to-cell molecular variability preceding irreversible commitment in a differentiation process. PLoS Biol. **14**(12), e1002585 (2016). https://doi.org/10.1371/journal.pbio.1002585
8. Sarkar, A., Stephens, M.: Separating measurement and expression models clarifies confusion in single-cell RNA sequencing analysis. Nat. Genet. **53**(6), 770–777 (2021). https://doi.org/10.1038/s41588-021-00873-4
9. Schwanhäusser, B., et al.: Global quantification of mammalian gene expression control. Nature **473**(7347), 337–342 (2011). https://doi.org/10.1038/nature10098
10. Semrau, S., Goldmann, J.E., Soumillon, M., Mikkelsen, T.S., Jaenisch, R., van Oudenaarden, A.: Dynamics of lineage commitment revealed by single-cell transcriptomics of differentiating embryonic stem cells. Nat. Commun. **8**(1), 1096 (2017). https://doi.org/10.1038/s41467-017-01076-4
11. Shahrezaei, V., Swain, P.S.: The stochastic nature of biochemical networks. Curr. Opin. Biotechnol. **19**(4), 369–374 (2008). https://doi.org/10.1016/j.copbio.2008.06.011
12. Stumpf, P.S., et al.: Stem cell differentiation as a non-Markov stochastic process. Cell Syst. **5**(3), 268–282 (2017). https://doi.org/10.1016/j.cels.2017.08.009
13. Tunnacliffe, E., Chubb, J.R.: What is a transcriptional burst? Trends Genet. **36**(4), 288–297 (2020). https://doi.org/10.1016/j.tig.2020.01.003
14. Ventre, E.: Reverse engineering of a mechanistic model of gene expression using metastability and temporal dynamics. In Silico Biol. **14**(3–4), 89–113 (2021). https://doi.org/10.3233/ISB-210226
15. Ventre, E., Espinasse, T., Bréhier, C.E., Calvez, V., Lepoutre, T., Gandrillon, O.: Reduction of a stochastic model of gene expression: lagrangian dynamics gives access to basins of attraction as cell types and metastabilty. J. Math. Biol. **83**(5), 59 (2021). https://doi.org/10.1007/s00285-021-01684-1
16. Ventre, E., Herbach, U., Espinasse, T., Benoit, G., Gandrillon, O.: One model fits all: combining inference and simulation of gene regulatory networks. PLoS Comput. Biol. **19**(3), e1010962 (2023). https://doi.org/10.1371/journal.pcbi.1010962

Approximate Constrained Lumping of Polynomial Differential Equations

Alexander Leguizamon-Robayo[1(✉)], Antonio Jiménez-Pastor[1], Micro Tribastone[2], Max Tschaikowski[1], and Andrea Vandin[3,4]

[1] Aalborg University, Aalborg, Denmark
alexanderlr@cs.aau.dk

[2] IMT School for Advanced Studies Lucca, Lucca, Italy

[3] Sant'Anna School of Advanced Studies, Pisa, Italy

[4] DTU Technical University of Denmark, Lyngby, Denmark

Abstract. In life sciences, deriving insights from dynamic models can be challenging due to the large number of state variables involved. To address this, model reduction techniques can be used to project the system onto a lower-dimensional state space. Constrained lumping can reduce systems of ordinary differential equations with polynomial derivatives up to linear combinations of the original variables while preserving specific output variables of interest. Exact reductions may be too restrictive in practice for biological systems since quantitative information is often uncertain or subject to estimations and measurement errors. This might come at the cost of limiting the actual aggregation power of exact reduction techniques. We propose an extension of exact constrained lumping which relaxes the exactness requirements up to a given tolerance parameter ε. We prove that the accuracy, i.e., the difference between the output variables in the original and reduced model, is in the order of ε. Furthermore, we provide a heuristic algorithm to find the smallest ε for a given maximal approximation error. Finally, we demonstrate the approach in biological models from the literature by providing coarser aggregations than exact lumping while accurately capturing the original system dynamics.

Keywords: Approximate reduction · Dynamical systems · Constrained lumping

1 Introduction

Dynamical models of biochemical systems help discover mechanistic principles in living organisms and predict their behavior under unseen circumstances. Realistic and accurate models, however, often require considerable detail that inevitably leads to large state spaces. This hinders both human intelligibility and numerical/computational analysis. For example, even the relatively primary mechanism of protein phosphorylation yields, in the worst case, a combinatorial state space as a function of the number of phosphorylation sites [37].

As a general way to cope with large state spaces, model reduction aims at providing a lower-dimensional representation of the system under study that

J. Pang and J. Niehren (Eds.): CMSB 2023, LNBI 14137, pp. 106–123, 2023.
https://doi.org/10.1007/978-3-031-42697-1_8

retains some dynamical properties of interest to the modeler. In applications to systems biology, it is beneficial that the reduced state space keeps physical interpretability, mainly when the model is used to validate mechanistic hypotheses [3,38,41]. There is a variety of approaches in this context, such as those exploiting time-scale separation properties [32,51], quasi-steady-state approximation [35,39], heuristic fitness functions [42], spatial regularity [48], sensitivity analysis [40] and, at last, conservation analysis, which detects linear combinations of variables that are constant at all times [49].

By lumping, one generally refers to the latter class, with a self-consistent system of dynamical equations comprised of a set of macro-variables, each given in terms of combination of the original ones [1,6,22,32,40]. In linear lumping, this reduction is expressed as a linear transformation of the original state variables. Since this can destroy physical intelligibility in general, *constrained lumping* allows restricting to only part of the state, allowing defining linear combinations of state variables that ought to be preserved in the reduction [30]. Lumping techniques of Markov chains date back to early nineties [27], were later extended to stochastic process algebra [11,24,44,46,52] and have expanded recently to efficient algorithmic approaches for stochastic chemical reaction networks [12] and deterministic models of biological systems [15,17,19]. In these work, reduction mappings are linear mappings induced by a partition (i.e., an equivalence relation) of state variables; in the aggregated system each macro-variable represents the sum of the original variables of a partition block.

In this paper we are concerned with systems of ordinary differential equations (ODEs) with polynomial derivatives, subsuming mass-action kinetic models (e.g., [45,50]) which is however also rich enough to cover electric circuits [13]. For these, CLUE has been presented as an algorithm that efficiently computes constrained linear lumping [33], avoiding the computational shortcomings of previous work due to the symbolic computation of the eigenvalues of a nonconstant matrix [28,30], enabling the analysis of models with several thousands of equations on standard hardware. In particular, CLUE can compute *the smallest* linear dimensional reduction that preserves the dynamics of arbitrary linear combinations of original state variables given by the user.

The aforementioned lumping approaches are *exact*, in that the reduced model does not incur any approximation error (but only loss of information because, in general, the aggregation map is not invertible). Approximate lumping is a natural extension that has been studied for a long time (e.g., [29]). Indeed, although exact reduction methods have been experimentally proved successful in a large variety of biological systems (e.g., [34]), approximate reductions can be more robust to parametric uncertainty—which notoriously affects systems biology models (e.g., [4,7])—and offer a flexible trade-off between the aggressiveness of the reduction and its precision. For partition-based lumping algorithms, this has been recently explored. In [18], the authors present an algorithm for approximate aggregation parameterized by a tolerance parameter ε, which, informally, relaxes an underlying criterion of equality for two variables to be exactly lumped into the same block.

In this paper, we present for the first time a polynomial time algorithm for the computation of approximate constrained lumpings. It is an extension of CLUE that relaxes its conditions for exact lumping using a *lumping tolerance* parameter that is roughly related to how close a lumping matrix is to an exact lumping. Moreover, we show that the actual error is proportional to the lumping tolerance. Using a prototype implementation, we evaluate our approximate constrained lumping on three representative case studies from the literature. Overall, the numerical results show that our approximate extension can lead to substantially smaller reduced models while introducing limited errors in the dynamics.

Further Related Work. We are complementary to classic works [29,30] which do not address the efficient algorithmic computation of lumping. Moreover, we are more general than [18,26,47] which consider so-called "proper" lumping (i.e., [40]) where each state variable appears in exactly one aggregated variable. Likewise, we are complementary to rule-based reduction techniques [21] which are independent of kinetic parameters [14]. As less closely related abstraction techniques, we mention (bisimulation) distance approaches for the approximate reduction of Markov chains [5,20] and proper orthogonal projection [2], which, however, apply to linear systems. Abstraction of chemical reaction networks by means of learning [10,36] and simulation [23] are complementary to lumping methods.

Outline. Section 2 provides necessary preliminary notations, Sect. 3 introduces approximate constrained lumping, while Sect. 4 discusses how to compute it and how to choose the lumping tolerance value. Section 5 evaluates our proposal on models from the literature, while Sect. 6 concludes the paper.

Notation. The derivative with respect to time of a function $x : [0; T] \to \mathbb{R}^m$ is denoted by $\dot{x}$. We denote a dynamical system by $\dot{x} = f(x)$, and denote its initial condition by $x(0) = x^0$. For $f : \mathbb{R}^m \to \mathbb{R}^n$, we denote the Jacobian of f at x by $J(x)$. For any vector $x \in \mathbb{R}^m$, we denote by $\mathbb{R}[x]$ the rings of polynomial functions of x with real coefficients respectively. Given a matrix $L \in \mathbb{R}^{m \times n}$, the rowspace of L or the vector space generated by the rows of L is denoted by $\mathrm{rowsp}(L)$. We denote by $\bar{L} \in \mathbb{R}^{n \times m}$ a right pseudoinverse of L (i.e. $L\bar{L} = I_n$ where $I_n \in \mathbb{R}^{n \times n}$ is the identity matrix). The term *pseudoinverse* refers to a right pseudoinverse.

2 Preliminaries

In this paper we study systems of ODEs with polynomial derivatives of the form:

$$\dot{x} = f(x), \tag{1}$$

where $f : \mathbb{R}^m \to \mathbb{R}^m, x \mapsto (f_1(x), \ldots, f_m(x))^T$ and $f_i \in \mathbb{R}[x]$, for $i = 1, \ldots, m$.

Definition 1. Given a system of polynomial ODEs, and a full rank matrix $L \in \mathbb{R}^{l \times m}$ with $l < m$, we say that L is an *exact lumping of dimension* l (or that the system is *exactly lumpable* by L) if there exists a function $g : \mathbb{R}^l \to \mathbb{R}^l$ with polynomial entries such that $L \circ f = g \circ L$.

Example 1. Consider the following system

$$\dot{x_1} = x_2^2 + 4x_2x_3 + 4x_3^2, \qquad \dot{x_2} = 2x_1 - 4x_3, \qquad \dot{x_3} = -x_1 - x_2. \quad (2)$$

Then, the matrix $L = (1\ 0\ 0, 0\ 1\ 2)^T$ is an exact lumping of dimension 2. To see this we compute

$$\begin{pmatrix} \dot{y}_1 \\ \dot{y}_2 \end{pmatrix} = \begin{pmatrix} \dot{x}_1 \\ \dot{x}_2 + 2\dot{x}_3 \end{pmatrix} = \begin{pmatrix} (x_2 + 2x_3)^2 \\ -2x_2 - 4x_3 \end{pmatrix} = \begin{pmatrix} y_2^2 \\ -2y_2 \end{pmatrix}.$$

In this case, we have $g(y) = (y_2^2, -2y_2)^T$. ■

Definition 2. Given a system of polynomial ODEs, and an exact lumping $L \in \mathbb{R}^{l \times m}$, we say that $y = Lx$ are the *reduced (or lumped) variables*, and their evolution is given by the *reduced system* $\dot{y} = g(y)$.

Given an initial condition $x^0 \in \mathbb{R}^n$ and an exact lumping L, there is a corresponding initial condition y^0 in the lumped variables given by $y^0 = Lx^0$. Similarly, since $y = Lx$ and $g \circ L = Lf$, we have that

$$L\dot{x} = Lf(x) = g(Lx) = g(y) = \dot{y}\ .$$

This means that we can study the evolution of the lumped variables $y(t)$ by solving the (smaller) reduced system rather than the (larger) original one.

Suppose that we want now to recover the evolution of some linear combination of state variables. To answer whether this is possible in the context of our example, we introduce the notion of constrained lumping. Roughly speaking, this is an exact lumping able to preserve given observations of interest.

Definition 3. Let $x_{obs} = Mx$ for some matrix $M \in \mathbb{R}^{p \times m}$, for $p < m$. We say that a lumping L is a *constrained lumping* with *observables* x_{obs} if $\text{rowsp}(M) \subseteq \text{rowsp}(L)$. This means that each entry of x_{obs} is a linear combination of the reduced variables y.

Example 2. Consider the system in Eq. (2) and suppose we are interested in observing the quantity $2x_1 + x_2 + 2x_3$. In this case $x_{obs} = Mx$ with $M = (2\ 1\ 2)$. We can see that L from Example 1 is a constrained lumping as we can recover the observable from the reduced system, i.e., $x_{obs} = 2y_1 + y_2$. Suppose now that we want to observe the quantity $x_1 + x_2 + x_3$. This will be described by $x_{obs} = Mx$ with $M = (1\ 1\ 1)$. In this case, the matrix L is not a constrained lumping as there is no way to obtain x_{obs} as a linear combination of y_1 and y_2. ■

To understand how constrained lumping can be computed, we first review the following known characterization of lumping.

Theorem 1 (Characterization of Exact Lumping [43]) *Given a system $\dot{x} = f(x)$ of m polynomial ODEs and a matrix $L \in \mathbb{R}^{l \times m}$ with rank l, the following are equivalent.*

1. *The system is exactly lumpable by L.*
2. *For any pseudoinverse $\bar{L}$ of L, $Lf = (Lf) \circ \bar{L}L$.*
3. *The row space of L is invariant under $J(x)$ for all $x \in \mathbb{R}^m$, where $J(x)$ is the total derivative of f at x, known also as the Jacobian. More formally,* $\text{rowsp}(LJ(x)) \subseteq \text{rowsp}(L)$ *for all $x \in \mathbb{R}^m$.*

The characterization of exact lumpings in Theorem 1 provides us with a way to compute constrained lumpings L. This is because thanks to Point 3, the problem of computing a lumping is equivalent to the problem of finding a $J(x)$-invariant subspace of $\mathbb{R}^m$ for all $x \in \mathbb{R}^m$. This gives rise to Algorithm 1.

Algorithm 1. Constrained lumping [33]

Input: A system $\dot{x} = f(x)$ of m polynomial ODEs; A $p \times m$ matrix M with row rank p.
1: **compute** $J(x)$, the Jacobian of $f(x)$
2: **compute** a representation $J(x) = \sum_{i=1}^{N} J_i \mu_i(x)$ (Theorem 2);
3: **compute** L (Algorithm 2)
4: **return** Constrained lumped ODE system $\dot{y} = Lf(\bar{L}y)$.

Algorithm 2. Computation of L

Input: Matrices J_i, $i = 1, \ldots, N$, such that $J(x) = \sum_{i=1}^{N} J_i \mu_i(x)$ (Theorem 2); A $p \times m$ matrix M with row rank p.
1: **set** $L := M$
2: **repeat**
3: **for all** $1 \leq i \leq \kappa$ **and rows** r **of** L **do**
4: **compute** rJ_i
5: **if** $rJ_i \notin \text{rowsp}(L)$ **then**
6: **append** row rJ_i to L
7: **end if**
8: **end for**
9: **until** no rows are appended to L
10: **return** Lumping matrix L.

We can see that Algorithm 1 uses a secondary algorithm, Algorithm 2, to actually compute constrained lumpings L. We explain next how this is done efficiently. In particular, the next theorem shows how to verify Point 3 of Theorem 1 by performing a finite number of numerical checks. The number is proportional to the number of monomials present in the polynomial vector field f.

Theorem 2 ([33, Lemma I.1]). *Consider a system of polynomial ODEs as in Eq.* (1). *Let $J(x)$ be the Jacobian matrix of f. We have that $J(x)$ can be represented as*

$$J(x) = \sum_{i=0}^{N} J_i \mu_i(x), \tag{3}$$

where $\{\mu_i(x) : i \in \{0, \ldots, N\}\}$ is the set of monomials in $J(x)$. Then, for all rows r of L, $rJ(x) \in \text{rowsp}(L)$ for all x if and only if $rJ_i \in \text{rowsp}(L)$ for $i = 1, \ldots, N$.

Example 3. The Jacobian for the system of Example 1 is given by

$$J(x) = \begin{pmatrix} 0 & 2x_2 + 4x_3 & 4x_2 + 8x_3 \\ 2 & 0 & -4 \\ -1 & -1 & 0 \end{pmatrix}.$$

Using Theorem 2, $J(x)$ can be represented as

$$\begin{aligned} J(x) &= J_1\mu_1(x) + J_2\mu_2(x) + J_3\mu_3(x) \\ &= \begin{pmatrix} 0 & 0 & 0 \\ 2 & 0 & -4 \\ -1 & -1 & 0 \end{pmatrix} 1 + \begin{pmatrix} 0 & 2 & 4 \\ 0 & 0 & 0 \\ 0 & 0 & 0 \end{pmatrix} x_2 + \begin{pmatrix} 0 & 4 & 8 \\ 0 & 0 & 0 \\ 0 & 0 & 0 \end{pmatrix} x_3. \end{aligned}$$

We note that, in this example, the μ_i functions are $\mu_1(x) = 1, \mu_2(x) = x_2$ and $\mu_3(x) = x_3$. Thanks to the fact that the vector field is polynomial, monomials μ_i can be computed in polynomial time [33]. ■

Theorem 2 gives a computational way to verify Point 3 in Theorem 1, see Algorithm 1–2 whose polynomial complexity is addressed in [33].

3 Approximate Constrained Lumping

In this section we provide an estimation for the error underlying an approximate lumping. We begin by demonstrating why the exact lumping condition can be too restrictive for finding reductions. Consider the following system of ODEs:

$$\dot{x}_1 = x_2^2 + 4.05x_2x_3 + 4x_3^2, \quad \dot{x}_2 = 2x_1 - 4x_3, \quad \dot{x}_3 = -x_1 - x_2. \tag{4}$$

Notice that the system given by Eq. (4) is similar to that of Eq. (2), as we have just added a term $0.05x_2x_3$ to its first equation. This new system is not exactly lumpable by the matrix L of Example 1. However, we would like to know if it is still possible to use the matrix L to obtain a *useful* reduced system which approximates well the original one. To do so, we first want to evaluate how *close* the matrix L is to be (or how much it *deviates* from) an exact lumping.

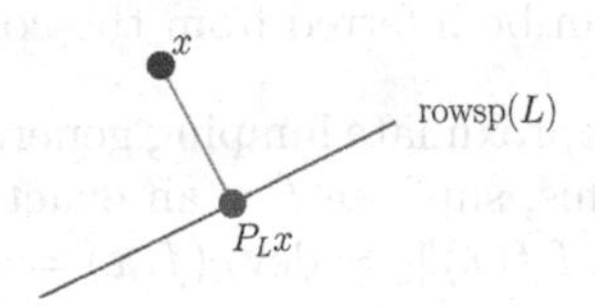

Fig. 1. Points projected to rowsp(L).

By Theorem 1, Point 2, a full rank matrix L is a lumping if and only if the equality $Lf(\bar{L}Lx) - Lf(x) = 0$ is satisfied for all $x \in \mathbb{R}^m$. In our proposal of approximate lumping, we relax this requirement by asking it to be satisfied up to a certain tolerance.

Definition 4. Consider the system of ODEs as in Eq. (1). Let $L \in \mathbb{R}^{l\times m}$ be full rank matrix with $l < m$ and denote by $\bar{L}$ the Moore-Penrose right pseudo inverse of L. We define the projection onto rowsp(L) as $P_L = \bar{L}L$ and the *deviation of* L *at* x by

$$\text{dev}_L(f, x) := \|Lf(\bar{L}Lx) - Lf(x)\|_2. \tag{5}$$

To understand the intuition behind Definition 4, consider the point x in Fig. 1. Then the deviation computes the difference between the images under $L \circ f$ of x (purple) and $P_L x$ (green). The deviation is identically zero if and only if L is a lumping.

Example 4. Consider the matrix L given in Example 1. A pseudoinverse of L is given by $\bar{L} = (1\ 0\ 0, 0\ 0.2\ 0.4)^T$. Writing f^1 for the corresponding vector field, we obtain that $dev_L(f^1, (1,1,1)^T) = 0$, as L is an exact lumping. Now consider the system given by Eq. (4) and use f^2 to denote the corresponding vector field. Then, we get that $\text{dev}_L(f^2, (1,1,1)^T) = 0.014$. Therefore, the matrix L is not an exact lumping of Eq. (4). ■

From a modelling perspective, a reduced model is meaningful insofar as its predictions for a set of initial conditions of interest are *close enough* to those of the original model throughout a given finite time horizon of interest. On the other hand, the theory in [33] guarantees that reduced models provide exact predictions. This might restrict the actual aggregation power of the technique. Here, we aim at relaxing the theory by considering reductions that do not need to be exact or tight. Having this in mind, we introduce the following notion.

Definition 5. Consider the system of ODEs given by Eq. (1) and a set of initial conditions S. Let $L \in \mathbb{R}^{l \times m}$ be a full rank matrix with $l < m$. Given $\eta > 0$ and a time horizon $T > 0$, we say that Eq. (1) is *approximately lumpable* by L with *deviation tolerance* η if

$$\text{dev}_L(f, x(t)) \leq \eta, \tag{6}$$

for all $t \in [0, T]$ and all initial conditions $x(0) \in S$, where $x(t)$ is the solution of the system in Eq. (1). We say that L is an *approximate lumping* for the set S, time horizon T, and deviation tolerance η (or (S, T, η)-lumping). We will omit S, T or η whenever they can be inferred from the context.

Remark 1. The notion of approximate lumping generalizes that of exact lumping from Definition 1. To see this, suppose L is an exact lumping. Then by Point 2 of Theorem 1, $\|Lf(\bar{L}Lx) - Lf(x)\|_2 = \text{dev}_L(f, x) = 0$, for all $x \in \mathbb{R}^m$. It follows that L is a $(\mathbb{R}^m, \infty, 0)$-lumping.

A common type of study is to see the evolution of a system $\dot{x}(t) = f(x(t))$ until it reaches a steady-state for $t \to \infty$. In practice, modelers equip biological models with finite time horizon T, thus implicitly assuming that the system will reach steady-state at T. We shall adhere to this heuristic but point out that a rigorous steady-state analysis would require to ensure asymptotic convergence using, e.g., Lyapunov functions [18].

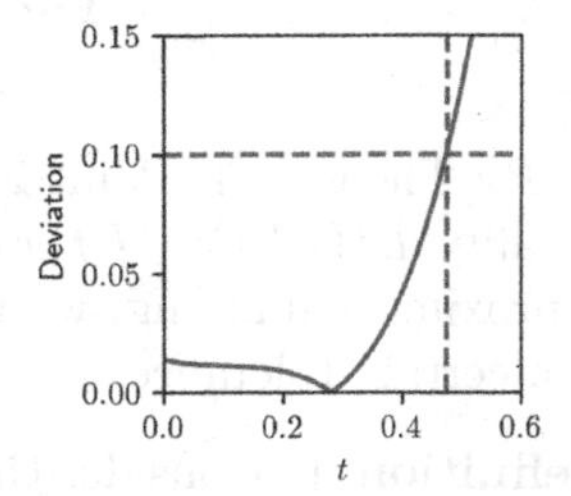

Fig. 2. Example 5: evolution of $\text{dev}_L(f(x(t)))$.

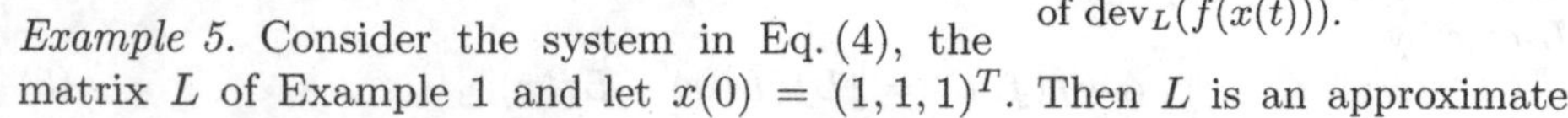

Example 5. Consider the system in Eq. (4), the matrix L of Example 1 and let $x(0) = (1,1,1)^T$. Then L is an approximate

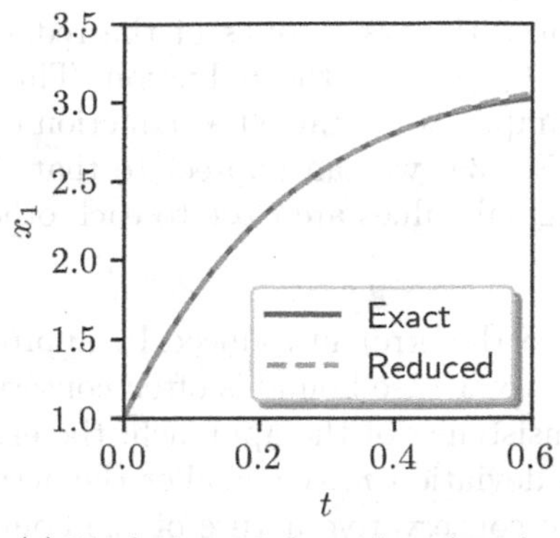

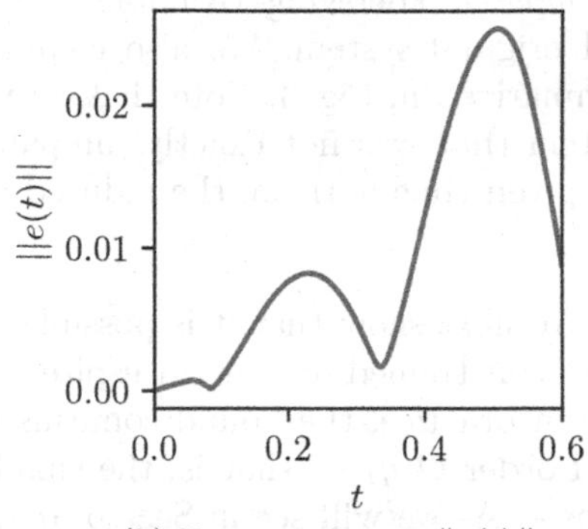

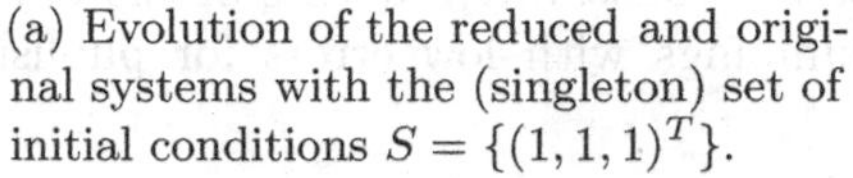

(a) Evolution of the reduced and original systems with the (singleton) set of initial conditions $S = \{(1, 1, 1)^T\}$.

(b) Error evolution $\|e(t)\|_2$

Fig. 3. Reduced system and error computation for Example 6 using the matrix L of Example 1 on Eq. (4).

lumping for $x(0)$, time $T = 0.475$, and deviation tolerance 0.1, i.e., L is a $(\{x(0)\}, 0.475, 0.1)$-lumping. To see this, note that in Fig. 2 the deviation of the dynamics is bounded by 0.1. ■

After having generalized the notion of lumping in that of approximate lumping, we next introduce approximate constrained lumping in the obvious manner.

Definition 6. Let $x_{obs} = Mx$ for some matrix $M \in \mathbb{R}^{p \times m}$ with $p < m$. We say that an approximate lumping L of Eq. (1) is an *approximate constrained lumping* with *observables* x_{obs} if $\text{rowsp}(M) \subseteq \text{rowsp}(L)$.

Similarly to Definition 2, Definition 6 means that the observables x_{obs} can be recovered as a linear combination of the reduced variables $y = Lx$.

Definition 7. The *reduced system* induced by an approximate lumping L is given by $\dot{y} = Lf(\bar{L}y)$, where $y \in \mathbb{R}^l$.

Example 6. Consider the system in Eq. (4) and the matrix L given in Example 1. We can compute the reduced system as follows

$$\begin{pmatrix} \dot{y}_1 \\ \dot{y}_2 \end{pmatrix} = Lf\left(\bar{L}\begin{pmatrix} y_1 \\ y_2 \end{pmatrix}\right) = Lf\left(\begin{pmatrix} y_1 & 0 \\ 0 & 0.2y_2 \\ 0 & 0.4y_2 \end{pmatrix}\right) = L\begin{pmatrix} 1.004y_2^2 \\ 2y_1 - 1.6y_2 \\ -y_1 - 0.2y_2 \end{pmatrix} = \begin{pmatrix} 1.004y_2^2 \\ -2y_2 \end{pmatrix}.$$

We can see that L induces the reduced system $\dot{y} = (1.004y_2^2, -2.0y_2)^T$. ■

Definition 8. Given the system in Eq. (1) and an approximate lumping L, we define the *error* of the reduction by $e(t) := y(t) - Lx(t)$, where y is the solution to the reduced system given by Definition 7. The dynamics of the approximation error, instead, is given by $\dot{e} = \dot{y} - L\dot{x}$, with $e(0) = 0$.

Example 7. Following Example 6, we compute the trajectories of the reduced and original system. We also compute the L_2-error of the reduction. This is summarized in Fig. 3. Note that, in this example, we obtained a reduction of a system that was not exactly lumpable. In Fig. 3a, we can appreciate that, for the given time horizon, the reduced and original values are close to each other. ■

We next show that it is possible to bound the error introduced by approximate constrained lumping. Despite that our worst case bound is often conservative in practice, the bound confirms the consistency of the approach: the error is of order $\mathcal{O}(\eta)$ — that is, the smaller the deviation η, the smaller the actual error e. As we will see in Sect. 5, despite the conservative nature of the bound, our framework can find approximate lumpings with low errors for published biological models.

Theorem 3 (Error Bounds). *Fix a bounded set of initial conditions S, a finite time horizon T and assume that the respective reachable set of the m-dimensional polynomial ODE system $\dot{x} = f(x)$ is bounded on $[0;T]$. Then, there exists a constant $C \geq 0$ such that for any $\eta > 0$ for which L is a (S,T,η)–lumping, it holds that $\|e(t)\|_2 \leq \eta \cdot K_{C,L,T}$, where*

$$K_{C,L,T} = \frac{1}{C\|L\|_2\|\bar{L}\|_2}\left(e^{C\|L\|_2\|\bar{L}\|_2 T} - 1\right)$$

and C is the Lipschitz constant of f over the set of initial conditions S.

4 Computing Constrained Approximated Lumping

Armed with the estimation of the error underlying approximate lumpings, we next focus on their computation. In Sect. 4.1, we introduce a numerical lumping tolerance ε to compute a lumping matrix L, opposed to the *deviation tolerance* η. Moreover, we show how the lumping tolerance ε and the deviation tolerance η relate to each other. In Sect. 4.2 instead we discuss how to pick the best value for the lumping tolerance ε.

4.1 Lumping Algorithm

In this section, we relax the condition in Line 5 of Algorithm 2, thus allowing to find approximate reductions. Intuitively, we add a new row rJ_i to the matrix L only if it is far enough from rowsp(L). We make this rigorous by fixing a lumping tolerance ε and adding a row rJ_i only if $\|rJ_i - \pi_i\|_2 > \varepsilon$, where $\pi_i := rJ_iP_L$ is the orthogonal projection of rJ_i onto rowsp(L). Thus, we propose Algorithm 3.

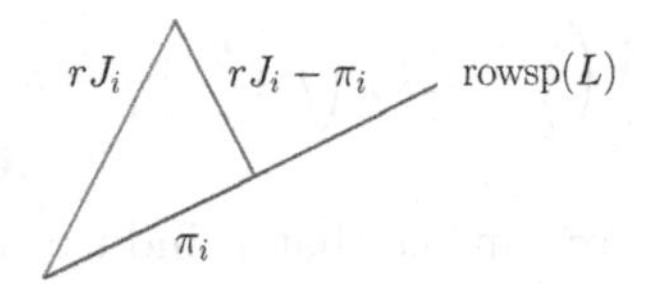

Fig. 4. Decomposition of rJ_i into rowsp(L) and rowsp$(L)^{\perp}$.

Algorithm 3. Approximate Constrained Lumping Algorithm

Input: numerical threshold $\varepsilon \geq 0$;
set of matrices J_i, $i = 1, \ldots, N$ such that $J(x) = \sum_{i=1}^{N} J_i \mu_i(x)$ (Theorem 2)
a $p \times m$ matrix of observables M with row rank p

1: **compute** orthonormal rows spanning the row space of M and store them as M
2: **set** $L := M$
3: **repeat**
4: **for all** $1 \leq i \leq N$ and rows r of L **do**
5: **compute** $\pi_i := rJ_iL^TL$
6: **if** $\|rJ_i - \pi_i\|_2 > \varepsilon$ **then**
7: **append** row $(rJ_i - \pi_i)/\|rJ_i - \pi_i\|_2$ to L
8: **end if**
9: **end for**
10: **until** no rows have been appended to L
11: **return** lumping matrix L.

Remark 2. In Line 7 of Algorithm 3 we do not append rJ_i, given by the orange vector in Fig. 4. Rather, we append the normalized component of rJ_i in the orthogonal direction to $\text{rowsp}(L)$, which is $rJ_i - \pi_i$, the green vector in Fig. 4. This ensures that the matrix L is orthonormal, thus we can use the fact that the pseudoinverse of an orthonormal matrix is its transpose to obtain that $\bar{L} = L^T$.

We now provide a detailed example of the use of Algorithm 3.

Example 8. We apply Algorithm 3 to Eq. (4) for $\varepsilon = 0.1$ and $M = (1, 0, 0)$. In other words, we are observing the component x_1 from the original system. The Jacobian of $f(x)$ can be represented as $J_1(1) + J_2(x_2) + J_3(x_3)$, where

$$J_1 = \begin{pmatrix} 0 & 0 & 0 \\ 2 & 0 & -4 \\ -1 & -1 & 0 \end{pmatrix}, \quad J_2 = \begin{pmatrix} 0 & 2 & 4.05 \\ 0 & 0 & 0 \\ 0 & 0 & 0 \end{pmatrix}, \quad J_3 = \begin{pmatrix} 0 & 4.05 & 8 \\ 0 & 0 & 0 \\ 0 & 0 & 0 \end{pmatrix}.$$

We begin the computation in Line 2 by setting $L = M = (1, 0, 0)$ and begin the main loop in Line 4 with $r = (1, 0, 0)$. To carry out the computation of Line 5, we have that $rJ_1 = (0, 0, 0)$, $rJ_2 = (0, 2, 4.05)$, $rJ_3 = (0, 4.05, 8)$. Since $rJ_1 = \pi_1 = (0, 0, 0)$, by the check of Line 6, we do not append any new row to L. As next, we compute π_2 following Line 5. As $\pi_2 = 0$, it follows that $\|rJ_2 - \pi_2\|_2 = \|rJ_2\|_2 = 4.52 > 0.1$, which, by Line 6, means that we need to append a new row to L. By Line 7, we add normalized $rJ_2 - \pi_2$ as a row of L (Line 7), thus obtaining

$$L = \begin{pmatrix} 1 & 0 & 0 \\ 0 & 0.443 & 0.897 \end{pmatrix}. \tag{7}$$

Going back to Line 5, we compute $\pi_3 = rJ_3L^TL = (0, 3.970, 8.040)$. Following Line 6, since $\|rJ_3 - \pi_3\|_2 = 0.09 < 0.1$, we do not add any additional row to L. As we have only checked one of the rows of L so far, we follow the main loop (Line 4) by setting $r = (0, 0.443, 0.897)$. We compute

$$rJ_1 = (-0.011, -0.900, -1.771), \; rJ_2 = (0, 0, 0), \; rJ_3 = (0, 0, 0).$$

We skip to the check of Line 6, where we have that $\|rJ_1 - \pi_1\|_2 = 0.02 < 0.1$. Similarly, we have that $\|rJ_2 - \pi_2\|_2 = \|rJ_3 - \pi_3\|_2 = 0$, meaning that no more rows should be added to L, terminating the algorithm. The output of Algorithm 3 is then the matrix L of Eq. (7). ■

Being a generalization of Algorithm 2, the polynomial complexity of Algorithm 3 follows from [33] and the fact that linear projections can be computed in polynomial time. Likewise, its result is unique and minimal, provided that the input matrix M has full row-rank.

The next result relies on Theorem 2 and ensures that the approximate reductions found by Algorithm 3 admit a deviation tolerance of order $\mathcal{O}(\varepsilon)$. That is, the deviation tolerance η is in the order of the lumping tolerance ε.

Theorem 4. (Correctness of Algorithm *3***).** *Consider a system of ODEs of the form* (1) *of dimension* m *and a set of initial conditions* S*, such that none of the solutions explode in a finite time* $t \leq T$*, where* T *is a given time horizon. Let* L *be the matrix computed via Algorithm 3 with lumping tolerance* ε *and let* Ω *be an open ball centered in the origin of diameter* K *such that for all* $x(0) \in S$*, the solution* $x(t)$ *of* (1) *remains in* Ω *for all* $t \in [0, T]$*. Then,* L *is a* $(T, S, \sqrt{m}C'K\varepsilon)$*-lumping, where* $C' = \sup_{i,x\in\Omega} |\mu_i(x)|$*.*

4.2 Heuristic Search of Lumping Tolerance

In order to apply Algorithm 3, it is important to choose an appropriate value for ε. While Theorem 3 together with Theorem 4 allows for an estimation of the error in terms of ε, in practice, this bound is not tight enough. For this reason, we introduce a heuristic approach to appropriately choose ε.

Intuitively, by increasing ε, one can lump more variables together at the price of incurring larger approximation errors. We would like to find the largest admissible ε such that the approximation error is small enough. Note that the minimum value that ε can have is 0, corresponding to an exact reduction. A naive idea would be to set $\varepsilon = 0$ and add small increments until the reduction is satisfying. However, this requires an appropriate choice of increment which in turn depends on the model. Instead, Lemma 1 gives us an upper bound for ε which can be computed for each model.

Lemma 1. (Upper bound on ε**).** *Consider a matrix of observables* $M \in \mathbb{R}^{p\times m}$ *of rank* p *with* $j-$*th row denoted by* r_j *and a decomposition of the Jacobian* $J(x) = \sum_{i=1}^{n} J_i\mu_i(x)$*. Then there is value* $\varepsilon_{\max}$ *given by*

$$\varepsilon_{\max} = \max_{j,i} \|r_j J_i\|_2, \tag{8}$$

for $i = 1, \ldots, n$ *and each* $j = 1, \ldots, l$*, such that for any* $\varepsilon \geq \varepsilon_{\max}$ *the output of Algorithm 3 will be a matrix of orthonormal rows spanning the row space of* M*.*

In other words, any $\varepsilon \geq \varepsilon_{\max}$ will collapse the dynamics of the system onto M. Note also that the error e and the deviation dev_L of the reduction underlying an approximate lumping L increase as the lumping tolerance ε increases. To find the largest admissible ε, we need a way to approximate the error without having to simulate the original and reduced systems. To this aim, we estimate the deviation of lumping matrix L by sampling and averaging the deviation for N randomly chosen points over $[0; \|x^0\|_2]^m$, where x^0 is the center of the initial set. The computation corresponds to the Monte Carlo approximation of the L_1 norm of dev_L over $[0; \|x^0\|_2]^m$. The choice of the compactum is motivated by the fact that the diameter of the reachable set will be of the order of the initial set.

Algorithm 4. Finding the largest acceptable ε for Algorithm 3

Input: set of matrices J_i, $i = 1, \dots, N$ such that $J(x) = \sum_{i=1}^{N} J_i \mu_i(x)$ (Theorem 2)
  maximum deviation $\eta_{\max}$
  minimal difference $d_{\min}$
  set of initial conditions S and its center x^0
  number of random samples $M \geq 1$

1: **set** $\varepsilon_{\min} = 0$
2: **compute** $\varepsilon_{\max} = \max_{j,i} \|r_j J_i\|_2$
3: **repeat**
4: **compute** $\varepsilon = (\varepsilon_{\max} + \varepsilon_{\min})/2$
5: **compute** L using Algorithm 3 with ε
6: **compute** $\mathrm{dev}_L = \frac{1}{M}\sum_{i=1}^{M} \|Lf(L^T Lx^i) - Lf(x^i)\|_2$ for random $x^i \in [0; \|x^0\|_2]^m$
7: **if** $\mathrm{dev}_L > \eta_{\max}$ **then**
8: **set** $\varepsilon_{\max} = \varepsilon$
9: **else**
10: **set** $\varepsilon_{\min} = \varepsilon$
11: **end if**
12: **compute** $d = \varepsilon_{\max} - \varepsilon_{\min}$
13: **until** $d < d_{\min}$
14: **return** $\varepsilon_{\min}$.

To find the largest admissible ε that leads to a reduction close to a given maximal deviation tolerance $\eta_{\max}$, it suffices to search over $\varepsilon \in [0, \varepsilon_{\max}]$ thanks to Lemma 1. We find such an ε by performing a binary search over $[0, \varepsilon_{\max}]$, relying on the intuition that $\mathrm{dev}_{L(\varepsilon_1)} \leq \mathrm{dev}_{L(\varepsilon_2)}$ where $\varepsilon_1 \leq \varepsilon_2$ and $L(\varepsilon_i)$ is the result of Algorithm 3 for input ε_i. The corresponding realization is provided in Algorithm 4, where the maximal acceptable deviation tolerance $\eta_{\max}$ is chosen as a percentage of $\|f(x^0)\|_2$, with x^0 being again the center of the initial set. This is motivated by the fact that dev_L is stated in terms of f, see Definition 4.

5 Evaluation

In this section, we evaluate our method by demonstrating its higher reduction power over exact constrained lumping. We use the following selection of models from the literature, with observables taken from the cited papers (For models with more than one observable we arbitrarily selected one.)

- **BioNetGen_CCP** [31]: the central carbon pathway of E. coli is a metabolic process that converts glucose into energy and building blocks for cell growth. **Observable:** concentration of 1,3-diphosphoglycerate (D13PG). This corresponds to the sum of 2 out of the 87 variables in the model.
- **NIHMS80246_S6** [9]: FceRI-like network of a cell-surface receptor of the receptor tyrosine kinase (RTK) family.
  Observable: total amount of phosphorylation at Y2 (S2P). This corresponds to the sum of 10 out of the 24 variables in the model.
- **pcbi_1003544_s006** [25]: trivalent-ligand bivalent-receptor (TLBR) model, which is a simplified representation of receptor aggregation following multivalent ligand binding.
  Observable: total concentration of membrane receptor (RC). This corresponds to the sum of 18 out of the 62 variables in the model.

These are chemical reaction networks written in the input language of the tool BioNetGen [8] as available in the corresponding cited papers. We use MATLAB to implement all algorithms presented in this paper, as well as to carry out the simulations of the reduced and original models. The Jacobians for all models were automatically computed using ERODE [16], using the importing capabilities of ERODE for the `.net` format generated using BioNetGen version 2.2.5-stable. The time horizon T and initial conditions for all models were taken from their original papers. For reproducibility, we provide all the generated MATLAB files, including the specifications of the models, at https://www.erode.eu/models/CMSB2023.zip.

Overall, these models showcase three exemplary situations: (1) the suitable ε computed by Algorithm 4 gives good reductions with low errors, while an excessively large ε completely destroys the observed dynamics; (2) our approach can only provide limited reduction improvements; (3) a model that cannot be reduced by exact constrained lumping, whereas approximate constrainted lumping is able to halve the number of variables while adding limited errors.

Table 1 presents the detailed results of these three experiments. Using Algorithm 4, we seek to find the largest lumping tolerance ε that does not exceed a given maximal deviation tolerance $\eta_{\max}$, where the latter is described as a percentage of the initial slope via $\eta_{\max} = \texttt{SlopePercentage}\|f(x_0)\|_2$. Similarly to the choice of the compact set for the evaluation of dev_L, the choice of $\eta_{\max}$ is justified by the fact that the slopes at points contained in the reachable set are roughly of the same order to the slope at the center of the initial set $f(x_0)$. For difference percentages, we then find the largest admissible lumping tolerance ε using Algorithm 4. Table 1 summarizes the results of the experiments. The column *Red. size* shows the size of the reduced model obtained by Algorithm 3. We display the absolute error and the relative error at the time horizon, respectively, in the columns $e(T)$ and $e(T)_{Rel}$ (where the relative error is given as the absolute error divided by the value of the observable of the exact reduction). We also display the number of iterations and the average computation time of Algorithm 4 over 5 runs in the *Time* column. All experiments were carried out on a 4.7 GHz Intel Core i7 computer with 32 GB of RAM. For each model, we plot in Fig. 5 the corresponding simulations of the observables for the considered values of ε.

Table 1. Validation: greater reduction power on original models.

Red. size	$e(T)$	$e_{Rel}(T)[\%]$	ε	*SlopePercentage*	*Iterations*	*Time[s]*
Experiment 1) Model: BioNetGen_CCP, size: 87, exact lumping: 30						
30	6.210E–06	5.960E–05	0.13313	10%	22	2.098
23	3.199E+00	3.070E+01	0.34426	30%	22	1.459
9	4.835E+00	4.641E+01	0.71208	50%	22	0.719
5	1.042E+01	1.000E+02	0.71720	70%	22	0.593
4	1.042E+01	1.000E+02	1.32714	90%	22	0.395
Experiment 2) Model: NIHMS80246_S6, size: 24, exact lumping: 19						
19	2.706E–08	2.706E–08	0.01696	10%	19	0.607
18	4.793E–04	4.793E–04	0.03000	30%-70%	19	0.895
8	9.999E+01	1.000E+02	0.21033	90%	19	0.441
Experiment 3) Model: pcbi_1003544_s006, size: 62, exact lumping: 62						
28	3.200E–09	2.366E–09	0.00079	10%-90%	22	2.762

In Experiment 1, we can see that, while for `SlopePercentage` 10% we get no reduction improvements, for `SlopePercentage` 30% and 50% get substantial reduction improvements going down to 23 and 9 variables, respectively, while preserving the shape of the dynamics of the observable (middle lines in Fig. 5a). Furthermore, it is interesting to note that aggressive reductions come at the price of larger errors. In fact, by setting `SlopePercentage` to 70% we lump the system so much that the observable in question degenerates to a constant (bottom flat line in Fig. 5a).

In Experiment 2 we can see that by setting `SlopePercentage` to 10% we do not get reduction improvements, while by allowing it to go up to 70% we only get very limited improvements (we reduce to 18 variables rather than to 19), without adding noticeable errors (this can be seen in Fig. 5b, were the exact and approximate reductions are indistinguishable). More aggressive reductions, e.g., obtaining 8 variables for `SlopePercentage` 90%, destroy the dynamics (the bottom flat line in Fig. 5b).

Finally, Experiment 3 shows a model that is not exactly lumpable. That is, exact lumping does not allow to lump its 62 variables. Instead, our approximate variant allows to reduce the model to less than 50% of the original variables (45%). This happens for any `SlopePercentage` from 10% to 90%. Most importantly, this reduction comes at very limited cost: as we can see in Fig. 5c, the lines can be hardly distinguished.

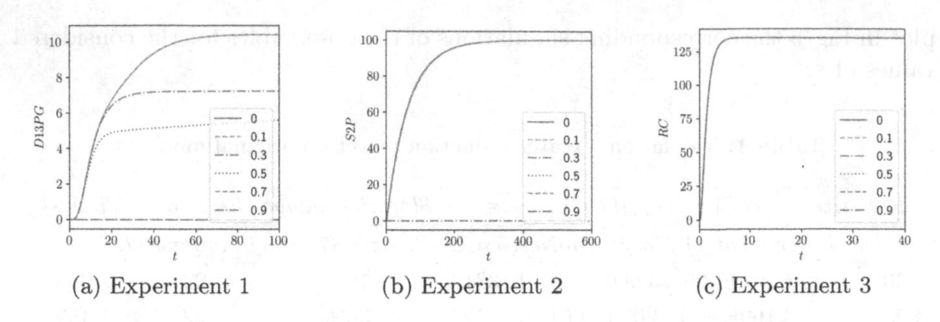

(a) Experiment 1 (b) Experiment 2 (c) Experiment 3

Fig. 5. Simulation of exact and approximately lumped models from Table 1, for varying `SlopePercentage`. The 0-line refers to exact constrained lumping.

6 Conclusion

In this work we introduced approximate constrained lumping, a framework for the reduction of systems of polynomial differential equations. While approximate lumping has been studied before, to the best of our knowledge no polynomial time algorithm for its computation has been provided before. The proposed algorithm takes as input a numerical tolerance and computes an approximate constrained lumping whose error is guaranteed to be proportional to the numerical threshold (with the proportional constant being not dependent on the threshold). The applicability of the framework is demonstrated by finding approximate lumpings of published models with low errors. Future work will consider the extension to more general vector fields, where the challenge is to express the derivative of the vector field as a sum of computationally amenable functions. In the case of polynomial vector fields considered here, for instance, these are monomials.

Acknowledgment. The work was partially supported by the DFF project REDUCTO 9040-00224B, the Poul Due Jensen Grant 883901, the Villum Investigator Grant S4OS, the PRIN project SEDUCE 2017TWRCNB and the co-funding of European Union - Next Generation EU, in the context of The National Recovery and Resilience Plan, Investment 1.5 Ecosystems of Innovation, Project Tuscany Health Ecosystem (THE), CUP: B83C22003920001.

References

1. Abate, A., Andriushchenko, R., Ceska, M., Kwiatkowska, M.: Adaptive formal approximations of Markov chains. Perform. Evaluation **148** (2021)
2. Antoulas, A.: Approximation of Large-Scale Dynamical Systems. Advances in Design and Control. SIAM (2005)
3. Apri, M., de Gee, M., Molenaar, J.: Complexity reduction preserving dynamical behavior of biochemical networks. J. Theor. Biol. **304**, 16–26 (2012)
4. Babtie, A., Stumpf, M.: How to deal with parameters for whole-cell modelling. J. Roy. Soc. Interface **14**(133), 20170237 (2017)

5. Bacci, G., Bacci, G., Larsen, K.G., Mardare, R.: On-the-fly exact computation of bisimilarity distances. In: N. Piterman and S. A. Smolka, editors, TACAS, vol. 7795. LNCS, pp. 1–15 (2013)
6. Backenköhler, M., Bortolussi, L., Großmann, G., Wolf, V.: Abstraction-guided truncations for stationary distributions of Markov population models. In: QEST, pp. 351–371 (2021)
7. Barnat, J., Beneš, N., Brim, L., Demko, M., Hajnal, M., Pastva, S., Šafránek, D.: Detecting attractors in biological models with uncertain parameters. In: Feret, J., Koeppl, H. (eds.) CMSB 2017. LNCS, vol. 10545, pp. 40–56. Springer, Cham (2017). https://doi.org/10.1007/978-3-319-67471-1_3
8. Blinov, M.L., Faeder, J.R., Goldstein, B., Hlavacek, W.S.: BioNetGen: software for rule-based modeling of signal transduction based on the interactions of molecular domains. Bioinformatics **20**(17), 3289–3291 (2004)
9. Borisov, N.M., Chistopolsky, A.S., Faeder, J.R., Kholodenko, B.N.: Domain-oriented reduction of rule-based network models. IET Syst. Biol. **2**(5), 342–351 (2008)
10. Cairoli, F., Carbone, G., Bortolussi, L.: Abstraction of Markov population dynamics via generative adversarial nets. In: Cinquemani, E., Paulevé, L. (eds.) CMSB 2021. LNCS, vol. 12881, pp. 19–35. Springer, Cham (2021). https://doi.org/10.1007/978-3-030-85633-5_2
11. Cardelli, L.: From processes to odes by chemistry. In: Ausiello, G., Karhumäki, J., Mauri, G., Ong, L. (eds.) Fifth Ifip International Conference on Theoretical Computer Science - Tcs 2008 (2008)
12. Cardelli, L., Pérez-Verona, I.C., Tribastone, M., Tschaikowski, M., Vandin, A., Waizmann, T.: Exact maximal reduction of stochastic reaction networks by species lumping. Bioinform. **37**(15), 2175–2182 (2021)
13. Cardelli, L., Tribastone, M., Tschaikowski, M.: From electric circuits to chemical networks. Nat. Comput. **19**(1), 237–248 (2020)
14. Cardelli, L., Tribastone, M., Tschaikowski, M., Vandin, A.: Forward and backward bisimulations for chemical reaction networks. In: CONCUR, pp. 226–239 (2015)
15. Cardelli, L., Tribastone, M., Tschaikowski, M., Vandin, A.: Comparing chemical reaction networks: a categorical and algorithmic perspective. In: Grohe, M., Koskinen, E., Shankar, N. (eds.) Proceedings of the 31st Annual ACM/IEEE Symposium on Logic in Computer Science, LICS 2016, July 5–8, 2016, pp. 485–494. ACM, New York (2016)
16. Cardelli, L., Tribastone, M., Tschaikowski, M., Vandin, A.: ERODE: a tool for the evaluation and reduction of ordinary differential equations. In: Legay, A., Margaria, T. (eds.) TACAS 2017. LNCS, vol. 10206, pp. 310–328. Springer, Heidelberg (2017). https://doi.org/10.1007/978-3-662-54580-5_19
17. Cardelli, L., Tribastone, M., Tschaikowski, M., Vandin, A.: Maximal aggregation of polynomial dynamical systems. PNAS **114**(38), 10029–10034 (2017)
18. Cardelli, L., Tribastone, M., Tschaikowski, M., Vandin, A.: Guaranteed error bounds on approximate model abstractions through reachability analysis. In: McIver, A., Horvath, A. (eds.) QEST 2018. LNCS, vol. 11024, pp. 104–121. Springer, Cham (2018). https://doi.org/10.1007/978-3-319-99154-2_7
19. Cardelli, L., Tribastone, M., Tschaikowski, M., Vandin, A.: Symbolic computation of differential equivalences. Theoret. Comput. Sci. **777**, 132–154 (2019)
20. Daca, P., Henzinger, T.A., Kretínský, J., Petrov, T.: Linear distances between Markov chains. In: Desharnais, J., Jagadeesan, R. (eds.) CONCUR, vol. 59. LIPIcs, pp. 20:1–20:15 (2016)

21. Feret, J., Danos, V., Krivine, J., Harmer, R., Fontana, W.: Internal coarse-graining of molecular systems. PNAS **106**(16), 6453–6458 (2009)
22. Großmann, G., Kyriakopoulos, C., Bortolussi, L., Wolf, V.: Lumping the approximate master equation for multistate processes on complex networks. In: McIver, A., Horvath, A. (eds.) QEST 2018. LNCS, vol. 11024, pp. 157–172. Springer, Cham (2018). https://doi.org/10.1007/978-3-319-99154-2_10
23. Helfrich, M., Ceska, M., Kretínský, J., Marticek, S.: Abstraction-based segmental simulation of chemical reaction networks. In: Petre, I., Paun, A. (eds.) CMSB, vol. 13447, pp. 41–60 (2022)
24. Hillston, J., Tribastone, M., Gilmore, S.: Stochastic process algebras: from individuals to populations. Comput. J. **55**(7), 866–881 (2011)
25. Hogg, J.S., Harris, L.A., Stover, L.J., Nair, N.S., Faeder, J.R.: Exact hybrid particle/population simulation of rule-based models of biochemical systems. PLOS Comput. Biol. **10**(4), e1003544, April 2014. Publisher: Public Library of Science
26. Iacobelli, G., Tribastone, M.: Lumpability of fluid models with heterogeneous agent types. In: 2013 43rd Annual IEEE/IFIP International Conference on Dependable Systems and Networks (DSN), pp. 1–11, June 2013. ISSN: 2158-3927
27. Larsen, K.G., Skou, A.: Bisimulation through probabilistic testing. Inf. Comput. **94**(1), 1–28 (1991)
28. Li, G., Rabitz, H.: A general analysis of exact lumping in chemical kinetics. Chem. Eng. Sci. **44**(6), 1413–1430 (1989)
29. Li, G., Rabitz, H.: A general analysis of approximate lumping in chemical kinetics. Chem. Eng. Sci. **45**(4), 977–1002 (1990)
30. Li, G., Rabitz, H.: New approaches to determination of constrained lumping schemes for a reaction system in the whole composition space. Chem. Eng. Sci. **46**(1), 95–111 (1991)
31. Mu, F., Williams, R.F., Unkefer, C.J., Unkefer, P.J., Faeder, J.R., Hlavacek, W.S.: Carbon-fate maps for metabolic reactions. Bioinformatics (Oxford, England) **23**(23), 3193–3199 (2007)
32. Okino, M., Mavrovouniotis, M.: Simplification of mathematical models of chemical reaction systems. Chem. Rev. **2**(98), 391–408 (1998)
33. Ovchinnikov, A., Pérez Verona, I., Pogudin, G., Tribastone, M.: CLUE: exact maximal reduction of kinetic models by constrained lumping of differential equations. Bioinformatics **37**(12), 1732–1738, June 2021
34. Pérez-Verona, I.C., Tribastone, M., Vandin, A.: A large-scale assessment of exact model reduction in the BioModels repository. In: Bortolussi, L., Sanguinetti, G. (eds.) CMSB 2019. LNCS, vol. 11773, pp. 248–265. Springer, Cham (2019). https://doi.org/10.1007/978-3-030-31304-3_13
35. Radulescu, O., Gorban, A.N., Zinovyev, A., Noel, V.: Reduction of dynamical biochemical reactions networks in computational biology. Front. Genet. **3**, 131 (2012)
36. Repin, D., Petrov, T.: Automated deep abstractions for stochastic chemical reaction networks. Inf. Comput. **281**, 104788 (2021)
37. Salazar, C., Höfer, T.: Multisite protein phosphorylation – from molecular mechanisms to kinetic models. FEBS J. **276**(12), 3177–3198 (2009)
38. Schmidt, H., Madsen, M., Danø, S., Cedersund, G.: Complexity reduction of biochemical rate expressions. Bioinformatics **24**(6), 848–854 (2008)
39. Segel, L., Slemrod, M.: The quasi-steady-state assumption: a case study in perturbation. SIAM Rev. **31**(3), 446–477 (1989)

40. Snowden, T., van der Graaf, P., Tindall, M.: Methods of model reduction for large-scale biological systems: a survey of current methods and trends. Bull. Math. Biol. **79**(7), 1449–1486 (2017)
41. Sunnaker, M., Cedersund, G., Jirstrand, M.: A method for zooming of nonlinear models of biochemical systems. BMC Syst. Biol. **5**(1), 140 (2011)
42. Tognazzi, S., Tribastone, M., Tschaikowski, M., Vandin, A.: Egac: a genetic algorithm to compare chemical reaction networks. In: GECCO, GECCO 2017, p. 833–840 (2017)
43. Tomlin, A.S., Li, G., Rabitz, H., Tóth, J.: The effect of lumping and expanding on kinetic differential equations. SIAM J. Appl. Math. **57**(6), 1531–1556 (1997). Publisher: Society for Industrial and Applied Mathematics
44. Tribastone, M.: Behavioral relations in a process algebra for variants. In: Gnesi, S., Fantechi, A., Heymans, P., Rubin, J., Czarnecki, K., Dhungana, D. (eds.) SPLC, pp. 82–91. ACM (2014)
45. Troják, M., Šafránek, D., Pastva, S., Brim, L.: Rule-based modelling of biological systems using regulated rewriting. Biosyst. **225**, 104843 (2023)
46. Tschaikowski, M., Tribastone, M.: Exact fluid lumpability in Markovian process algebra. Theoret. Comput. Sci. **538**, 140–166 (2014)
47. Tschaikowski, M., Tribastone, M.: Approximate reduction of heterogeneous nonlinear models with differential hulls. IEEE TAC (2016)
48. Tschaikowski, M., Tribastone, M.: Spatial fluid limits for stochastic mobile networks. Perform. Evaluation **109**, 52–76 (2017)
49. Vallabhajosyula, R., Chickarmane, V., Sauro, H.: Conservation analysis of large biochemical networks. Bioinformatics **22**(3), 346–353 (2005)
50. Voit, E.O.: Biochemical systems theory: a review. ISRN Biomathematics **2013**, 53 (2013)
51. Whitby, M., Cardelli, L., Kwiatkowska, M., Laurenti, L., Tribastone, M., Tschaikowski, M.: PID control of biochemical reaction networks. IEEE Trans. Autom. Control **67**(2), 1023–1030 (2022)
52. Wirsing, M., et al.: Sensoria patterns: augmenting service engineering with formal analysis, transformation and dynamicity. In: Margaria, T., Steffen, B. (eds.) Leveraging Applications of Formal Methods, Verification and Validation, pp. 170–190 (2008)

Core SBML and Its Formal Semantics

Joachim Niehren[1,2(✉)], Cédric Lhoussaine[1], and Athénaïs Vaginay[3]

[1] BioComputing Team of CRIStAL Lab, Université de Lille, Lille, France
joachim.niehren@inria.fr
[2] Inria Center of the University of Lille, Lille, France
[3] Université de Lorraine, CNRS, CRAN & LORIA, 54000 Nancy, France

Abstract. The systems biology markup language (SBML) permits to represent biological models mixing reaction networks, algebraic equations, differential equations, and events. Its main objective is to exchange biological models between various tools for simulation and analysis. The specification of SBML, however, lacks a formal semantics. This makes it often difficult to understand SMBL models and to design correct and general interfaces with SBML. In the present paper, we propose Core SBML, a novel language covering a large subset of SBML with clear formal semantics. We present a compiler of the delay-free fragment of SBML to Core SBML (without any formal correctness guarantees). We then show how to compile Core SBML further to BioCham while preserving the semantics. We implemented and applied our compilers to the more than 500 SBML models from the curated part of the BioModels database.

1 Introduction

The three most prominent formalisms to model the quantitative dynamical evolution of biological systems over time are differential equations, reaction networks [9], and hybrid networks [11]. If all parameters and initial values are known, then such models can be simulated numerically over time by some variant of the Euler's algorithm for deterministic simulation [6]. In practice, this is supported by various tools, including Copasi [12] and BioCham [3,8].

All the formalisms mentioned above can be integrated into a unified framework, as provided by the Systems Biology Markup Language (SBML) [19]. Each of these formalisms has variables for denoting real-valued functions. These values of these variables can be constrained in different manners, by chemical reactions, differential equations, algebraic equations, and events. For reaction networks, the *species* of chemical reactions are considered as variables for the species' concentrations. These must satisfy the ordinary differential equations (ODEs) inferred from the reaction network. These ODEs specify how the concentrations of the species evolve continuously over real time according to the kinetic laws of the reactions. In hybrid networks, the values of variables can be changed by assignments triggered by some events. An event is triggered at the first time points when some logical conditions becomes valid. In mathematical models, variables

J. Pang and J. Niehren (Eds.): CMSB 2023, LNBI 14137, pp. 124–143, 2023.
https://doi.org/10.1007/978-3-031-42697-1_9

may be constrained by algebraic equations and differential equations respectively. When allowed to combine the four kinds of constraints for variables (reactions, events, algebraic equations, differential equations) and to use not only species but all kinds of variables in kinetic laws of reactions, then the integrated framework is obtained.

The concrete *syntax* of SBML is an instance of the Extendable Markup Language (XML) that is formally defined by an XML schema. However, the *semantics* of SBML models lacks a formal definition. In particular, its advanced concepts such as boundaryConditions, assignment rules, rate rules, and events with and delays are often difficult to interpret unambiguously. In such cases, most users (even experts) have to consult the equations that are inferred from by some tool (e.g. Copasi [22]) in order to understand the meaning of a given model. The lack of a formal semantics that is independent of any implementation also makes it difficult to design correct interfaces with large coverage between various tools in systems biology, while using SBML as an exchange format. It is also not clear how to specify fragments of SBML that can be compiled to other languages such as BioCham. Furthermore, the correctness of such compilers cannot even be stated formally, given that SBML doesn't come with a formal semantics.

In the present paper, we propose Core SBML, a novel language covering a large subset of SBML, while coming with a clear formal semantics. We base Core SBML on an abstract term syntax, enabling the definition of the formal semantics. We also provide a concrete XML syntax of Core SBML following its abstract syntax. As a consequence the concrete syntax of Core SBML becomes different from that of SBML. We define the XML syntax of Core SBML by an XML schema that is composed of a DTD and a Schematron [15] (rather than in RelaxNG [31] or XML Schema [28] as for SBML on level 3 and respectively level 2). All Core SBML networks that we produced were validated with respect to our XML schema.

Second, we present a compiler of a large fragment of SBML to Core SBML. Besides some tedious semantic aspects, a trick is needed to deal with r¡compartment volumes in full generality: we introduce algebraic variables for concentrations of species, and link them to variables for the amount of species via an algebraic equation. A formal correctness result cannot be expected, given that SBML lacks a formal semantics. We have implemented our compiler and applied it to all SBML models of the curated part of the BioModels database [21]. It turns out that the only aspects of SBML models that cannot be translated by our compiler are delays in differential equations and events. These occur in only 19 out of the 547 SBML networks, though. Adding delays to Core SBML is not difficult syntactically, but would require further efforts in semantics and implementation.

The third contribution is a compiler from Core SBML to BioCham. The abstract syntax of BioCham is a restriction of that of Core SBML. In particular, BioCham does neither support delays, nor in differential equation nor in events. BioCham's concrete syntax is not based on XML. Its conciseness increases the human readability but makes automatic processing more laborious. One concept missing in BioCham compared to Core SBML are differential variables, but these can be easily converted to species variables. Another difference is that BioCham does not permit to specify initial values of species by arithmetic expressions in dependence of the initial values of other species. Therefore, our compiler has to evaluate the arithmetic expressions for initial values in Core SBML networks statically. Finally, some arithmetic functions such as tanh are built in (Core) SBML, but need to be encoded explicitely in BioCham. In this way, our compiler from Core SBML to BioCham preserves the differential algebraic equations and the initial values up to logical equivalence. It also preserves the events. This implies its correctness with respect to the semantics of both languages. It should be noted that BioCham has its own function for importing SBML models and converting them into reaction networks in BioCham's format. A major difference to our compiler is that BioCham's import ignores compartment volumes all over, which is correct only if all volumes are equal to 1.

We implemented both compilers in the XML Stylesheet Transformation Language (XSLT) [32] with the Saxon programming tool [18], and started testing it by comparing numerical simulations. We made all Core SBML and BioCham networks produced by our compilers from the SBML networks in the curated part of the BioModels database available at https://biocomputing.univ-lille.fr/core-sbml.

Related Work. Reaction networks lay the foundation of all quantitative modeling frameworks in systems biology. They may be given different semantics besides the deterministic continuous based on ODEs [7]. The non-deterministic rewrite semantics [20] ignores the kinetic expressions, while the stochastic semantics uses them differently, for computing the probability of a reaction to happen. Reaction networks also have a Boolean semantics which abstracts from the rewrite semantics (rather than the ODE semantics [24]). This Boolean semantics serves in BioCham for model verification.

Qualitative logical reasoning in systems biology is usually supported by Boolean networks [10,23,26,27]. Compilers from reaction networks to Boolean networks with correctness guarantees with respect to the ODE semantics were considered in [24] and without in [29]. Alternatively, reaction networks with partial kinetic information on reactions on inhibitors and activators were considered for quantitative reasoning [1,17,25]. SBML level 3 models can support some kinds of partial kinetic information too. In the present paper, however, we consider only models with complete kinetic information, as supported by SBML level 2 models. Neither do we consider models with missing kinetic parameters or missing initial concentrations.

Outline. After preliminaries on arithmetic expressions and algebraic differential equations in Sect. 2, we discuss SMBL informally in Sect. 3. The language of Core SBML is contributed in Sect. 4. A compiler of a large fragment of SBML to Core SBML is described in Sect. 5. The compiler to BioCham is discussed in Sect. 6. The appendix of the long version [16] contains supplementary material, including the Core SBML and BioCham networks obtained by our compilers from three SBML models in the BioModels database.

2 Preliminaries

Let $\mathbb{B} = \{0, 1\}$ be the set of Booleans, $\mathbb{N}$ the set of natural numbers including 0, $\mathbb{R}$ the set of real numbers, and $\mathbb{R}_+$ the set of positive real numbers including 0. Note that $\mathbb{B} \subseteq \mathbb{N} \subseteq \mathbb{R}_+ \subseteq \mathbb{R}$.

A partial function $g : S \hookrightarrow T$ is a binary relation $g \subseteq S \times T$ that is functional, i.e., for all $s \in S$ there exists at most one $t \in T$ such that $(s,t) \in g$. For any partial function g, we write $g(s) = t$ if and only if $(s,t) \in g$. The domain of a partial function is $dom(g) = \{s \in S \mid \exists t \in T.\ g(s) = t\}$. A total function $g : S \to T$ is a partial function $g : S \hookrightarrow T$ with $dom(g) = S$. The restriction of g to some $S' \subseteq dom(g)$ is written as $g_{|S'}$.

A substitution $g = [s_1/t_1, \ldots, s_n/t_n]$ is the finite function with domain $\{s_1, \ldots, s_n\}$ such that $g(s_i) = t_i$ for all $1 \leq i \leq n$. We also write $g' = g[s/t]$ for the function with $g'(s) = t$ and $g'(s') = g(s')$ for all $s' \neq s$. If $s_1, \ldots, s_n \in \mathbb{R}_+$ then we sometimes write the substitution g as a formal sum: $g = s_1t_1 + \ldots + s_nt_n$. Such functions can be seen as chemical solutions that contain species t_i at concentration s_i for all $1 \leq i \leq n$ and all other species at concentration 0. If $n = 0$, then g is the empty chemical solution that we denote by $\mathbf{0}$.

Arithmetic and Boolean Expressions. We introduce arithmetic and Boolean expressions. They depend mutually recursively on each other, since Boolean expressions may occur in the conditionals of arithmetic expressions, while arithmetic expressions may occur in the comparisons of Boolean expressions. In SBML and BioCham, arithmetic expressions are used in kinetic laws of reactions, while Boolean expressions serve as conditions of events. Both kinds of expressions are also well-known from various programming languages, so their scope is more general than just systems biology.

We assume a set of variables $x \in \mathcal{V}$ and a collection $F = (F^{(n)})_{n \in \mathbb{N}}$ with sets of function symbols $f \in F^{(n)}$ each requiring n arguments. The set of arithmetic expressions $e \in \mathcal{E}_{\mathcal{V}}$ and boolean expressions $b \in \mathcal{B}_{\mathcal{V}}$ are the set of all terms with the following abstract syntax:

$$
\begin{array}{rcll}
e, e', e_1, \ldots, e_n \in \mathcal{E}_{\mathcal{V}} & ::= & x \mid k \mid & \text{where } x \in \mathcal{V},\ k \in \mathbb{R}, \\
& \mid & e + e' \mid e - e' \mid ee' \mid e/e' & \\
& \mid & f(e_1, \ldots, e_n) & f \in F^{(n)}, n \in \mathbb{N} \\
& \mid & \textbf{if}\ b\ \textbf{then}\ e\ \textbf{else}\ e' & \\
b, b' \in \mathcal{B}_{\mathcal{V}} & ::= & e \leq e' \mid \neg b \mid b \wedge b' &
\end{array}
$$

The set of free variables $fv(e)$ is the set of all those variables that occur in e. The operators of addition $+$, multiplication ee', and division/can be interpreted over the relational structure of the real numbers as usual. We will formally do it in Fig. 1. Note that division by zero is undefined. For any symbols $f \in F^{(n)}$ we assume a real-valued function $f^{\mathbb{R}^n \to \mathbb{R}} : \mathbb{R}^n \to \mathbb{R}$ defined elsewhere.

$$[\![x]\!]^\eta = \eta(x) \qquad [\![e + e']\!]^\eta = [\![e]\!]^\eta +^{\mathbb{R}} [\![e']\!]^\eta$$

$$[\![k]\!]^\eta = k \qquad [\![e - e']\!]^\eta = [\![e]\!]^\eta -^{\mathbb{R}} [\![e']\!]^\eta$$

$$[\![ee']\!]^\eta = [\![e]\!]^\eta *^{\mathbb{R}} [\![e']\!]^\eta \qquad [\![e/e']\!]^\eta = \begin{cases} [\![e]\!]^\eta /^{\mathbb{R}} [\![e']\!]^\eta & \text{if } [\![e']\!]^\eta \neq 0 \\ \mathit{undef} & \text{otherwise} \end{cases}$$

$$[\![f(e_1, \ldots, e_n)]\!]^\eta = f^{\mathbb{R}^n \to \mathbb{R}}([\![e_1]\!]^\eta, \ldots, [\![e_n]\!]^\eta)$$

$$[\![\textbf{if } b \textbf{ then } e \textbf{ else } e']\!]^\eta = \begin{cases} [\![e]\!]^\eta & \text{if } [\![b]\!]^\eta = 1 \\ [\![e']\!]^\eta & \text{if } [\![b]\!]^\eta = 0 \\ \mathit{undef} & \text{otherwise} \end{cases}$$

$$[\![e \leq e']\!]^\eta = \begin{cases} 1 & \text{if } [\![e]\!]^\eta \leq^{\mathbb{R}} [\![e']\!]^\eta \\ 0 & \text{if } [\![e]\!]^\eta >^{\mathbb{R}} [\![e']\!]^\eta \\ \mathit{undef} & \text{otherwise} \end{cases}$$

$$[\![b \wedge b']\!]^\eta = [\![b]\!]^\eta \wedge^{\mathbb{B}} [\![b']\!]^\eta \qquad [\![\neg b]\!]^\eta = \neg^{\mathbb{B}}([\![b]\!]^\eta)$$

Fig. 1. Interpretation of arithmetic and boolean expressions over the reals in environments $\eta : \mathcal{V} \hookrightarrow \mathbb{R}$.

This permits to have other real-valued functions such as exponentiation $\sqrt{.}$, $.^2$, exp, sin, cos, etc. For instance, we can define for any arithmetic expression v_1, v_2, J_1, J_2 another arithmetic expression as follows:

$$\mathit{goldbeter\_koshland}(v_1, v_2, J_1, J_2) =_{\text{def}} \frac{2 v_1 J_2}{v_2 - v_1 + J_1 v_2 + J_2 \ v_1 + \sqrt{(v_2 - v_1 + J_1 v_2 + J_2 v_1)^2 - 4(v_2 - v_1) \ v_1 J_2}}$$

Arithmetic expressions support conditionals **if** b **then** e **else** e' which use a boolean expression b as condition and arithmetic subexpressions e and e' as branches. The set of free variables of boolean expression $fv(b)$ is the set of those variables that occur in b. A boolean expression can test two arithmetic expressions for inequality $e \leq e'$ while being closed under the boolean operators of conjunction $b \wedge b'$ and negation $\neg b$. Equality and strict inequality tests can be defined as boolean expressions too:

$$e \stackrel{\circ}{=} e' =_{\text{def}} e \leq e' \wedge e' \leq e \qquad \text{and} \qquad e < e' =_{\text{def}} e \leq e' \wedge \neg e \stackrel{\circ}{=} e'$$

Furthermore, we note that disjunctions $b \vee b'$ can be defined as usual by $\neg(\neg b \wedge \neg b')$.

Interpretation Over the Reals. Let $\eta : \mathcal{V} \hookrightarrow \mathbb{R}$ be a real-valued variable assignment. For any arithmetic expression e with $fv(e) \subseteq dom(\eta)$ we can define its interpretation $[\![e]\!]^\eta$ as a real number or undefined in Fig. 1

Undefineness comes from undefined operations such as the division by zero mentioned previously. When interpreting an expression, the value *undef* for undefinedness propagates recursively. In mutual recursion, we define in Fig. 1 for any boolean expression b with $fv(b) \subseteq dom(\eta)$ a boolean interpretation $[\![b]\!]^\eta$ which may also be undefined.

Interpretation as Real-Valued Functions over Time. Since dynamical systems evolve over time, we are interested in interpreting arithmetic expressions as real-valued functions over time. We now define such an interpretation.

Let $T \subseteq \mathbb{R}_+$ be a subset of time points. For any variable assignment $\alpha : \mathcal{V} \hookrightarrow (T \to \mathbb{R})$, we can interpret any arithmetic expression e with $fv(e) \subseteq dom(\alpha)$ to a real-valued function of type $[\![e]\!]^\alpha : T \to \mathbb{R}$ or remain undefined. The interpretation of boolean expressions b as functions $[\![e]\!]^\alpha : T \to \mathbb{B}$ from time points to Booleans works in analogy.

For this, we define any time point $t \in T$ the real-valued assignments $\alpha_t : \mathcal{V} \hookrightarrow \mathbb{R}$ such that $\alpha_t(x) = \alpha(x)(t)$. The interpretations over real functions then satisfies for all time points $t \in T$:

$$[\![e]\!]^\alpha(t) = [\![e]\!]^{\alpha_t} \qquad \text{and} \qquad [\![b]\!]^\alpha(t) = [\![b]\!]^{\alpha_t}.$$

Suppose that we are given a variable *time* whose interpretation $\alpha(\mathit{time})$ is fixed as the identity function on $\mathbb{R}_+$, so that for all time points $t \in \mathbb{R}_+$:

$$\alpha(\mathit{time})(t) = t$$

Arithmetic expressions can be used to define piecewise real-valued functions such as:

$$\textbf{if } 0 \leq \mathit{time} \wedge \mathit{time} \leq 2.5 \textbf{ then } 1.3\ \mathit{time} \textbf{ else } 0$$

In MathML as used by SBML, the same arithmetic expression can be defined.

Differential Algebraic Equation Systems. We recall the syntax and semantics of systems of algebraic and differential equations. We use the standard framework of first-order logic. A differential algebraic system of equations is a first-order formula with the following abstract syntax, where $x \in \mathcal{V}$ and $e \in \mathcal{E}_\mathcal{V}$:

$$S, S' ::= x \stackrel{\circ}{=} e \mid \dot{x} \stackrel{\circ}{=} e \mid S \wedge S' \mid \exists x.S$$

We can define inequations $S \geq 0$ by the equation system $\exists x.\ S \stackrel{\circ}{=} xx$. An equation system is called algebraic if it contains no differential equation $\dot{x} \stackrel{\circ}{=} e$. It is called an ordinary differential equation (ODE) if it contains no algebraic equation $x \stackrel{\circ}{=} e$.

Let $T \subseteq \mathbb{R}_+$ and $\alpha : \mathcal{V} \hookrightarrow (T \to \mathbb{R})$. We call α a solution of the algebraic equation $x \stackrel{\circ}{=} e$ if $\alpha(x) = [\![e]\!]^\alpha$. We call α a solution of a differential equation $\dot{x} \stackrel{\circ}{=} e$ if $\dot{\alpha(x)} = [\![e]\!]^\alpha$. Note that the derivative of $\alpha(x)$ may be undefined, in which case the above equation does not hold. We call α a solution of $S \wedge S'$ if it is a solution of both S and S'. And finally, α is a solution of $\exists x.S$ if there exists a function $f : D \to \mathbb{R}$ such that $\alpha[x/f]$ is a solution of S.

For algebraic equations $x \stackrel{\circ}{=} e$, we can define real-valued solutions $\eta : \mathcal{V} \hookrightarrow \mathbb{R}$ such that $\eta(x) = [\![e]\!]^\eta$. Note however, that the structure of real numbers cannot give interpretation to the derivative operator, so real-valued solutions do not make sense for differential equations.

Adding Delays. For getting delay differential equations, it is sufficient to extend the set of arithmetic expressions $\mathcal{E}_{\mathcal{V}}$ with a delay operator. The abstract syntax becomes:

$$e \in \mathcal{E}_{\mathcal{V}}(delay) ::= \ldots \mid delay_k(e) \qquad \text{where } k \in \mathbb{R}_+$$

The semantics over real-valued functions is such that $[\![delay_k(e)]\!]^\alpha(t) = [\![e]\!]^\alpha(t-k)$.

3 SBML

The systems biology markup language (SBML) is widely used to represent models of dynamical systems in systems biology. SBML models subsume chemical reaction networks, algebraic equations, differential equations, and events. While systems biology is the main application domain of SBML, the scope of dynamical systems and thus SMBL is way larger (economics, weather forcast, etc.). The BioModels database [4] provides more than 500 curated SBML models for biological systems online, many of which were introduced in research papers. The prime tool for the numerical simulation of SBML models is Copasi [22] but many other tools in systems biology have SBML interfaces.

In the present paper, we focus on complete SBML models that can be simulated numerically. Such models are available from SBML level 2 [14]. SBML levels correspond to upward-compatible specifications that add features and expressive power. Partial descriptions of complete models were added on level 3 of SBML [13]. These are out of the scope of the present paper.

Inside a level, successive versions supersed one another. Each version has an XML syntax that is defined by some schema for XML documents. An SBML model is an XML document that is valid for the schema of the considered version. The schemas of the different versions of SBML on level 2 are defined in the language XML Schema [30], while the schemas of the versions of SBML on level 3 are defined in language RelaxNG [31].

We now present the concepts of the language SBML on level 2 informally, while focussing on their semantics. A formal semantic of SBML models, however, is missing. This is problematic since even experts often have difficulties to understand the precise meaning of an SBML model. The semantics is described by mixing algebraic and differential equations with events. Tools like Copasi [12] and SBML2LaTeX [5] permit to infer these equations from SBML models. These informal description of how this works may be insufficient to predict the precise equations manually.

SBML models describe the values of a finite set $\mathcal{V}$ of variables, such that each variable $x \in \mathcal{V}$ denotes some real valued function $[\![x]\!] : \mathbb{R}_+ \to \mathbb{R}$. The variables are constrained in different manners, by reactions, assignment rules, rate rules, and events.

Let e be an arithmetic expression in $\mathcal{E}_{\mathcal{V}}(delay)$. Assignment rules $x \overset{\circ}{=} e$ make a direct assignment to the variable values, while rate rules $\dot{x} \overset{\circ}{=} e$ make a direct assignment to the rate of change of the variable x, denoted $\dot{x}$. The simplest events have the form $b \Rightarrow x := e$ where $b \in \mathcal{B}_{\mathcal{V}}$, $x \in \mathcal{V}$ and $e \in \mathcal{E}_{\mathcal{V}}$. They change the value of a variable x to the current value of e at all earliest time point when condition b becomes true. But there are more complex forms of event with delays. Even events with priorities are pemitted, but these do not occur in the curated BioModels.

Variables may stand for various objects including kinetic parameters and compartment volumes. Some of the variables have a special role as they store the *amount* of the species of interest. In addition, for each such species variable $x \in \mathcal{V}$ there is a joined *concentration* variable denoted by $[x] \in \mathcal{V}$, whose semantics is defined implicitly, as the amount of the species divided by the volume c of the compartment in wich the species is located, so $[x] \stackrel{\circ}{=} x/c$. Reactions can only change the amount of species variables. They have the form:

$$k_1 x_1 + \ldots + k_n x_n \xrightarrow{e} k'_1 y_1 + \ldots + k'_m y_m$$

where $k_i, k'_i \in \mathbb{R}_+$, $x_i, y_i \in \mathcal{V}$, and $n, m \in \mathbb{N}$, while e is an arithmetic expression in $\mathcal{E}_{\mathcal{V}}(delay)$. Species on the left of the arrows are refered to as reactants and those on the right as products. Those species that appear in e but those amount is not changed by reaction are refered to as modifiers (of which some can be distinguished explicitely).

If the same variable is changed both by some rules and by some reactions, then the priority is given to the rules. Finally, all variables are given an initial values, either directly, or by the mean of an assignment rule.

In the present paper, SBML models are represented by graphically. The graphs are produced by converting SMBL to Core SBML and the applying BioComputing's Network-Graph tool. The newest version can chose x-y-coordinates automatically. But for producing nicer graphs in the paper, we have improved the layout semi-automatically for the examples with yet another tool of the BioComputing group.

In the graphs of SBML models (see e.g. Fig. 2) circles represent species variables and gray boxes represent reactions. Variables identifiers for species amount and concentration are in red, as well as the reaction name. Other parameters are in black. The kinetic expression of a reaction is annotated to its box. Its modifiers are indicated by dash edges. Plain directed edges from a species to a reaction, and from a reaction to a specie encode the reactants and the product of the reaction, respectively. These edges are annotated with the stoichiometric coefficients of the reaction k_i, k'_i if these are different from 1.

The products and reactants of reactions are drawn in yellow circles. A white circles indicates that the value of the corresponding species is defined by some assignment rules. The expression that is assigned is annotated to the circle. Similarly, a brown circle indicate that the rate of the variable is directly dependent of a rate rule. Here, the rate is annotated to the circle.

The SMBL model B309 of the BioModels database (called BIOMD0000000309 there) is presented graphically in Fig. 2. This model is a toy example of homeostasis with negative feedback. It involves four reactions *r0*, *r2*, *r3* and *r4*, ten parameters and four species (E, Ep, R and S). All species belong to the same compartment *env* with fixed volume 1. Only the value of the species R is actually modified by reactions. All the other species are in fact change by rules:

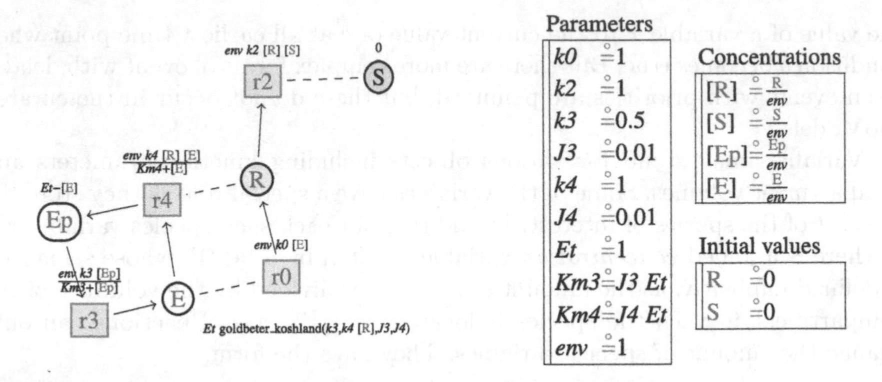

Fig. 2. A graphical illustration of the SBML model B309 of the BioModels database. The values of the species in the white circles are defined by algebraic equations $Ep \overset{\circ}{=} Et - [\mathrm{E}]$ and $E \overset{\circ}{=} goldbeter\_koshland(k3, k4\ conc_R, J3, J4)$ from assignment rules. Therefore, the species Ep and E are to be ignored as reactants and products of reactions $r3$ and $r4$. Species S has a default rate 0 yielding the ODE $\dot{S} \overset{\circ}{=} 0$. The kinetic law of reaction $r4$ is $\frac{env\ k3\ [\mathrm{Ep}]}{Km3+[\mathrm{Ep}]}$. The parameters there are defined by the algebraic equations on the right. The dashed lines indicate modifiers of a reaction. (Color figure online)

assignment rule for Ep and E and a rate rule for $\dot{S}$. These species are thus to be ignored as reactant and product of the reactions, even if they are listed such. Graphically, this is reflected by having a yellow circle for R, white circles for E and Ep, and a brown circle for S.

4 Core SBML

Reaction networks of Core SBML also contain reactions, equations and events. A main difference to SMBL is that each variable has a unique type, so that it cannot be constrained in contradictory manner a priori. So no priorities need to be formulated to rule out conflicting definitions.

In Core SBML it is optional to relate a species variable to some variable for its concentration and to another variable for its compartment volume. In SBML, in contrast, this is more mandatory. In case that the option is taken, the algebraic equation defining the concentration from the amount of the species and the volume must be stated explicitly in Core SBML. The consistency of this algebraic equation is then required by the schema of Core SBML networks.

Abstract Syntax and Semantics. We start the formal definition with a set $\mathcal{V} = \mathcal{S} \uplus \mathcal{V}_{\text{alg}} \uplus \mathcal{V}_{\text{diff}}$ with three disjoints sets of variables: species in $\mathcal{S}$, algebraic variables in $\mathcal{V}_{\text{alg}}$, and differential variables in $\mathcal{V}_{\text{diff}}$.

Definition 1. *Let $\mathcal{S} \subseteq \mathcal{V}$ be finite sets. A reaction of Core SBML with species in $\mathcal{S}$ and variables in $\mathcal{V}$ is a triple (R, P, e) where $R, P : \mathcal{S} \to \mathbb{R}_+$ and $e \in \mathcal{E}_{\mathcal{V}}$.*

We denote reactions as $R \xrightarrow{e} P$ instead of (R, P, e). The expression $e \in \mathcal{E}_\mathcal{V}$ is called the kinetic law of the reaction. The species $A \in \mathcal{S}$ with $R(A) > 0$ are called the reactants of the reaction and the species with $P(A) > 0$ are called its products. In contrast to SBML, the reactants and products in Core SBML do only permit variables in $\mathcal{S}$. Other variables in $\mathcal{V} \setminus \mathcal{S}$ are ruled out since these are constrained by equations or events. Furthermore, delays are not permitted in arithmetic expressions. An example for a Core SBML reaction with mass-action kinetics is:

$$2A + B \xrightarrow{2.1xA^2B} A + 2.1C$$

where $A, B, C \in \mathcal{S}$ and $x \in \mathcal{V} \setminus \mathcal{S}$. This reaction has the reactants A with coefficient 2 and B with coefficient 1. Its products are A with coefficient 1 and C with coefficient 2.1. Note that the kinetic expression does not only contain the reactants A and B but also the variable $x \in \mathcal{V} \setminus \mathcal{S}$, which can be used to model the volume of the compartment in which A and B reside.

Any species variable $x \in \mathcal{S}$ denotes a positive real numbers, which stands for the species' amount. All other kinds of variables $y \in \mathcal{V} \setminus \mathcal{S}$ denote arbitrary real numbers. Algebraic variables $z \in \mathcal{V}_{\mathrm{alg}}$ can be used to define compartment values and species concentrations.

Definition 2. *Let $\mathcal{V}_{\mathrm{alg}} \subseteq \mathcal{V}$. An event E with variables in $\mathcal{V}$ and algebraic variables in $\mathcal{V}_{\mathrm{alg}}$ has the following form:*

events	$E ::= b \Rightarrow u$	*where* $b \in \mathcal{B}_\mathcal{V}$
updates	$u, u' ::= x := e \mid u; u'$	*where* $e \in \mathcal{E}_\mathcal{V}$ *and* $x \in \mathcal{V} \setminus \mathcal{V}_{\mathrm{alg}}$

Intuitively, an event E is triggered at the earliest time point when its boolean condition b becomes true. In this case a sequence of updates u are executed from the left to the right. Any update $x := e$ may change the value of some variable $x \in \mathcal{V} \setminus \mathcal{V}_{\mathrm{alg}}$ to the value of the arithmetic expression e at the current time point.

Note that values of algebraic variables $x \in \mathcal{V}_{\mathrm{alg}}$ cannot be updated, since they always must satisfy $x \stackrel{\circ}{=} expr(x)$. In contrast, the values of differential variables $y \in \mathcal{V}_{\mathrm{diff}}$ may be updated while preserving its equation $\dot{y} = expr(y)$ except for the time point of the update. Also, the values of species can be updated by events.

Definition 3. *A Core SBML network is a tuple $N = (\mathcal{V}, Reacts, expr, init, Evts)$ where $\mathcal{V} = \mathcal{S} \uplus \mathcal{V}_{\mathrm{alg}} \uplus \mathcal{V}_{\mathrm{diff}}$ is a finite set of variables, Reacts is a finite set of Core SMBL reactions with species in $\mathcal{S}$ and variables in $\mathcal{V}$ and Evts a finite set of events with variables in $\mathcal{V}$ and algebraic variables in $\mathcal{V}_{\mathrm{alg}}$. Furthermore:*

$$expr : \mathcal{V} \setminus \mathcal{S} \to \mathcal{E}_\mathcal{V}, \qquad init : \mathcal{V} \setminus \mathcal{V}_{\mathrm{alg}} \to \mathcal{E}_\mathcal{V}.$$

Any reaction network N of Core SBML defines a system of algebraic differential equations that we introduce next. The value of reaction species $x \in \mathcal{S}$

evolves over time according to the following ODE induced by the set of reactions of N:

$$\dot{x} \stackrel{\circ}{=} \sum_{(R,P,e)\in Reacts} e(P(x) - R(x))$$

Furthermore $x \geq 0$ must hold for the trajectories of all species $x \in \mathcal{S}$, so at all time points. Note that the kinetic expression e may depend on any kind of variables in $\mathcal{V}$ not only on species. The values of the algebraic variables $x \in \mathcal{V}_{\text{alg}}$ are specified by the algebraic equation $x \stackrel{\circ}{=} expr(x)$. A differential variable $y \in \mathcal{V}_{\text{diff}}$ comes with a differential equation $\dot{y} \stackrel{\circ}{=} expr(y)$.

The differential-algebraic system of equations of a reaction network N, denoted $dae(N)$, is defined by:

$$\begin{aligned} dae(N) = & \bigwedge_{x\in\mathcal{S}} \dot{x} \stackrel{\circ}{=} \textstyle\sum_{(R,P,e)\in Reacts} e(P(x) - R(x)) \\ & \wedge \bigwedge_{x\in\mathcal{S}} x \geq 0 \\ & \wedge \bigwedge_{x\in\mathcal{V}_{\text{alg}}} x \stackrel{\circ}{=} expr(x) \\ & \wedge \bigwedge_{x\in\mathcal{V}_{\text{diff}}} \dot{x} \stackrel{\circ}{=} expr(x) \end{aligned}$$

The function *init* defines the initial value of all variables that are not algebraic. The initial values at time point 0 must satisfy the following algebraic equations now interpreted over the reals.

$$initState(N) = \bigwedge_{x\in\mathcal{V}\setminus\mathcal{V}_{\text{alg}}} x \stackrel{\circ}{=} init(x) \ \wedge \ \bigwedge_{y\in\mathcal{V}_{\text{alg}}} y \stackrel{\circ}{=} expr(y)$$

For instance, the reaction network with $init(x) = 2.1$ and $expr(y) = x+1$ impose the algebraic equations to define the initial values at time 0:

$$x \stackrel{\circ}{=} 2.1 \wedge y \stackrel{\circ}{=} x + 1$$

We make the additional assumption on all Core SBML networks that there is no cyclic dependency between algebraic equations and initializations. This makes it possible to evaluate $init(x)$ for all nonalgebraic variables x. We note that this acyclicity assumption of of Core SBML is consistent with the SBML specification too.

For instance, the reaction network with $init(y) = x$ and $expr(x) = y$ is ill-formed where $y \in \mathcal{V}\setminus\mathcal{V}_{\text{alg}}$ and $x \in \mathcal{V}_{\text{alg}}$ is an algebraic variable, since the initial values are required to have the following cyclic dependencies:

$$x \stackrel{\circ}{=} y \wedge y \stackrel{\circ}{=} x$$

We note that Core SBML networks subsume differential algebraic equations. These can be expressed by using algebraic and differential variables. Core SBML also subsumes chemical reaction networks [9]. Delays in differential equations or events are not supported though.

Core SBML networks may be enriched with extra information on top of Definition 3. The first kind concerns the definition of macros and the second how to represent compartments, volumes, and concentrations of species.

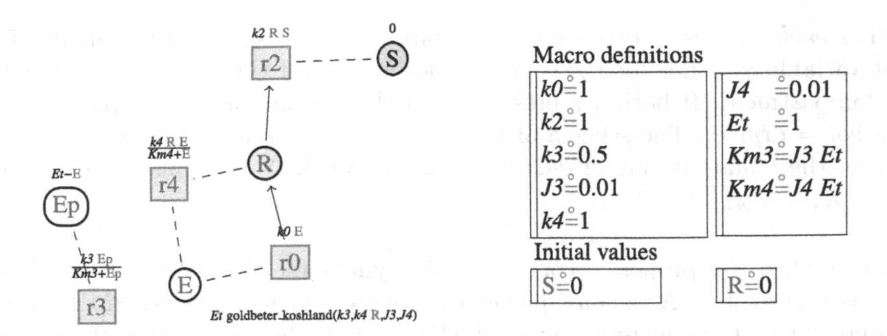

Fig. 3. Core SBML network capturing the essence of the SMBL model B309 in Fig. 2. Note that the variables E and Ep were removed from the reactants and products of reactions $r3$ and $r4$ since irrelevant. There are macro definitions for representing the SMBL parameters. Irrelevant variables for concentrations and compartment volumes were eliminated. Initial values are not changed.

Macros. A subset $\mathcal{V}[macros] \subseteq \mathcal{V}_{\text{alg}}$ of macros may be distinguished. Given a macro $x \in \mathcal{V}[macros]$ we call the $expr(x)$ the macros' definition. We assume that the definitions of macros are acyclic (in contast to more other algebraic equations). Therefore, all macros can be removed by exhaustively replacing them by their definitions. Macros can be used to represent SBML's parameters possibly with assignment rules. In graphs of Core SMBL networks, we show the macros definitions in a table while non-macro algebraic variables are drawn as nodes in white circles in the graphs of networks.

Example. We illustrate Core SBML at the networks in Fig. 3; it captures the essence of the SBML model B309 in Fig. 2. The Core SBML network has a single species $\mathcal{S} = \{R\}$, the algebraic variables $\mathcal{V}_{\text{alg}} = \{E, Ep\} \cup \mathcal{V}[macros]$, where $\mathcal{V}[macros] = \{Et, k_3, k_4, J_3, J_4, \ldots\}$, a single differential variable $\mathcal{V}_{\text{diff}} = \{S\}$. The definitions of the macros are given in Fig. 3. The initial values are given there too. Note that algebraic values don't have initial values.

In contrast to the SBML model in Fig. 2, there are no more variables for species' concentrations. This simplification makes sence given that the volume of compartment of all species was $env = 1$, so that concentrations and amounts did coincide anyway.

We note another differnce to the SBML model B309: the species Ep is no more a product of reaction $r3$ and the species E is no more a reactant of reaction $r4$. In SMBL they existed syntactically, but had no semantic effect on the equations. In Core SBML, such inconsistencies are forbidden a priori by Definition 1.

Concentrations, Compartments, and Volumes. The second kind of extra information of Core SBML networks serves for representing species's concentations, compartments, and volumes. This is done in such a way that all equations are made explicit, even though this may leads to redundant information. The consistency of this redudant information is verified by the schema of Core SBML.

For each species x, two algebraic variables may be specified optionally. The first variable y stands for the concentration of x, and the second variable c for its compartment. If both are present, then the volume of the compartment of x is $vol = expr(c)$. The schema of Core SBML then imposes that $expr(y) = \frac{x}{c}$. Hence, the equation system $dae(N)$ of the network N contains the equations $y \overset{\circ}{=} \frac{x}{c} \wedge c \overset{\circ}{=} vol$.

XML Syntax. We propose a concrete XML syntax for Core SBML that follows its abstract syntax. As a consequence the concrete syntax of Core SMBL is *not* a fragment of the concrete syntax of SBML. Still the idea is that it captures most of the expressiveness of SBML. This is consistent with the usual role of core languages in the context of programming languages.

Each variable of Core SBML is equipped with its type as "species", "algebraic", "differential", or "control", in contrast to the untyped identifiers such as <ci> Ep </ci> of SBML inherited from MathML. For instance, we can define an algebraic variable *Ep* for the SBML species with an assignment rule, a macro [*Ep*] for its concentration, and a macro *env* for its compartment as follows.

```
<variable type="algebraic" id="Ep"
   concentration="conc_Ep" compartment="env">
  <kinetic-expression>
    <minus>
      <expr id="Et"/>
      <expr id="conc_E"/>
    </minus>
  </kinetic-expression>
</variable>
<expression id="conc_Ep" latex-look="[Ep]">
    <divide>
      <var type="algebraic" id="Ep"/>
      <expr id="env"/>
    </divide>
</expression>
<expression id="env"> <constant value="1"/> </expression>
```

Variable references such as <var type="algebraic" id="Ep"/> are typed too. Note that the attribute latex-look permits to display species more nicely while keeping their identifiers simple. Mathematical operators such as divide are named such as in MathML, but can be applied directly (without requiring an additional apply element).

We defined an XML schema for Core SBML documents based on a document type descriptor (DTD) and a Schematron [15]. Compared to XML Schema [28] or RelaxNG [31], our approach has the advantage to nicely localize errors in invalid documents. This enables informative error messages. The DTD defines the hierarchical structure of Core SBML networks. The Schematron ensures the consistency of concentrations, amounts, and compartment volumes. It also verifies that there are no dangling references, i.e. that there is a variable definition for any `var` references with the same identifier and the same type. The Schematron also rules out empty identifiers or forbidden symbols in identifiers.

We developed some additional tools for Core SBML networks by using XML technology. We developped a graph drawing tool mapping Core SBML networks to the format of BioComputing's NetworkGraph tool. We also have a script computing the algebraic differential equations of a Core SBML network in XML format. Last not least, we implemented an inverse compiler from Core SBML to SBML.

5 Compiler from a SBML Fragment to Core SBML

Many concepts of SBML and Core SBML correspond in a direct manner. The main semantic difference are the permission of delays in ODEs and events of SMBL. These cannot be translated to Core SBML. Apart of this the syntax of Core SMBL (abstract and XML) is different from that of SMBL in order to guarantee the consistency of the equations. The compiler needs to identify the information in SBML models that is to be ignored in order to resolve inconsistencies based on SMBL's priorities.

Resolving inconsistencies is tedious given the informal character of the SBML specifications. In particular, it requires to precisely understand the relevance of species' attributes such as `@boundaryCondition` and `@constant` which interact in a complex manner. Selecting the right values from attributes `@initialConcentration`, `@initialAmount` and element `<initialAssignment>` is another issue. The details on these engineering aspects of the compiler cannot be presented in detail here.

Compartments, Volumes, and Concentrations. The proper treatment of compartments, volumes and concentration raises a non-obvious difficulty, given that Core SMBL does not reserve any extra treatment for them besides the consistency validation.

In order to explain this difficulty, let us first recall how compartment volumes and concentrations are treated in SBML. Each SBML species A is mapped to some compartment variable c, whose value is equal to that of some arithmetic expression vol standing for the compartments volume. The species A itself is used as a variable for the amount of A. SBML models may refer to a second variable $[A]$ that stands for A's concentration. The ODE for a species A – as infered by Copasi [22] or SBML2LaTeX [5] as used by the BioModels database [21] – then has the following form, where some other concentration variables such as $[B]$ may occur on the right:

$$\dot{A} \stackrel{\circ}{=} \ldots [A] \ldots [B] \ldots .$$

The variables for the amount and the concentration are coupled via the volume vol of the compartment c of species A, as expressed by the following equation:

$$[A] \stackrel{\circ}{=} A/c \wedge c \stackrel{\circ}{=} vol$$

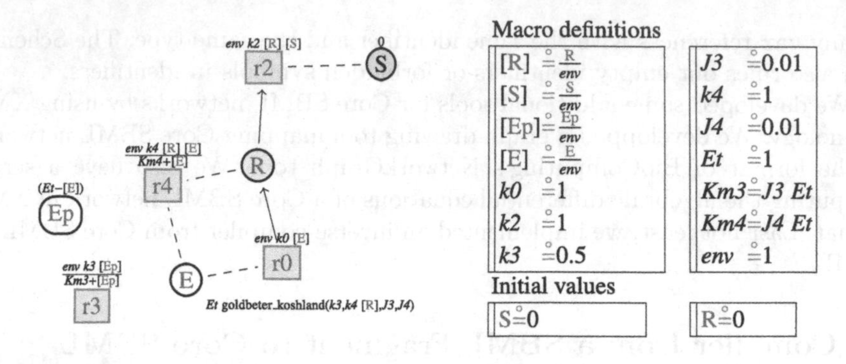

Fig. 4. The Core SBML network compiled from B309 in Fig. 2.

If $vol = 1$ then $[A] \stackrel{\circ}{=} A$ so we can identify A and $[A]$ so that a single variable is enough. That is why a single variable per species was enough in the Core SBML network for the essence of the SMBL model B309 in Fig. 3.

Our idea for correctly compiling SBML models to Core SBML networks without restrictions on compartment volumes follows the two variable strategy of SBML. The compiler will introduce for any species variable A an algebraic variable $[A]$ for its concentration and another algebraic variable c for its compartment, such that $expr([A]) = A/c$ and $expr(c) = vol$ where vol is the arithmetic expression for the volume of c. We can then use the variable $[A]$ in the kinetic expressions of the reactions of Core SBML as done by SBML. The Core SBML networks produced by the compiler will use the option to store for each species A the relationship to the concentration variable $[A]$, and the compartment variable c, so that $expr([A]) = A/c$ is guaranteed by Core SMBL's schema.

Examples for the Compiler. The compiler applied to SBML model B309 yields the Core SBML network with the graph in Fig. 4. By replacing the algebraic variables for the concentration by the amount variables (given that the volume of compartment *env* is equal to 1), we obtain the previous Core SBML network in Fig. 3. We note that the compartment structure present in the SMBL network is made fully explicit by the Core SBML network produced by our compiler. This structure is consistent with the equations in Fig. 4:

Amount	concentration	compartment
R	$[R]$	*env*
S	$[S]$	*env*
Ep	$[Ep]$	*env*
E	$[E]$	*env*

In the appendix of the long version, we present the Core SMBL networks for the BioModels B001 and B111, which contain events and respectively nontrivial compartment volumes.

Implementation. We implemented our compiler from a large fragment of SBmML to Core SBML in the XML stylesheet transformation language (XSLT) [32]. We also implemented a straightforward inverse compiler to SBML. Based thereon, we can obtain simulation trajectories for Core SBML networks via Copasi. This permits us to compare the trajectories of the Core SBML networks obtained by our compiler to those of the original SBML model. On few tests on SBML models with dynamic compartment changes, we observed equal trajectories, confirming the correctness of our treatment of compartment volumes.

We applied our compiler to the 548 SBML models in the curated BioModels database. The compilation results are available at https://biocomputing.univ-lille.fr/core-sbml. Out of the 548 curated SBML models, all but 19 could be compiled to Core SBML. Of these 13 contain delays in events and 6 delays in differential equations. Adding these two features to our formal treatment of Core SBML is easy, but practically it would require 2 more days of implementation. We also note that 56 curated BioModels contain piecewise linear functions, that we compiled to conditionals while using the constant *time*.

Limitations. Another issue is that the SMBL specification supports events with priorities. On the one hand side, the semantics of event priorities is quite complicated, on the other hand side they are not used in any SBML model of curated part of the BioModels database. Therefore we decided to not add any event priorities to Core SBML.

A limitation of the current compiler is that it ignores most aspects of SBML irrelevant for the dynamic semantics. In particular, this applies to units like mol per liter, but also to the information on the reversibility of reactions (which only talks about the motivation of the kinetic law of the reaction). Furthermore, all XML elements for annotation and notes are ignored too.

6 Compiler of Core SBML to BioCham

BioCham supports pure chemical reaction networks with kinetic expressions and events, while ruling out arithmetic and differential variables. When ignoring its analysis and verification facilities, the abstract syntax of BioCham's modeling language is basically the same as that of Core SBML without algebraic and algebraic variable. Macros for expressions and function definitions are equally available.

Note that the import of SBML models by BioCham 4 follows the single variable approach where $[A] \stackrel{\circ}{=} A$ is valid for all species A. Compartment volumes are ignored all over. This is correct for compartments with fixed volume equal to 1, but incorrect otherwise. The compilation of SBML to Core SBML, in contrast, introduces two different variables for A and $[A]$ and then imposed the expected relationship. In this way, even dynamical changes of compartment volumes can be modeled correctly.

Compiler. We next show how to map Core SBML networks to BioCham 4. In combination with our compiler from SMBL to Core SBML, this can be used to improve on BioCham current SBML import.

First, we notice that our set of arithmetic expression $\mathcal{E}_{\mathcal{V}}$ coincides with that of BioCham 4. In particular, conditionals are supported there but no delays. However, not all build-in functions of SBML coming via MathML are natively supported by BioCham 4. A counter example is *tanh*. But we found BioCham definitions for all those used in the SBML models of the Curated part of the BioModels database.

Second notice, that BioCham can express any chemical reactions $R \xrightarrow{e} P$ of Core SBML by the statement: `e for R => P`. Third, BioCham 4 can express any algebraic equation $x \overset{\circ}{=} e$ by the statement `function(x=e)`. Fourth, differential variables of Core SBML can be eliminated at beforehand: It is sufficient to turn any differential variable x into a species, that is produced by an artificial reaction with kinetic expression $expr(x)$ and not consumed by any other reaction. Fifth, the events of Core SBML coincide with those of BioCham 4 up to details of their concrete syntax.

Sixth, initial values in BioCham must be specified by reals: they cannot be specified by arithmetic expressions, so that BioCham could compute them from other initial values. Therefore, the compiler has to evaluate the arithmetic expressions $init(x)$ to some real number for any $x \in \mathcal{V} \setminus \mathcal{V}_{\text{alg}}$. The evaluation cannot loop, since we assumed that there are no cyclic dependency between algebraic equations and initializations.

Our compiler from Core SBML to BioCham is correct in that maintains the differential algebraic equations of the network and its initial conditions up to equivalence, and also the events of the network. The correctness could be stated and proven formally. We note that BioCham cannot represent compartment structures. Our compiler can safely ignore Core SBML's logical information relating amounts, compartments, and concentrations, since consistency with the equations is garanteed anyway.

Implementation. We implemented our compiler from Core SBML to BioCham 4 in XSLT. We notice that the concrete syntax varies with the version of BioCham. The concrete syntax of the BioCham 4 networks for the SBML models B309, B001, and B111 can be found in the appendix of the long version [16]. The full collection of BioCham networks for all curated BioModels is available online at https://biocomputing.univ-lille.fr/core-sbml. In this way, we obtained a second method to simulate Core SBML networks via BioCham. In particular, we could sucessfully simulate some SMBL models with dynamic compartment volumes via BioCham.

7 Conclusion

We presented the Core SBML, an exchange format for systems biology, which, in contrast to SBML, has a clear formal semantics based on differential algebraic equations. We argued that Core SBML covers a large part of the curated

BioModels. In order to cover all of them, it is sufficient to add delays in differential equations and events. Conceptually this is not difficult, but it requires some more implementation efforts. In any case, delays are shown to make the only true difference in expressiveness between SBML and BioCham. We believe that Core SBML can help to reduce the difficulties to bridge the various tools in systems biology. A good next step could be to use Core SBML to revise the SMBL import of the scientific computing language Julia [2]. On this way, we hope that Core SBML will find a large acceptance for tools in systems biology eventually.

Acknowledgements. We thank Sylvain Soliman and François Fages for their unfailable and timely support with BioCham questions. We also enjoyed efficient support from the Copasi people. Finally, this work was supported by the French National Research Agency (ANR), by funding the project MIGAD (ANR- 21-CE45-0017).

References

1. Allart, E., Niehren, J., Versari, C.: Computing difference abstractions of metabolic networks under kinetic constraints. In: Bortolussi, L., Sanguinetti, G. (eds.) CMSB 2019. LNCS, vol. 11773, pp. 266–285. Springer, Cham (2019). https://doi.org/10.1007/978-3-030-31304-3_14
2. Bezanson, J., Edelman, A., Karpinski, S., Shah, V.B.: Julia: a fresh approach to numerical computing. SIAM Rev. **59**(1), 65–98 (2017). https://doi.org/10.1137/141000671
3. Calzone, L., Fages, F., Soliman, S.: BIOCHAM: an environment for modeling biological systems and formalizing experimental knowledge. Bioinformatics **22**(14), 1805–1807 (2006). https://doi.org/10.1093/bioinformatics/btl172
4. Chelliah, V., et al.: BioModels: ten-year anniversary. Nucl. Acids Res. (2015). https://doi.org/10.1093/nar/gku1181
5. Dräger, A., et al.: Sbml2latex: conversion of sbml files into human-readable reports. Bioinformatics **25**(11), 1455–1456 (2009)
6. Euler, L.: Institutionum Calculi Integralis. No. vol. 1 in Institutionum Calculi Integralis, imp. Acad. imp. Saènt. (1768)
7. Fages, F., Soliman, S.: Abstract interpretation and types for systems biology. Theor. Comput. Sci. **403**(1), 52–70 (2008)
8. Fages, F., Gay, S., Soliman, S.: Inferring reaction systems from ordinary differential equations. Theor. Comput. Sci. **599**, 64–78 (2015). https://doi.org/10.1016/j.tcs.2014.07.032
9. Feinberg, M.: Chemical reaction network structure and the stability of complex isothermal reactors-I. the deficiency zero and deficiency one theorems. Chem. Eng. Sci. **42**(10), 2229–2268 (1987). https://doi.org/10.1016/0009-2509(87)80099-4. http://www.sciencedirect.com/science/article/pii/0009250987800994
10. Glass, L., Kauffman, S.A.: The logical analysis of continuous, non-linear biochemical control networks. J. Theor. Biol. **39**(1), 103–129 (1973). https://doi.org/10.1016/0022-5193(73)90208-7. https://www.sciencedirect.com/science/article/pii/0022519373902087
11. Harel, D.: Statecharts: a visual formalism for complex systems. Sci. Comput. Program. **8**(3), 231–274 (1987). https://doi.org/10.1016/0167-6423(87)90035-9

12. Hoops, S., et al.: Copasi-a complex pathway simulator. Bioinformatics **22**(24), 3067–3074 (2006)
13. Hucka, M., et al.: The systems biology markup language (SBML): language specification for level 3 version 2 core release 2. J. Integrat. Bioinf. **16**(2), 20190021 (2019). https://doi.org/10.1515/jib-2019-0021. https://www.degruyter.com/view/j/jib.ahead-of-print/jib-2019-0021/jib-2019-0021.xml
14. Hucka, M., et al.: Systems biology markup language (SBML) level 2 version 5: structures and facilities for model definitions. J. Integrat. Bioinf. **12**(2), 271 (2015). https://doi.org/10.2390/biecoll-jib-2015-271
15. Jelliffe, R.: Schematron (2006). iSO/IEC 19757–3
16. Joachim, N., Lhoussaine, C., Ahténaïs, V.: Core SBML and its formal semantics. In: CMSB: 21th International Conference on Formal Methods in Systems Biology, Luxembourg, Luxembourg (2023). https://inria.hal.science/hal-04125922
17. John, M., Nebut, M., Niehren, J.: Knockout prediction for reaction networks with partial kinetic information. In: 14th International Conference on Verification, Model Checking, and Abstract Interpretation, Rome, Italy, pp. 355–374 (2013). http://hal.inria.fr/hal-00692499
18. Kay, M.: The saxon xslt and xquery processor (2004). https://www.saxonica.com
19. Keating, S.M., et al.: SBML level 3 community members: SBML level 3: an extensible format for the exchange and reuse of biological models. Molec. Syst. Biol. **16**(8), e9110 (2020). https://doi.org/10.15252/msb.20199110
20. Madelaine, G., Lhoussaine, C., Niehren, J.: Attractor equivalence: an observational semantics for reaction networks. In: Fages, F., Piazza, C. (eds.) FMMB 2014. LNCS, vol. 8738, pp. 82–101. Springer, Cham (2014). https://doi.org/10.1007/978-3-319-10398-3_7
21. Malik-Sheriff, R.S., et al.: BioModels-15 years of sharing computational models in life science. Nucl. Acids Res. **48**(D1), D407–D415 (2020). https://doi.org/10.1093/nar/gkz1055
22. Mendes, P., Hoops, S., Sahle, S., Gauges, R., Dada, J., Kummer, U.: Computational modeling of biochemical networks using copasi. Methods Molec. Biol. (Clifton, N.J.) **500**, 17–59 (2009)
23. Mizera, A., Pang, J., Qu, H., Yuan, Q.: Taming asynchrony for attractor detection in large Boolean networks. IEEE/ACM Trans. Comput. Biol. Bioinf. **16**(1), 31–42 (2019). https://doi.org/10.1109/TCBB.2018.2850901
24. Niehren, J., Vaginay, A., Versari, C.: Abstract simulation of reaction networks via boolean networks. In: Petre, I., Paun, A. (eds.) Computational Methods in Systems Biology - 20th International Conference, CMSB 2022, Bucharest, Romania, 14–16 September 2022, Proceedings. Lecture Notes in Computer Science, vol. 13447, pp. 21–40. Springer, Heidelberg (2022). https://doi.org/10.1007/978-3-031-15034-0_2
25. Niehren, J., Versari, C., John, M., Coutte, F., Jacques, P.: Predicting changes of reaction networks with partial kinetic information. BioSystems **149**, 113–124 (2016). https://hal.inria.fr/hal-01239198
26. Paulevé, L., Kolçà, J., Chatain, T., Haar, S.: Reconciling qualitative, abstract, and scalable modeling of biological networks. Nat. Commun. **11** (2020). https://doi.org/10.1038/s41467-020-18112-5. https://hal.archives-ouvertes.fr/hal-02518582
27. Thomas, R.: Boolean formalization of genetic control circuits. J. Theor. Biol. **42**(3), 563–585 (1973). https://doi.org/10.1016/0022-5193(73)90247-6. https://www.sciencedirect.com/science/article/pii/0022519373902476
28. Thompson, H.S., Beech, D., Maloney, M., Mendelsohn, N.: Xml schema part 1: Structures, 2nd edn. (2004). http://www.w3.org/TR/xmlschema-1/

29. Vaginay, A., Boukhobza, T., Smaïl-Tabbone, M.: From quantitative SBML models to Boolean networks. Appl. Netw. Sci. **7**(1), 73 (2022). https://doi.org/10.1007/s41109-022-00505-8
30. van der Vlist, E.: XML Schema. O'Reilly, Beijing (2003)
31. Van der Vlist, E.: RELAX NG: A Simpler Schema Language for XML, 1. aufl edn.. O'Reilly & Assoc. (2004)
32. W3C: XSL transformations (XSLT) version 3.0 (2017). https://www.w3.org/TR/xslt-30

Average Sensitivity of Nested Canalizing Multivalued Functions

Élisabeth Remy[1(✉)] and Paul Ruet[2]

[1] Aix Marseille Univ, CNRS, I2M, Marseille, France
elisabeth.remy@univ-amu.fr
[2] CNRS, Université Paris Cité, Paris, France
ruet@irif.fr

Abstract. The canalizing properties of biological functions have been mainly studied in the context of Boolean modelling of gene regulatory networks. An important mathematical consequence of canalization is a low average sensitivity, which ensures in particular the expected robustness to noise. In certain situations, the Boolean description is too crude, and it may be necessary to consider functions involving more than two levels of expression. We investigate here the properties of nested canalization for these multivalued functions. We prove that the average sensitivity of nested canalizing multivalued functions is bounded above by a constant. In doing so, we introduce a generalization of nested canalizing multivalued functions, which we call weakly nested canalizing, for which this upper bound holds.

Keywords: nested canalizing functions · multivalued functions · average sensitivity · regulatory network modelling

1 Introduction

The concept of canalization in biology was proposed by Waddington in the early 1940s [26]. It corresponds to the property of a biological process of being able to produce a relatively stable phenotype despite the presence of variability, as a kind of noise filter inherent in the process [2,15]. This phenomenon is observed in natural systems, for example in [22] it has been shown that, at the molecular level, the development process of the Drosophilia embryo is canalized: the expression of genes controlling embryo segmentation is concentrated in a small area of state space.

Canalizing Boolean functions form a class of Boolean functions introduced by Kauffman [7,8] that formalize this canalizing behaviour observed in gene regulatory networks. In short, letting $\mathbb{Z}/2\mathbb{Z}$ denote the 2-element ring of integers modulo 2, *canalizing* Boolean functions are functions f from $(\mathbb{Z}/2\mathbb{Z})^n$ to $\mathbb{Z}/2\mathbb{Z}$ (or possibly to $\mathbb{R}$) such that at least one input variable, say x_i ($1 \leqslant i \leqslant n$), has a value $a = 0$ or 1 which determines the value of $f(x)$. *Nested canalizing* (NC) functions provide a "recursive" version of canalizing functions: an NC function f is canalizing and, moreover, its restriction $f\restriction_{x_i \neq a}$ is itself NC.

J. Pang and J. Niehren (Eds.): CMSB 2023, LNBI 14137, pp. 144–156, 2023.
https://doi.org/10.1007/978-3-031-42697-1_10

It is worth noting that NC Boolean functions are both rare among the set of Boolean functions and frequent among Boolean functions modeling gene networks. Indeed, on the one hand, they form a sparse set of Boolean functions [5]: the fraction of NC functions of arity n among all Boolean functions of arity n tends to 0 as $n \to \infty$. On the other hand, [21] gives evidence that gene regulatory networks, which are built from biological data and knowledge from the literature [24], are far from random: in particular, NC functions are predominant in Boolean gene networks. This canalizing property can be, and has been [4,27], used as a guide to filter through the large number of candidate functions to parameterise the regulatory graph, a critical point in the process of modelling networks.

On the mathematical side, a striking feature of NC functions is their low *average sensitivity*. The average sensitivity $\mathbf{AS}(f)$ of a Boolean function f is a measure of its "complexity" in the sense of Boolean functions analysis [14]: roughly speaking, for $f : (\mathbb{Z}/2\mathbb{Z})^n \to \mathbb{Z}/2\mathbb{Z}$, $\mathbf{AS}(f)$ measures how scattered the frontier between 0's and 1's is. For arbitrary Boolean functions, $\mathbf{AS}(f) = \mathcal{O}(n)$, but some functions have significantly lower average sensitivity. For NC functions, $\mathbf{AS}(f)$ is bounded above by a constant [10,11]. This low average sensitivity has several consequences. Most importantly, it entails noise stability (noise in inputs is not amplified [14,20]), a robustness property observed indeed in gene networks [7,9]. Functions with low **AS** also depend on few coordinates [3]. And more theoretically, the Fourier-Walsh spectrum of functions with low **AS** is concentrated on low degrees, and the function can be more easily learned from examples [14].

If in most cases, Boolean variables are sufficient to capture the main and key characteristics of the system under study, in some situations this description is too crude, and it may be necessary to consider more levels. To model such a situation correctly, multivalued variables have been introduced [25]. Then it is necessary to consider multivalued functions $f : (\mathbb{Z}/k\mathbb{Z})^n \to \mathbb{Z}/k\mathbb{Z}$ or $\mathbb{R}$ for some $k \geqslant 2$, where $\mathbb{Z}/k\mathbb{Z}$ denotes the ring of integers modulo k. The notion of average sensitivity generalizes to the multivalued setting [14], and multivalued NC functions are defined in [12,13]. Very little is known about their spectral properties. In [6], a variant of average sensitivity, the normalized average c-sensitivity, is defined for multivalued functions, and used to measure the stability of networks based on NC functions.

A natural question is whether the average sensitivity of NC multivalued functions is bounded above by a constant, too. We prove in Theorem 3 that this is the case. We actually show that the upper bound holds for a more general class of functions, which we call *weakly nested canalizing* (WNC), and at the same time this enables us to establish the upper bound in a simpler way than in [11] for NC Boolean functions.

The paper is organized as follows. In Sect. 2, we recall the definition of nested canalizing multivalued functions from [12,13] and illustrate it with example functions inspired from the logical modelling of the phage lambda. In Sect. 3, we define weakly nested canalizing multivalued functions, provide examples of multivalued functions which are WNC but not NC, including a function arising from the phage lambda modelling. We also give an alternative characterization of WNC functions and prove that NC functions are indeed WNC. In Sects. 4 and 5, we recall the definition of average sensitivity and prove Theorem 3. Section 6 concludes with possible perspectives for further research.

2 Nested Canalizing Multivalued Functions

2.1 Definition

Let k, n be positive integers, $k \geqslant 2$. $\mathbb{Z}/k\mathbb{Z}$ is the ring of integers modulo k.

Following [6,12,13], we shall say that $f : (\mathbb{Z}/k\mathbb{Z})^n \to \mathbb{Z}/k\mathbb{Z}$ is *canalizing with respect to coordinate* i *and* $(a,b) \in \mathbb{Z}/k\mathbb{Z} \times \mathbb{Z}/k\mathbb{Z}$ if there exists a function $g : (\mathbb{Z}/k\mathbb{Z})^n \to \mathbb{Z}/k\mathbb{Z}$ different from the constant b such that

$$f(x) = \begin{cases} b & \text{if } x_i = a \\ g(x) & \text{if } x_i \neq a. \end{cases}$$

We shall simply say that f is *canalizing* if it is canalizing with respect to some i, a, b.

A *segment* is a (proper, nonempty) subset of $\mathbb{Z}/k\mathbb{Z}$ of the form $\{0, \dots, i\}$ or $\{i, \dots, k-1\}$, with $0 \leqslant i \leqslant k-1$.

Let $\sigma \in \mathfrak{S}_n$ be a permutation, $A_1, \dots, A_n$ be segments, and $c_1, \dots, c_{n+1} \in \mathbb{Z}/k\mathbb{Z}$ be such that $c_n \neq c_{n+1}$. Then f is said to be *nested canalizing (NC) with respect to* σ, $A_1, \dots, A_n$, $c_1, \dots, c_{n+1}$ if

$$f(x) = \begin{cases} c_1 & \text{if } x_{\sigma(1)} \in A_1 \\ c_2 & \text{if } x_{\sigma(1)} \notin A_1, x_{\sigma(2)} \in A_2 \\ \vdots & \vdots \\ c_n & \text{if } x_{\sigma(1)} \notin A_1, \dots, x_{\sigma(n-1)} \notin A_{n-1}, x_{\sigma(n)} \in A_n \\ c_{n+1} & \text{if } x_{\sigma(1)} \notin A_1, \dots, x_{\sigma(n-1)} \notin A_{n-1}, x_{\sigma(n)} \notin A_n. \end{cases}$$

We shall simply say that f is *NC* if it is NC with respect to some σ, $A_1, \dots, A_n$, $c_1, \dots, c_{n+1}$.

2.2 Example: Logical Modelling of the Phage Lambda

The following example is inspired from the logical modelling of the phage lambda, a model system whose study revealed the basic concepts and mechanistic details of gene regulation. This regulator model that has been widely studied to understand the decision between lysis and lysogenization [1,16,17] is described by NC functions.

The model involves two genes, CI and Cro. CI is either expressed or not, and its expression level is therefore modelled by a Boolean variable, while Cro can take 3 values $\{0, 1, 2\}$ [23]. This simple model is sufficient to display both multistability (representing lysis and lysogeny fates) and oscillations (lysogeny state) [18,19].

In state $x = (x_{CI}, x_{Cro}) \in \mathbb{Z}/2\mathbb{Z} \times \mathbb{Z}/3\mathbb{Z}$, the next target value of CI is given by the following function $f_{CI} : \mathbb{Z}/2\mathbb{Z} \times \mathbb{Z}/3\mathbb{Z} \to \mathbb{Z}/2\mathbb{Z}$:

$$f_{CI}(x) = \begin{cases} 0 & \text{if } x_{Cro} \geqslant 1 \\ 1 & \text{otherwise.} \end{cases}$$

For instance, in state $(1, 2)$, the next value of CI can be 0 because $f_{CI}(1, 2) = 0$, and in state $(0, 2)$, the value of CI cannot change because $f_{CI}(0, 2) = 0$. Similarly, the target value of Cro is given by a function $f_{Cro} : \mathbb{Z}/2\mathbb{Z} \times \mathbb{Z}/3\mathbb{Z} \to \mathbb{Z}/3\mathbb{Z}$. For instance,

$$f_{Cro}(x) = \begin{cases} 0 & \text{if } x_{CI} = 1 \\ 1 & \text{if } x_{CI} = 0 \text{ and } x_{Cro} = 2 \\ 2 & \text{otherwise.} \end{cases}$$

x_{Cro} \ x_{CI}	0	1
2	1	0
1	2	0
0	2	0

In the above table to the right, the $x = (x_{CI}, x_{Cro})$ entry is the value of $f_{Cro}(x)$. In state $(1, 2)$, the target value of Cro is 0 because $f_{Cro}(1, 2) = 0$, so the value of Cro can decrease.

In this context of discrete dynamics, to represent the trajectories of the dynamics of the whole system, we need to choose the update rule for the pair of functions (f_{CI}, f_{Cro}). We choose the *asynchronous* setting, which means that at each time step, the level of at most one gene can change. So, in state $(1, 2)$, since the levels of both genes can decrease ($f_{CI}(1, 2) = 0 < 1 = x_{CI}$ and $f_{Cro}(1, 2) = 0 < 2 = x_{Cro}$), the system can (non-deterministically) reach either state $(1, 1)$ or state $(0, 2)$. Note that there is no direct transition from $(1, 2)$ to $(1, 0)$ because we limit the length step to 1. The set of asynchronous trajectories is summarized in the following state transition graph, where the vertices are the system states (x_{CI}, x_{Cro}) and the arrows connect two consecutive states:

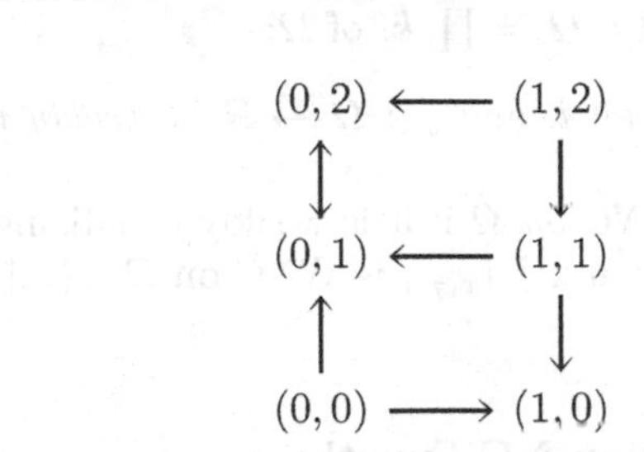

Fig. 1. State transition graph: asynchronous trajectories of the system (f_{CI}, f_{Cro}). It contains two attractors: the stable state $(1, 0)$ and the cyclical attractor $\{(0, 1), (0, 2)\}$.

The presence of two attractors (terminal strongly connected components, see Fig. 1) reflects multistability: the stable state $(1, 0)$ representing lysis fate and the cyclical attractor lysogeny fate.

Clearly, f_{CI} and f_{Cro} are both NC. To see from the above definition of f_{CI} that it is NC, it suffices to let the segment for x_{Cro} be $\{1,2\} \subset \{0,1,2\}$. It is again easy to see that f_{Cro} is NC: the segment corresponding to x_{CI} is $\{1\} \subset \{0,1\}$, and then the segment corresponding to x_{Cro} is $\{2\} \subset \{0,1,2\}$.

3 Weakly Nested Canalizing Multivalued Functions

In Theorem 3, we shall give an upper bound on average sensitivity which holds not only for NC functions, but for the more general class of weakly nested canalizing functions, which we define now.

3.1 Definition

Let n be a positive integer. For each $i \in \{1,\ldots,n\}$, Ω_i is a finite set of cardinality $k_i > 0$, $\Omega = \prod_i \Omega_i$, and $f : \Omega \to \mathbb{R}$. Note that we do not require $k_i \geqslant 2$ for all i. If $k_j = 1$ for some j, f could be viewed as a function with one less variable, *i.e.* as a function on $\prod_{i \neq j} \Omega_i$, but we still consider it as a function defined on $\prod_i \Omega_i$.

We shall say that f is *weakly canalizing with respect to coordinate i and $(a,b) \in \Omega_i \times \mathbb{R}$* if $f(x) = b$ whenever $x_i = a$, and simply that it is *weakly canalizing* if it is weakly canalizing with respect to some i, a, b.

Note that this definition differs slightly from the usual definition by the absence of condition on the values of f for $x_i \neq a$: we do not require the existence of some x such that $x_i \neq a$ and $f(x) \neq b$. In particular, constant functions are weakly canalizing, though not canalizing.

If f is canalizing with respect to i, a, b and $k_i \geqslant 2$, we shall consider

$$f\restriction_{x_i \neq a} : \Omega \cap \{x \mid x_i \neq a\} \to \mathbb{R},$$

the restriction of f to the set of $x \in \Omega$ such that $x_i \neq a$.

The class of weakly nested canalizing functions on $\Omega = \prod_i \Omega_i$ is then defined by induction on the cardinality $|\Omega| = \prod_i k_i$ of Ω:

- If $|\Omega| = 1$, *i.e.* $k_i = 1$ for all i, any $f : \Omega \to \mathbb{R}$ is *weakly nested canalizing (WNC) on Ω*.
- If $|\Omega| > 1$, $f : \Omega \to \mathbb{R}$ is *WNC on Ω* if it is weakly canalizing with respect to some i, a, b such that $k_i \geqslant 2$ and $f\restriction_{x_i \neq a}$ is WNC on $\Omega \cap \{x \mid x_i \neq a\}$, a strict subset of Ω.

3.2 Examples of WNC Non NC Functions

As we shall see in Proposition 2, the class of WNC functions contains the class of NC functions. We give here a few examples of simple multivalued functions which are WNC but not NC.

- As we have already observed, constant functions from $(\mathbb{Z}/k\mathbb{Z})^n$ to $\mathbb{Z}/k\mathbb{Z}$ are WNC but not NC.
- In decomposing a WNC function $f : (\mathbb{Z}/k\mathbb{Z})^n \to \mathbb{Z}/k\mathbb{Z}$, it is possible to "peel" a coordinate hyperplane defined on some coordinate i (*i.e.* by some equation $x_i = a$), then a coordinate hyperplane defined on j, and later a coordinate hyperplane defined on i again. This is because of the recursive nature of the definition of WNC functions, and gives more freedom in the construction of WNC functions than in the construction of NC functions.
  For instance, the functions min and max : $(\mathbb{Z}/k\mathbb{Z})^2 \to \mathbb{Z}/k\mathbb{Z}$ are not NC, as observed in [6]. However, an easy induction on k shows that they are WNC. For instance, $\min = \min_k : \{0,\dots,k-1\}^2 \to \{0,\dots,k-1\}$ is weakly canalizing with respect to $1,0,0$, $\min_k \upharpoonright_{x_1\neq 0}$ is weakly canalizing with respect to $2,0,0$, and $\min_k \upharpoonright_{x_1\neq 0, x_2\neq 0}$ is identical to the function $\min_{k-1} : \{1,\dots,k-1\}^2 \to \{1,\dots,k-1\}$, which is WNC.
- Also, in constructing a WNC function $f : (\mathbb{Z}/k\mathbb{Z})^n \to \mathbb{Z}/k\mathbb{Z}$, the values a used to define $f(x)$ for $x_i = a$ need not be extremal values (initially 0 or $k-1$), they can be intermediate values: $0 < a < k-1$.
  For instance, the function from $\mathbb{Z}/3\mathbb{Z}$ to $\mathbb{Z}/3\mathbb{Z}$ defined by $0 \mapsto 0, 1 \mapsto 1, 2 \mapsto 0$ is not NC because it is canalizing with respect to either the intermediate value 1 (for its unique variable), or the values 0 and 2 (which do not form a segment). But any function from $\mathbb{Z}/k\mathbb{Z}$ to $\mathbb{Z}/k\mathbb{Z}$ is WNC.

3.3 Back to the Phage Lambda Example

Getting back to the model of the phage lambda presented in Sect. 2.2, since f_{CI} and f_{Cro} are NC, they are also WNC by Proposition 2 below.

Let us consider the following function:

$$f'_{Cro}(x) = \begin{cases} 1 & \text{if } x_{Cro} = 2 \\ 0 & \text{if } x_{Cro} \neq 2 \text{ and } x_{CI} = 1 \\ 2 & \text{if } x_{Cro} \neq 2, x_{CI} \neq 1 \text{ and } x_{Cro} = 1 \\ 1 & \text{otherwise} \end{cases}$$

x_{Cro} \ x_{CI}	0	1
2	1	1
1	2	0
0	1	0

In the above table to the right, the $x = (x_{CI}, x_{Cro})$ entry is the value of $f'_{Cro}(x)$. By the same argument as developed in Sect. 2.2, we can see that (f_{CI}, f'_{Cro}) gives rise to the same state transition graph as (f_{CI}, f_{Cro}) (represented in Fig. 1).

It is interesting to remark that f_{Cro} and f'_{Cro} do not have the same canalizing property: f_{Cro} is NC, while f'_{Cro} is clearly WNC but not NC: Indeed, the table shows that the first canalizing step has to be $x_{Cro} = 2$ (unique choice which fixes the value of f'_{Cro}), the resulting restriction $f'_{Cro} \upharpoonright_{x_{Cro}\neq 2}$ is canalizing only with $x_{CI} = 1$, and the restriction $f'_{Cro} \upharpoonright_{x_{Cro}\neq 2, x_{CI}\neq 1}$ is not constant. Thus, in this example two functions that represent the same asynchronous dynamics do not have the same canalizing properties.

3.4 Properties of WNC Functions

Intuitively, a function $f : \Omega \to \mathbb{R}$ is WNC if its domain Ω can be "peeled" by successively removing coordinate hyperplanes (defined by equations of the form $x_i = a$) whose points are mapped by f to the same value, whence the following characterization:

Proposition 1. (Characterization) *Letting* $K = \sum_i k_i$, f *is WNC if and only if there exist a function* $v : \{1, \ldots, K\} \to \{1, \ldots, n\}$ *and numbers* $a_i \in \Omega_{v(i)}$ *and* $b_i \in \mathbb{R}$ *for each* $i \in \{1, \ldots, K\}$ *such that:*

$$f(x) = \begin{cases} b_1 & \text{if } x_{v(1)} = a_1 \\ b_2 & \text{if } x_{v(1)} \neq a_1, x_{v(2)} = a_2 \\ \vdots & \vdots \\ b_K & \text{if } x_{v(1)} \neq a_1, \ldots, x_{v(K-1)} \neq a_{K-1}, x_{v(K)} = a_K. \end{cases}$$

Proof. By induction on K.

The base case corresponds to $|\Omega| = 1$. Then for all i, Ω_i is a singleton $\{a_i\}$, and $K = n$. In that case, a function $f : \Omega \to \mathbb{R}$ is simply a number $b \in \mathbb{R}$ and can be defined for instance by

$$f(x) = \begin{cases} b & \text{if } x_1 = a_1 \\ b_2 & \text{if } x_1 \neq a_1, x_2 = a_2 \\ \vdots & \vdots \\ b_n & \text{if } x_1 \neq a_1, \ldots, x_{n-1} \neq a_{n-1}, x_n = a_n, \end{cases}$$

where $b_2, \ldots, b_n$ are arbitrary numbers because the conditions of lines 2 to n are obviously not matched.

We now assume that $|\Omega| > 1$ and that $f : \Omega \to \mathbb{R}$ is WNC on Ω: this means that f is weakly canalizing with respect to some i, a, b such that $k_i \geqslant 2$ and that $f' = f \restriction_{x_i \neq a}$ is WNC on $\Omega' = \Omega \cap \{x \mid x_i \neq a\}$. Therefore

$$f(x) = \begin{cases} b & \text{if } x_i = a \\ f'(x) & \text{if } x_i \neq a. \end{cases}$$

Moreover, for $f' : \Omega' \to \mathbb{R}$, we have $K' = K - 1$, hence by the induction hypothesis, there exist a function $v : \{2, \ldots, K\} \to \{1, \ldots, n\}$ and numbers $a_i \in \Omega_{v(i)}$ and $b_i \in \mathbb{R}$ indexed by $i \in \{2, \ldots, K\}$ (a set of cardinality K') such that

$$f'(x) = \begin{cases} b_2 & \text{if } x_{v(2)} = a_2 \\ b_3 & \text{if } x_{v(2)} \neq a_2, x_{v(3)} = a_3 \\ \vdots & \vdots \\ b_K & \text{if } x_{v(2)} \neq a_2, \ldots, x_{v(K-1)} \neq a_{K-1}, x_{v(K)} = a_K. \end{cases}$$

This entails the following expression for f:

$$f(x) = \begin{cases} b & \text{if } x_i = a \\ b_2 & \text{if } x_i \neq a, x_{v(2)} = a_2 \\ b_3 & \text{if } x_i \neq a, x_{v(2)} \neq a_2, x_{v(3)} = a_3 \\ \vdots & \quad \vdots \\ b_K & \text{if } x_i \neq a, x_{v(2)} \neq a_2, \ldots, x_{v(K-1)} \neq a_{K-1}, x_{v(K)} = a_K, \end{cases}$$

which is of the expected form by letting $a_1 = a, b_1 = b$ and extending v to $v : \{1, 2, \ldots, K\} \to \{1, \ldots, n\}$ with $v(1) = i$. □

In decomposing an NC function $f : (\mathbb{Z}/k\mathbb{Z})^n \to \mathbb{Z}/k\mathbb{Z}$, each coordinate $i \in \{1, \ldots, n\}$ is considered exactly once (in some order prescribed by a permutation σ) and the value of f is fixed for $x_{\sigma(i)}$ in some segment A_i. This can be realized by successively fixing the value of f for each $\alpha \in A_i$, and therefore, the class of WNC functions contains the class of NC functions, as stated in the following Proposition:

Proposition 2. (NC $\Rightarrow$ WNC) *If $f : (\mathbb{Z}/k\mathbb{Z})^n \to \mathbb{Z}/k\mathbb{Z}$ is NC, then it is WNC.*

Proof. Assume f is NC with respect to σ, $A_1, \ldots, A_n$, $c_1, \ldots, c_{n+1}$. For each $i \in \{1, \ldots, n\}$, let

$$A_i = \{\alpha_i^1, \ldots, \alpha_i^{|A_i|}\}$$
$$(\mathbb{Z}/k\mathbb{Z}) \setminus A_i = \{\alpha_i^{1+|A_i|}, \ldots, \alpha_i^k\}$$

with $\alpha_i^1 < \cdots < \alpha_i^{|A_i|}$ and $\alpha_i^{1+|A_i|} < \cdots < \alpha_i^k$. This defines $K = nk$ numbers $\alpha_i^j \in \mathbb{Z}/k\mathbb{Z}$. For each $i \in \{1, \ldots, n\}$ and $j \in \{1, \ldots, k\}$, let

$$\beta_i^j = \begin{cases} c_i & \text{if } j \leqslant |A_i| \\ c_{n+1} & \text{otherwise.} \end{cases}$$

To comply with the characterization of WNC functions (Proposition 1), we relabel the numbers α_i^j, β_i^j by identifying the list

$$\alpha_1^1, \ldots, \alpha_1^{|A_1|}, \ldots, \alpha_n^1, \ldots, \alpha_n^{|A_n|}, \alpha_1^{1+|A_1|}, \ldots, \alpha_1^k, \ldots, \alpha_n^{1+|A_n|}, \ldots, \alpha_n^k$$

as the list $a_1, \ldots, a_K$, and by identifying similarly the list

$$\beta_1^1, \ldots, \beta_1^{|A_1|}, \ldots, \beta_n^1, \ldots, \beta_n^{|A_n|}, \beta_1^{1+|A_1|}, \ldots, \beta_1^k, \ldots, \beta_n^{1+|A_n|}, \ldots, \beta_n^k$$

as the list $b_1, \ldots, b_K$. Call φ this relabelling, which maps $r \in \{1, \ldots, K\}$ to the pair $\varphi(r) = (i, j)$ such that $a_r = \alpha_i^j$ and $b_r = \beta_i^j$. For instance, $\varphi(1) = (1, 1)$ and $\varphi(K) = (n, k)$. Then finally, a function $v : \{1, \ldots, K\} \to \{1, \ldots, n\}$ is defined by $v(r) = \sigma(i)$ if $\varphi(r) = (i, j)$. Then f clearly enjoys the characterization of WNC functions, with the choice of function v and numbers a_r, b_r. □

4 Average Sensitivity

In Sect. 5 we shall be interested in the average sensitivity of (WNC) multivalued functions, but before giving the (more technical) definition of average sensitivity for multivalued functions, we start by recalling the more intuitive definition for Boolean functions.

4.1 Boolean-Valued Boolean Functions

The average sensitivity of a Boolean function $f : (\mathbb{Z}/2\mathbb{Z})^n \to \mathbb{Z}/2\mathbb{Z}$ is the probability that the ith variable affects the outcome. It can be defined as follows. For $x \in (\mathbb{Z}/2\mathbb{Z})^n$, let $\overline{x}^i$ denote the vector obtained from x by changing the value of x_i. Then the *influence of the ith variable* is

$$\mathbf{Inf}_i[f] = \mathbf{Prob}_x[f(x) \neq f(\overline{x}^i)] \in [0,1],$$

where the probability is taken for the uniform distribution, and the *average sensitivity* (also called *influence* or *total influence*) is

$$\mathbf{AS}[f] = \sum_i \mathbf{Inf}_i[f] \in [0,n].$$

This can be reformulated in terms of boundary edges. Let an *edge* in the Hamming cube $(\mathbb{Z}/2\mathbb{Z})^n$ be a pair (x,y) such that the Hamming distance between x and y is 1, and let a *boundary edge* be an edge (x,y) such that $f(x) \neq f(y)$. Then the average sensitivity $\mathbf{AS}[f]$ is such that the fraction of edges which are boundary edges equals $\mathbf{AS}[f]/n$.

4.2 Multivalued Functions

This definition can be generalized to multivalued functions. Following [14, Chapter 8], we shall take the following definition.

First, Fourier decomposition is generalized to non Boolean domains. Let $\Omega = \prod_{i=1}^n \Omega_i$ be as above, with $|\Omega_i| = k_i$. On the vector space of real-valued functions defined on Ω, an inner product is given by $\langle f,g \rangle = \mathbf{E}_x[f(x)g(x)]$, where $\mathbf{E}$ denotes the expectation. Here, $x \in \Omega$ and we assume independent uniform probability distributions on the Ω_i. A *Fourier basis* is an orthonormal basis $(\varphi_\alpha)_{\alpha \in \prod_i \{0,\ldots,k_i\}}$ such that $\varphi_{(0,\ldots,0)} = 1$. It is not difficult to see that a Fourier basis always exists, although it is not unique.

Then, fix a Fourier basis (φ_α). The *Fourier coefficients of* $f : \Omega \to \mathbb{R}$ are $\widehat{f}(\alpha) = \langle f, \varphi_\alpha \rangle$, and $E_i f = \sum_{\alpha | \alpha_i = 0} \widehat{f}(\alpha)\varphi_\alpha$ turns out to be independent of the basis. The notation $\sum_{\alpha|\alpha_i=0}$ denotes the sum for all α such that $\alpha_i = 0$. For all $i \in \{1,\ldots,n\}$, let the *ith coordinate Laplacian operator* L_i be the linear operator defined by $L_i f = f - E_i f$.

Finally, the *influence of coordinate i on f* is defined by $\mathbf{Inf}_i[f] = \langle f, L_i f \rangle$, and the *average sensitivity* of f is then $\mathbf{AS}[f] = \sum_i \mathbf{Inf}_i[f]$.

By Plancherel's theorem (see [14]), we have

$$\mathbf{Inf}_i[f] = \sum_{\alpha_i \neq 0} \widehat{f}(\alpha)^2 = \mathbf{E}_x[\mathbf{Var}_{y_i}[f(x_1, \ldots, x_{i-1}, y_i, x_{i+1}, \ldots, x_n)]],$$

where **Var** denotes the variance ($\mathbf{Var}[g] = \mathbf{E}[g^2] - \mathbf{E}[g]^2$) and $y_i \in \Omega_i$. The above equality makes clear that the definition of influence of multivalued functions generalizes the Boolean case.

5 Upper Bound on the Average Sensitivity of WNC Multivalued Functions

For an arbitrary $f : \Omega \to [0, M]$, we have $\mathbf{Var}_i[f](x) \leqslant (M/2)^2$ for all i, therefore $\mathbf{Inf}_i[f] \leqslant M^2/4$ for all i and

$$\mathbf{AS}[f] \leqslant n \cdot M^2/4 = \mathcal{O}(n).$$

For WNC functions, this upper bound can be greatly improved. In the Boolean case, [11] proves (by a different method from ours) that $\mathbf{AS}[f] \leqslant 2$ for NC $\{-1, +1\}$-valued functions. This bound is improved in [10], where it is proved that $\mathbf{AS}[f] \leqslant 4/3$. For NC functions $f : \{0,1\}^n \to \{0,1\}$, the result in [11] means $\mathbf{AS}[f] \leqslant 1/2$.

Theorem 3 generalizes this result, by establishing that, in the more general multivalued case, the average sensitivity of WNC functions is bounded above by a constant.

Theorem 3. *Let* $\Omega = \prod_{i=1}^n \Omega_i$ *where each* Ω_i *has cardinality* $k_i > 0$. *Let* $f : \Omega \to [0, M]$ *and* $\kappa = \max_i (k_i - 1)/k_i < 1$. *If* f *is WNC (in particular if it is NC), then*

$$\mathbf{AS}[f] \leqslant \frac{M^2}{4(1-\kappa)}.$$

Proof. We prove this by induction on $\sum_i k_i$. If $\sum_i k_i = n$, *i.e.* $k_i = 1$ for all i, the inequality holds trivially: actually $\mathbf{AS}[f] = 0$.

Now assume $\sum_i k_i > n$ and f is WNC. This means that f is weakly canalizing with respect to some j, a, b such that $k_j \geqslant 2$, and we let $f' = f \restriction_{x_j \neq a}$. Let Ω' be the set of $x \in \Omega$ such that $x_j \neq a$, so that $f' : \Omega' \to [0, M]$. The induction hypothesis applied to f' reads

$$\mathbf{AS}[f'] \leqslant \frac{M'^2}{4(1-\kappa')}$$

with

$$M' = \max_{x \in \Omega'} f'(x) = \max_{x \in \Omega'} f(x) \leqslant M$$

$$\kappa' = \max\left\{\frac{k_j - 2}{k_j - 1}, \max_{i \neq j} \frac{k_i - 1}{k_i}\right\} \leqslant \kappa.$$

Note that the induction hypothesis implies $\mathbf{AS}[f'] \leqslant M^2/(4(1-\kappa))$. We shall use the notation

$$\mathbf{Var}_i[f](x) = \mathbf{Var}_{y_i}[f(x_1, \ldots, x_{i-1}, y_i, x_{i+1}, \ldots, x_n)].$$

Then $\mathbf{AS}[f] = \mathbf{E}_x[\sum_i \mathbf{Var}_i[f](x)]$ and

$$\begin{aligned}\mathbf{AS}[f] \cdot \prod_i k_i &= \sum_x \sum_i \mathbf{Var}_i[f](x) \\ &= \sum_{x_j = a} \mathbf{Var}_j[f](x) + \sum_{x_j \neq a} \Big(\mathbf{Var}_j[f](x) + \sum_{i \neq j} \mathbf{Var}_i[f](x) \Big)\end{aligned}$$

since $f(x)$ is constant when $x_j = a$, so that $\mathbf{Var}_i[f](x) = 0$ for $i \neq j$. Here and below, the notations $\sum_{x_j = a}$ and $\sum_{x_j \neq a}$ denote the sums for all x such that $x_j = a$ (resp. $x_j \neq a$). Furthermore, $\mathbf{Var}_j[f](x)$ is independent of x_j, and on the other hand, $\mathbf{Var}_i[f](x) = \mathbf{Var}_i[f'](x)$ when $x_j \neq a$ and $i \neq j$. Thus

$$\mathbf{AS}[f] \cdot \prod_i k_i = k_j \cdot \sum_{x_j = a} \mathbf{Var}_j[f](x) + \sum_{x_j \neq a} \sum_{i \neq j} \mathbf{Var}_i[f'](x).$$

Since $0 \leqslant f(x) \leqslant M$ for all x, we have $\mathbf{Var}_j[f](x) \leqslant M^2/4$, and on the other hand, $x_j \neq a \Leftrightarrow x \in \Omega'$. Therefore

$$\begin{aligned}\mathbf{AS}[f] \cdot \prod_i k_i &\leqslant k_j \cdot \prod_{i \neq j} k_i \cdot M^2/4 + \sum_{x \in \Omega'} \sum_{i=1}^{n} \mathbf{Var}_i[f'](x) \\ &= \prod_i k_i \cdot M^2/4 + \mathbf{AS}[f'] \cdot (k_j - 1) \cdot \prod_{i \neq j} k_i\end{aligned}$$

and

$$\mathbf{AS}[f] \leqslant \frac{M^2}{4} + \mathbf{AS}[f'] \cdot \frac{k_j - 1}{k_j} \leqslant \frac{M^2}{4} + \kappa \cdot \mathbf{AS}[f'].$$

To conclude the proof, it suffices to observe that $\mathbf{AS}[f'] \leqslant M^2/(4(1-\kappa))$ implies $\mathbf{AS}[f] \leqslant M^2/(4(1-\kappa))$ because

$$\begin{aligned}\mathbf{AS}[f] &\leqslant \frac{M^2}{4} + \kappa \cdot \mathbf{AS}[f'] \\ &\leqslant \frac{M^2}{4} + \kappa \cdot \frac{M^2}{4(1-\kappa)} \\ &= \frac{M^2}{4} \left(1 + \frac{\kappa}{1-\kappa} \right) \\ &\leqslant M^2/(4(1-\kappa)).\end{aligned}$$

□

In the Boolean case, $\kappa = 1/2$ and $M = 1$, so that the upper bound $M^2/(4(1-\kappa))$ equals 1/2 and the above result is a generalization of the result in [11].

The above proof is also significantly simpler than the one in [11]. It can be easily checked that in the Boolean case, our argument on variance essentially amounts to compute the fraction of edges in the Hamming cube which are boundary edges.

6 Concluding Remarks

In the context of modelling biological systems, the choice of a relevant function with respect to the biological application is a critical and difficult step in the modelling process, notably because of the large (exponential in n) amount of functions compatible with the state transition graph. A current challenge is thus to find good selection criteria. We have already mentioned that canalizing functions are significantly predominant in gene network modelling [21], and that networks with NC rules are stable [9]. While many works focus on the NC property for Boolean functions [9,21], we did not find any study of the canalizing properties for biological multivalued functions in the literature.

We have mentioned that in the process of modelling gene networks, canalization has been used as a guide to select candidate Boolean functions to parameterise the regulatory graph [4,27]. With the extensions proposed in this paper, such an approach could be considered for multivalued functions as well. Moreover, the simple example of phage lambda modelling (Sects. 2.2 and 3.3) suggests that it would be worth taking into account, in the sensitivity analysis, the updating rules and the notions of both NC and WNC functions.

On the theoretical side, an obvious question is whether the bound $M^2/(4(1-\kappa))$ in Theorem 3 can be improved for multivalued WNC, or at least NC, functions, along the lines of [10].

References

1. Danchin, A.: Biological innovation in the functional landscape of a model regulator, or the lactose operon repressor. C.R. Biol. **344**(2), 111–126 (2021)
2. Debat, V., Le Rouzic, A.: Canalization, a central concept in biology. Semin. Cell Dev. Biol. **88**, 1–3 (2019)
3. Friedgut, E.: Boolean functions with low average sensitivity depend on few coordinates. Combinatorica **18**, 27–35 (1998)
4. Hinkelmann, F., Jarrah, S.: Inferring biologically relevant models: nested canalyzing functions. ISRN Biomathematics **3**, 2012 (2012)
5. Just, W., Shmulevich, I., Konvalina, J.: The number and probability of canalyzing functions. Physica D **197**(3–4), 211–221 (2004)
6. Kadelka, C., Li, Y., Kuipers, J., Adeyeye, J.O., Laubenbacher, R.: Multistate nested canalizing functions and their networks. Theoret. Comput. Sci. **675**, 1–14 (2017)
7. Kauffman, S.A.: The origins of order: Self organization and selection in evolution. Oxford University Press (1993)

8. Kauffman, S., Peterson, C., Samuelsson, B., Troein, C.: Random Boolean network models and the yeast transcriptional network. Proc. Natl. Acad. Sci. **100**(25) (2003)
9. Kauffman, S., Peterson, C., Samuelsson, B., Troein, C.: Genetic networks with canalyzing Boolean rules are always stable. Proc. Natl. Acad. Sci. **101**(49) (2004)
10. Klotz, J.G., Heckel, R., Schober, S.: Bounds on the average sensitivity of nested canalizing functions. Plos One **8**(5) (2013)
11. Li, Y., Adeyeye, J.O., Murrugarra, D., Aguilar, B., Laubenbacher, R.: Boolean nested canalizing functions: a comprehensive analysis. Theoret. Comput. Sci. **481**, 24–36 (2013)
12. Murrugarra, D., Laubenbacher, R.: Regulatory patterns in molecular interaction networks. J. Theor. Biol. **288**, 66–72 (2011)
13. Murrugarra, D., Laubenbacher, R.: The number of multistate nested canalyzing functions. Physica D **241**(10), 929–938 (2012)
14. O'Donnell, R.: Analysis of Boolean functions. Cambridge University Press (2014)
15. Podlipniak, P.: The role of canalization and plasticity in the evolution of musical creativity. Front. Neurosci. **15**, 607887 (2021)
16. Ptachne, M.: A genetic switch. Blackwell Science, Phage lambda and higher organisms (1992)
17. Remy, É., Ruet, P.: From minimal signed circuits to the dynamics of Boolean regulatory networks. Bioinformatics **24**, i220–i226 (2008)
18. Remy, É., Ruet, P., Thieffry, D.: Graphic requirements for multistability and attractive cycles in a Boolean dynamical framework. Adv. Appl. Math. **41**(3), 335–350 (2008)
19. Ruet, P.: Local cycles and dynamical properties of Boolean networks. Math. Struct. Comput. Sci. **26**(4), 702–718 (2016)
20. Schober, S.: About Boolean networks with noisy inputs. In: Proceedings of Fifth International Workshop on Computational Systems Biology, pp. 173–176 (2008)
21. Subbaroyan, A., Martin, O.C., Samal, A.: Minimum complexity drives regulatory logic in Boolean models of living systems. PNAS Nexus **1**, 1–12 (2022)
22. Surkova, S., et al.: Characterization of the Drosophila segment determination morphome. Dev Biol. **313**(2), 844–862 (2008)
23. Thieffry, D., Thomas, R.: Dynamical behaviour of biological regulatory networks II. Immunity control in bacteriophage lambda. Bull. Math. Biol. **57**, 277–295 (1995)
24. Thomas, R.: Boolean formalization of genetic control circuits. J. Theor. Biol. **42**, 563–585 (1973)
25. Thomas, R.: Regulatory networks seen as asynchronous automata: a logical description. J. Theor. Biol. **153**, 1–23 (1991)
26. Waddington, C.H.: Canalization of development and the inheritance of acquired characters. Nature **150**, 563–565 (1942)
27. Zhou, J.X., Samal, A., d'Hérouël, A.F., Price, N.D., Huang, S.: Relative stability of network states in Boolean network models of gene regulation in development. Biosystems **142–143**, 15–24 (2016)

Tackling Universal Properties of Minimal Trap Spaces of Boolean Networks

Sara Riva[1(✉)], Jean-Marie Lagniez[2(✉)], Gustavo Magaña López[1(✉)], and Loïc Paulevé[1(✉)]

[1] Univ. Bordeaux, CNRS, Bordeaux INP, LaBRI, UMR 5800, 33400 Talence, France
{sara.riva,gustavo.magana,loic.pauleve}@labri.fr
[2] Univ. Artois, CNRS, CRIL, 62300 Lens, France
lagniez@cril.fr

Abstract. Minimal trap spaces (MTSs) capture subspaces in which the Boolean dynamics is trapped, whatever the update mode. They correspond to the attractors of the most permissive mode. Due to their versatility, the computation of MTSs has recently gained traction, essentially by focusing on their enumeration. In this paper, we address the logical reasoning on universal properties of MTSs in the scope of two problems: the reprogramming of Boolean networks for identifying the permanent freeze of Boolean variables that enforce a given property on all the MTSs, and the synthesis of Boolean networks from universal properties on their MTSs. Both problems reduce to solving the satisfiability of quantified propositional logic formula with 3 levels of quantifiers ($\exists\forall\exists$). In this paper, we introduce a Counter-Example Guided Refinement Abstraction (CEGAR) to efficiently solve these problems by coupling the resolution of two simpler formulas. We provide a prototype relying on Answer-Set Programming for each formula and show its tractability on a wide range of Boolean models of biological networks.

Keywords: Boolean networks · Attractors · Synthesis · QBF · CEGAR

1 Introduction

Since recent years, we observe a surge of successful applications of Boolean networks (BNs) in biology and medicine for the modeling and prediction of cellular dynamics in the case of cancer and cellular reprogramming [21, 29, 32]. Such applications face two main challenges: being able to design a qualitative Boolean model which is faithful to the behavior of the biological system and being able to compute predictions to control its (long-term) dynamics. From a computational point of view, the latter problem mostly depends on the complexity of the dynamical property to enforce, while the former additionally suffers from the combinatorics of candidate models.

In a BN, the state of interacting components are modeled with Boolean values $\mathbb{B} = \{0, 1\}$. Then, given n components, the dynamics evolve within the finite

J. Pang and J. Niehren (Eds.): CMSB 2023, LNBI 14137, pp. 157–174, 2023.
https://doi.org/10.1007/978-3-031-42697-1_11

discrete space of *configurations* $\mathbb{B}^n$. For each component i, the BN specifies a Boolean function $f_i : \mathbb{B}^n \rightarrow \mathbb{B}$ to compute the value towards which the component state evolves from a given configuration of the network. The transitions between the configurations are then computed according to an *update mode*. For instance, with the synchronous mode, a configuration $x \in \mathbb{B}^n$ has a (unique) transition to configuration $(f_1(x), \cdots, f_n(x))$, whereas, with the fully asynchronous mode, it has one transition for each component i such that $f_i(x) \neq x_i$ and going to $(x_1, \cdots, x_{i-1}, f_i(x), x_{i+1}, \cdots, x_n)$. There is a vast zoo of update modes defined in the literature. They reflect different modeling hypotheses on how the components evolve with respect to each other, and can have a great impact on the resulting dynamics [27]. These update modes can be compared using a simulation relation: an update mode simulates another if, for any transition $x \rightarrow y$ of the latter, there exists a trajectory from x to y with the former mode. This results in a hierarchy of update modes [27], where the *most permissive* [25,26] simulates all Boolean update modes. The most permissive mode captures any trajectory of any quantitative model which is a refinement of the BN (intuitively, a refinement adds quantitative information on interaction thresholds and state, while respecting the logic of state change). Hence, most permissive Boolean dynamics have formal connections with quantitative systems, contrary to (a)synchronous modes, which are known to preclude the prediction of actually feasible trajectories in biological systems [25].

Most applications of BNs to biological systems involve two types of dynamical properties: the trajectories between configurations, which model changes of the state of components over time, and the attractors, which capture the long-term dynamics of the system. An attractor can be characterized by a set of configurations from which there is no out-going transition, and such that there is a trajectory between any distinct pair of its configurations. When it is composed of a single configuration, the attractor is called a fixed point of the dynamics.

Capturing properties that are shared by *all* the attractors (or *all* attractors reachable from a given set of configurations) is, therefore, a fundamental task of BN modeling. In this paper, we focus on two problems related to these universal properties: the reprogramming of a given BN with the permanent freeze of components of the network, and the synthesis of a BN which matches with a given architecture while showing the desired universal property on its attractors. These are relevant problems since attractors usually capture biological phenotypes [11,18] and the reprogramming of a biological phenomena is linked to the search of treatments [21]. However, the computational complexity of these problems is stirred by the complexity of characterizing (all) the attractors of a BN. This complexity depends on the update mode. For (a)synchronous update modes, determining whether a configuration belongs to an attractor is an infamous PSPACE-complete problem, which largely impedes the tractability of analysis of networks with several hundreds of components. Indeed, attractors can have very different shapes with these modes.

The *minimal trap spaces* of BNs are related to their attractors but do not depend on the update mode. A *trap space* is a subcube of $\mathbb{B}^n$ which is closed by the local functions (the image by $f_1, \ldots, f_n$ of a vertex is one of its vertices). It is minimal whenever there is no other trap space within it. The fixed points are particular cases of minimal trap spaces. In some sense, a trap space delimits a portion of the space from where any trajectory with any update mode is trapped within. Thus, a minimal trap space encloses at least one attractor, with any update mode. However, an (a)synchronous (non-fixed point) attractor is not necessarily included in a minimal trap space. Nevertheless, in practice for biological models, minimal trap spaces have been observed to be good approximations of asynchronous attractors [19]. Moreover, it turns out that minimal trap spaces are exactly the attractors of the most permissive mode [25]. Back to our computational point of view, (minimal) trap spaces are more amenable objects thanks to a much lower complexity [23]. Recent approaches demonstrated the tractability of methods based on solving the satisfiability of logical formulas for enumerating minimal trap spaces in BNs with several thousands of components [25,30]. In large networks, however, their exhaustive enumeration can be intractable.

In this paper, we study the problem of reprogramming and synthesis of BNs from properties on their minimal trap spaces. In the case of the most permissive update mode, this gives exact methods to reason on the attractors. In the case of other update modes, this gives approximate methods to address their attractors. Specifically, we will consider *marker* properties of trap spaces: a marker is a partial map associating a subset of components with a Boolean value, e.g. $\{a \mapsto 1, c \mapsto 0\}$ where a, and c are components of the BN. A trap space matches with a marker if all its configurations match with it (e.g., a is always 1 and c always 0 in the trap space). The *marker reprogramming* [24] consists in permanently freezing a subset of its components to specific Boolean values so that all the minimal trap spaces of the resulting BN match with the marker. The *synthesis* consists in deriving a BN that matches with a given network architecture (influence graph) and such that all its minimal trap spaces match with a given marker. These problems can be expressed as a logical formula of the form "there is a permanent freeze P (resp. a BN matching with influence graph) such that all the minimal trap spaces of the BN perturbed by P (resp. of the BN) match with the given marker M". As we will explain, both problems boil down to solving the satisfiability of quantified propositional Boolean formulas (QBF) [8] with three levels of quantifiers ($\exists\forall\exists$, 3-QBF).

While modern SAT [6] and Answer-Set Programming (ASP) [3,15] solvers can address efficiently 1-QBF (NP) and 2-QBF problems respectively, the generic solving of higher order QBFs problems turns out to be very challenging. In [24], the marker reprogramming of minimal trap spaces is tackled by solving a complementary problem which is only 2-QBF. However, as we will show in experiments, this approach turns out to be intractable for large networks and for the synthesis problem. To the best of our knowledge, this is so far the only other method addressing universal properties over minimal trap spaces in BNs.

Instead of solving directly the 3-QBF problems, we introduce in Sect. 3 a logic approach based on a Counter-Example-Guided Abstraction Refinement

(CEGAR) of a simpler formula. Essentially, we extract candidate perturbations (resp. BNs) from an NP formula and verify using a 2-QBF formula whether they fulfill the universal property. If not, we extract a counter-example that we generalize and plug in the original NP formula. The procedure is repeated until either we prove that the 3-QBF problem is not satisfiable, or a candidate perturbation (resp. BN) verifies the universal property on the minimal trap spaces. We developed a prototype based on ASP and show in Sect. 4 its tractability for the reprogramming and synthesis of large BNs.

2 Preliminaries

2.1 Boolean Networks and Trap Spaces

A BN of dimension n is a function $f : \mathbb{B}^n \to \mathbb{B}^n$ where $\mathbb{B} = \{0, 1\}$. The vectors $x \in \mathbb{B}^n$ are its *configurations*, where, for each $i \in \{1, \ldots, n\}$, x_i denotes the *state* of *component* i. For each component i, $f_i : \mathbb{B}^n \to \mathbb{B}$ is called its *local function*. It can be specified using truth tables, Binary Decision Diagrams (BDDs) [14], or propositional formulas, to name but a few. Each f_i typically depends on a subset of components of the BN. The *influence graph* $G(f)$ captures these dependencies. It is the signed digraph $(\{1, \ldots, n\}, E_+ \cup E_-)$ such that there is a positive (resp. negative) influence of i on j, *i.e.*, $(i, j) \in E_+$ (resp. E_-) if and only if there exists at least one configuration x such that $f_j(x_1, \ldots, x_{i-1}, 1, x_{i+1}, \ldots, x_n) - f_j(x_1, \ldots, x_{i-1}, 0, x_{i+1}, \ldots, x_n) = 1$ (resp. -1). Remark that different BNs can have the same influence graph. A BN f is *locally monotone* whenever $E_+ \cap E_- = \emptyset$: a component i cannot influence positively and negatively a component j.

A vector $h \in \{0, 1, *\}^n$ denotes a subcube of $\mathbb{B}^n$ where dimensions with value $*$ are *free*, and others are *fixed*. Its vertices are the 2^k configurations $c(h) = \{x \in \mathbb{B}^n \mid h_i \neq * \Rightarrow x_i = h_i\}$ where k is the number of free dimensions. A subcube h is a *trap space* if it is closed by f, *i.e.*, $\forall x \in c(h)$, $f(x) \in c(h)$. Note that $(*)^n$ is always a trap space. Given two subcubes h, h', h is *smaller than* h', noted $h \preceq h'$, if and only if $c(h) \subseteq c(h')$. A trap space is *minimal* if it does not contain a smaller trap space. We denote by $TS_f(x) \in \{0, 1, *\}^n$ the smallest trap space of f containing the configuration x. $TS_f(x)$ always exists and is unique: if two subcubes h, h' are trap spaces, their intersection is a trap space.

Example 1. Consider the BN $f : \mathbb{B}^4 \to \mathbb{B}^4$ with $f_1(x) = x_2$, $f_2(x) = x_1$, $f_3(x) = \neg x_4 \wedge (x_1 \vee x_2)$, and $f_4(x) = \neg x_3$.

$G(f) =$ 1 + 2 + 3 4 + + − −

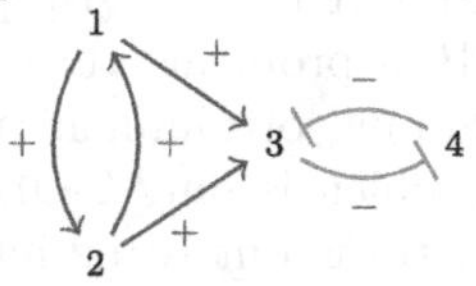

It is locally monotone and $h = 11**$ is a trap space since

$$\{f(1100), f(1101), f(1110), f(1111)\} \subseteq c(11**) = \{1100, 1101, 1110, 1111\}$$

but h is not minimal since it contains the (minimal) trap space 1101.

2.2 The Marker Reprogramming Problem

We assume the dimension n of BNs fixed. We denote by $\mathbb{M}$ the set of all partial maps from $\{1,\dots,n\}$ to $\mathbb{B}$. Given $k \in \mathbb{N}$, we write $\mathbb{M}^{\leq k}$ the partial maps with at most k associations. In the following, *markers* and *perturbations* are defined as partial maps. Notice that a partial map is equivalent to a subcube where the fixed dimensions are specified by the mapping. We say that a configuration $x \in \mathbb{B}^n$ *matches* with a marker M, denoted by $x \models M$, whenever for each $i \mapsto b \in M$, $x_i = b$. Similarly, given a subcube $h \in \{0,1,*\}^n$, $h \models M$ if and only if for each $i \mapsto b \in M$, $h_i = b$. Equivalently, $h \models M \Leftrightarrow \forall x \in c(h), x \models M$.

Given a BN f and a perturbation $P \in \mathbb{M}$, the *perturbed BN* f/P is obtained by replacing the corresponding local functions with constant values: for each component $i \in \{1,\dots,n\}$,

$$(f/P)_i(x) = \begin{cases} b & \text{if } i \mapsto b \in P, \\ f_i(x) & \text{otherwise.} \end{cases}$$

Intuitively, a perturbation permanently freezes involved components. It is important to remark that the minimal trap spaces of f/P and f can be very different.

Given a BN f and a marker $M \in \mathbb{M}$, the *marker reprogramming* problem consists in identifying perturbations $P \in \mathbb{M}$ such that *all* the minimal trap space of f/P match with M. Typically, we aim the complete identification of (subset)minimal perturbations only, *i.e.*, the perturbations so that no submap is a solution. Moreover, it is usual to limit the size of the perturbations to some constant k as many simultaneous perturbations can be difficult to implement experimentally. Similarly, some components may be *uncontrollable*, and the perturbations must not involve them. With either of these cases, the problem can be non-satisfiable (otherwise $P = M$ is a trivial solution), and deciding its satisfiability is already challenging. Finally, notice that if $P = \emptyset$ is a solution, then all the minimal trap spaces of f match with M.

Marker reprogramming generalizes the fixed point reprogramming considered in [5,22], limited to ensuring that all the fixed points only match with the marker. In [24], the problem over minimal trap spaces has been tackled for the first time in the scope of locally monotone BNs. The proposed approach relies on a modeling of the problem in QBF. Let us first consider the predicate $\text{IN\_MTS}_{f/P}(x)$ which is true if and only if x belongs to a minimal trap space of the BN f/P. The marker reprogramming problem can then be expressed as follows:

$$\exists P \in \mathbb{M}^{\leq k}, \forall x \in \mathbb{B}^n, \text{IN\_MTS}_{f/P}(x) \Rightarrow x \models M$$

In the equation, $\text{IN\_MTS}_{f/P}(x) \Rightarrow x \models M$ can be reformulated as $x \models M \vee \neg\text{IN\_MTS}_{f/P}(x)$. Then, one can remark that $\text{IN\_MTS}_{f/P}(x)$ is false if and only if $TS_{f/P}(x)$ is not minimal, *i.e.*, there exists a configuration $y \in TS_{f/P}(x)$ such that $TS_{f/P}(y) \subsetneq TS_{f/P}(x)$. In the case of locally monotone BNs, $TS_{f/P}(x)$ and $TS_{f/P}(y)$ can be computed in polynomial time [25]. Thus, the marker reprogramming boils down to the following 3-QBF:

$$\exists P \in \mathbb{M}^{\leq k}, \forall x \in \mathbb{B}^n : x \not\models M, \exists y \in TS_{f/P}(x), TS_{f/P}(y) \neq TS_{f/P}(x). \quad (1)$$

The approach of [24] relies on Answer-Set Programming (which is limited to 2-QBF problems) to solve the complementary problem of identifying all perturbations P such that at least one minimal trap space of f/P does not match with the marker ($\exists P \in \mathbb{M}^{\leq k}, \forall x \in \mathbb{B}^n, x \not\models M \wedge \forall y \in TS_{f/P}(x), TS_{f/P}(y) = TS_{f/P}(x)$). Then, the solutions are obtained by an ensemble difference with $\mathbb{M}^{\leq k}$. While $\mathbb{M}^{\leq k}$ is of polynomial size with n, this complementary problem becomes rapidly intractable with large networks having numerous wrong perturbations.

Example 2. Consider the BN f with $f_1(x) = \neg x_2$, $f_2(x) = \neg x_1$, $f_3(x) = x_1 \wedge \neg x_2 \wedge \neg x_4$, $f_4(x) = x_3 \vee x_5$ and $f_5(x) = \neg x_3 \wedge x_5$. It has two minimal trap spaces ($010**$ and $10***$). If we consider the marker $M = \{2 \to 1, 3 \to 1\}$ and $k = 2$, a possible solution is $P = \{3 \to 1, 1 \to 0\}$: the BN f/P has just a single minimal trap space, 01110, which matches with M.

2.3 The Synthesis Problem

The automatic design of BNs from specifications on their static and dynamical properties is another prime challenge for applications in biology [10,13,31].

There has been recent progress to address the synthesis from asynchronous dynamical properties, including attractors [4,17], but they still show limited tractability. Synthesis of BNs from specifications on their most permissive dynamics has been shown to have great scalability, thanks to a lower computational complexity, with applications to networks up to several thousands of components [9]. Nevertheless, prior work did not account for universal properties over minimal trap spaces.

Static properties of the network allow delimiting the domain of candidate BNs, which we denote by $\mathbb{F}$. It usually comes from a given signed influence graph $\mathcal{G} = (\{1, \ldots, n\}, \mathcal{E}_+ \cup \mathcal{E}_-)$. In that case, $\mathbb{F}$ could be, for instance, any BN f such that its influence graph $G(f) = (\{1, \ldots, n\}, E_+ \cup E_-)$ is equal to $\mathcal{G}$, or included in $\mathcal{G}$ (*i.e.*, $E_+ \subseteq \mathcal{E}_+$ and $E_- \subseteq \mathcal{E}_-$). Additionally, $\mathbb{F}$ could be restricted with already specified partial Boolean local functions.

Given a domain of BNs $\mathbb{F}$ and a marker $M \in \mathbb{M}$, the *synthesis problem* consists in identifying a BN $f \in \mathbb{F}$ such that *all* the minimal trap spaces of f match with M. It can be expressed as 3-QBF in a very similar fashion to the marker reprogramming problem (1):

$$\exists f \in \mathbb{F}, \forall x \in \mathbb{B}^n : x \not\models M, \exists y \in TS_f(x), TS_f(y) \neq TS_f(x). \quad (2)$$

Note this problem can be unsatisfiable. The main difference with marker reprogramming is the combinatorics of the domain $\mathbb{F}$. To our knowledge, there was no approach to tackle this synthesis problem efficiently.

Example 3. Consider $M = \{3 \to 1\}$ and $\mathcal{G}$ equal to the $G(f)$ presented in Example 1. A solution to the synthesis problem is $f_1(x) = 1$, $f_2(x) = x_1$, $f_3(x) = x_2 \vee \neg x_4$ and $f_4(x) = \neg x_3$. In this scenario, the influence graph is a sub-graph of $\mathcal{G}$ and the only minimal trap space (which also respects M) is 1110.

2.4 Counter-Example-Guided Abstraction Refinement (CEGAR)

CEGAR [12] is an incremental way to decide the satisfiability of a (possibly quantified) logic formula ϕ by the mean of a simpler formula ϕ_u (resp. ϕ_o) so that the models of ϕ_u subsume (resp. ϕ_o are subsumed by) the models of ϕ. Thus, $\phi_u \Leftarrow \phi$ (resp. $\phi_o \Rightarrow \phi$). The choice between ϕ_u and ϕ_o depends on whether one wants to *under-* or *over-approximate* ϕ.

We briefly explain the principle with the ϕ_u case. If ϕ_u is unsatisfiable, so is ϕ. Otherwise, a model $\vec{\mu}$ of ϕ_u is found ($\vec{\mu} \models \phi_u$), and one must verify whether $\vec{\mu} \models \phi$. If $\vec{\mu} \not\models \phi$, $\vec{\mu}$ is a *counter-example* and ϕ_u must be refined with some $\phi_r(\vec{\mu})$ so that $\phi_u \wedge \phi_r(\vec{\mu}) \Leftarrow \phi$. The process is repeated until the refined ϕ_u is either unsatisfiable, demonstrating the unsatisfiability of ϕ, or a model of the refined ϕ_u is a model of ϕ. The challenge is thus to design a refinement which makes this process converge rapidly, which is problem-specific.

3 A CEGAR for Minimal Trap Spaces

In this section, we introduce a CEGAR-based approach for addressing universal marker properties over minimal trap spaces. The refinement can be directly employed in solving the marker reprogramming (1) and synthesis (2) problems. In the case of locally monotone BNs, or BNs whose local functions are specified using propagation complete representations (such as BDDs or Petri net), this boils down to iteratively solving on the one hand an NP problem (ϕ_u) and on the other hand a 2-QBF problem for identifying counter-examples.

We first detail the CEGAR for the case of marker reprogramming before discussing its generalization to the synthesis problem.

3.1 Generalizing Counter-Examples for Refinement

Recall that the marker reprogramming problem consists in identifying perturbations of size at most k under which all the minimal trap spaces of the given BN f match with the given marker M. Let us consider the 3-QBF (1), that we refer to as ϕ in this section, following the notations of Sect. 2.4. Let us assume that we are given a candidate perturbation $\vec{P} \in \mathbb{M}^{\leq k}$, for instance, from a model of the NP formula $\phi_u = \exists P \in \mathbb{M}^{\leq k}$. To be a solution for ϕ, one must verify that all the minimal trap spaces of the perturbed BN $f/\vec{P}$ match with the marker. Thus, $\vec{P}$ and $\vec{x}$ are used to represent a specific perturbation and a specific configuration. A *counter-example* would be a configuration of a minimal trap space of $f/\vec{P}$ that does not match with the marker. Such counter-examples are models of the following QBF:

$$\phi_{ce}(\vec{P}) = \exists x \in \mathbb{B}^n \text{ s.t. } x \not\models M \text{ and } \forall y \in TS_{f/\vec{P}}(x), TS_{f/\vec{P}}(x) = TS_{f/\vec{P}}(y). \quad (3)$$

If $\phi_{ce}(\vec{P})$ is not satisfiable, then $\vec{P}$ is a model of ϕ and thus a solution to the marker reprogramming problem. Otherwise, let us denote by $\vec{x}$ a model of $\phi_{ce}(\vec{P})$,

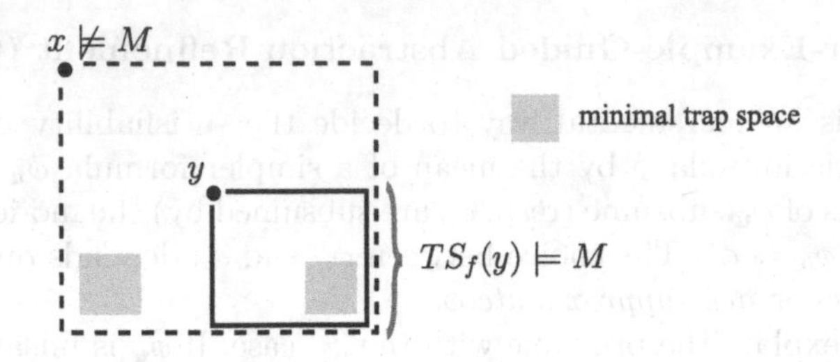

Fig. 1. Illustration of the refinement given an $x \in \mathbb{B}^n$ such that $x \not\models M$. If all the minimal trap spaces (gray squares) of f match with M, there must exist $y \in TS_f(x)$ (dashed line) so that all the configurations of $TS_f(y)$ (plain line) match with M.

i.e., a counter-example showing that $\vec{P}$ is not a model of ϕ: the configuration $\vec{x}$ belongs to a minimal trap space of $f/\vec{P}$ and does not match with the marker. Intuitively, $\vec{x}$ is a configuration that shows why $\vec{P}$ is not valid with respect to a marker M. The idea to move forward in the search for a valid perturbation is to avoid any other perturbation $P \in \mathbb{M}^{\leq k}$ such that $\vec{x}$ belongs to one of its minimal trap spaces. Let us now point out some useful properties concerning trap spaces and markers:

Property 1. For any BN $f : \mathbb{B}^n \to \mathbb{B}^n$, marker $M \in \mathbb{M}^{\leq n}$, perturbation $P \in \mathbb{M}^{\leq k}$ and configuration $x \in \mathbb{B}^n$:

1. if $TS_{f/P}(x) \models M$, all minimal trap spaces within $TS_{f/P}(x)$ match with M;
2. if $x \not\models M$, it holds that $TS_{f/P}(x) \not\models M$;
3. if $x \not\models M$, any perturbation P' such that x is in a minimal trap space of f/P' is not a model of ϕ;
4. if x is not in a minimal trap space of f/P, there exists a configuration $y \in TS_{f/P}(x)$ such that $TS_{f/P}(y) \subsetneq TS_{f/P}(x)$.

As illustrated by Fig. 1, these properties imply two constraints that must be verified by any perturbation P model of ϕ: (a) the trap space $TS_{f/P}(\vec{x})$ is not minimal, *i.e.*, there must exist a configuration $y \in TS_{f/P}(\vec{x})$ whose trap space is strictly smaller ($TS_{f/P}(y) \subsetneq TS_{f/P}(\vec{x})$); and (b) $TS_{f/P}(\vec{x})$ must contain a trap space which matches with the marker M. Combining (a) and (b), there must exists at least one configuration $y \in TS_{f/P}(\vec{x})$ such that $TS_{f/P}(y) \models M$. Therefore, we define the following refinement from the counter-example $\vec{x}$:

$$\phi_r(\vec{x}) = \exists y \in \mathbb{B}^n : TS_{f/P}(y) \subsetneq TS_{f/P}(\vec{x}) \wedge TS_{f/P}(y) \models M \ . \qquad (4)$$

Remarking that $\vec{P} \not\models \phi_u \wedge \phi_r(\vec{x})$, one can then apply the CEGAR approach by iterating the refinement until either no counter-example can be found (and thus $\vec{P}$ is a solution), or until the refined formula becomes non-satisfiable. The correctness is expressed by the following lemma:

Lemma 1. *Given $\vec{P} \models \phi_u$ and $\vec{x} \models \phi_{ce}(\vec{P})$, it holds that $\phi_u \wedge \phi_r(\vec{x}) \Leftarrow \phi$.*

Proof. Let us show that, starting from an under-approximation ϕ_u and adding ϕ_r, we get a new under-approximation of ϕ. In other words, we want to show that the set of solutions is shrunk by removing only invalid solutions. Initially, $\phi_u = \exists P \in \mathbb{M}^{\leq k}$. Then, $\phi_u \Leftarrow \phi$. Considering a model $\vec{P}$ of ϕ_u that is not a model of ϕ, one can find a model $\vec{x}$ of $\phi_{ce}(\vec{P})$. The identification of $\vec{x}$ is interesting since all P in $\mathbb{M}^{\leq k}$ such that $\text{IN\_MTS}_{f/P}(\vec{x})$ are not valid solutions (point 3 of Property 1). By initially defining $\phi_r(\vec{x}) = \exists y \in \mathbb{B}^n : TS_{f/P}(y) \subsetneq TS_{f/P}(\vec{x})$ we remove all candidates P for which $\vec{x}$ would be in a minimal trap space (point 4 of Property 1). The solution space is reduced by removing only invalid perturbations. Adding $TS_{f/P}(y) \models M$, we impose that at least one minimal trap space within $TS_{f/P}(\vec{x})$ matches the marker (point 1 of Property 1). This constraint must indeed be satisfied in the solutions of ϕ. It allows to eliminate all P for which $\nexists y \in \mathbb{B}^n : TS_{f/P}(y) \subsetneq TS_{f/P}(\vec{x}) \wedge TS_{f/P}(y) \models M$. In other words, all P where we fail to ensure that all minimal trap spaces in $TS_{f/P}(y)$ match the marker. Recall that the configuration y is needed to ensure $\vec{x}$ outside minimal trap spaces. Again only invalid solutions are removed and thus a new under-approximation is obtained. In a generic step, we have $\phi_u = \exists P \in \mathbb{M}^{\leq k} \wedge \phi_r(\vec{x}^{(1)}) \wedge \ldots \wedge \phi_r(\vec{x}^{(q)})$. The solution space is further reduced with $\phi_r(\vec{x}^{(q+1)})$, requiring that $\vec{x}^{(q+1)}$ is not within a minimal trap space and that $TS_{f/P}(\vec{x}^{(q+1)})$ contains a trap space matching with the marker. □

The actual complexity of derived QBF formulas largely depends on the encoding of the TS predicate, and is discussed in Sect. 3.4.

3.2 Necessary Condition on Perturbations

In the previous section, the initial perturbation candidate is a model of $\phi_u = \exists P \in \mathbb{M}^{\leq k}$. This can be already refined by remarking that there always exists at least one minimal trap space in any BN. Thus, one can already impose that at least one minimal trap space of f/P matches with the marker. By Prop. 1, this is ensured by the existence of a configuration w such that $TS_{f/P}(w) \models M$. This leads to the following formula:

$$\phi_u = \exists P \in \mathbb{M}^{\leq k}, \exists w \in \mathbb{B}^n, TS_{f/P}(w) \models M \ . \quad (5)$$

Remark that with this version of ϕ_u, only the first iteration of the CEGAR can be affected as the existence of w is subsumed by the refinement ϕ_r. Therefore, Lemma 1 still holds with this ϕ_u. Algorithm 1 summarizes the overall CEGAR procedure for solving the marker reprogramming problem.

Example 4. Consider the BN of Example 1. According to ϕ_u (5), a first candidate solution is $\vec{P} = \emptyset$. To verify if all minimal trap spaces of $f/\vec{P}$ (*i.e.*, f) match with $M = \{1 \to 1, 3 \to 1\}$, we verify the existence of a counter-example (with $\phi_{ce}(\vec{P})$) and identify its model $\vec{x} = 1101$. In fact, 1101 is a fixed point of $f/\vec{P}$. Then, the new under-approximation is $\phi_u \wedge \phi_r(1101)$ and we can identify a new candidate perturbation $\vec{P} = \{4 \to 0\}$. In this case, a counter-example is

Algorithm 1. CEGAR-based Reprogramming

Input: $f : \mathbb{B}^n \to \mathbb{B}^n, M \in \mathbb{M}^{\leq n}, k \in \{0, \ldots, n\}$
Output: $P \in \mathbb{M}^{\leq k}$ such that $\forall x \in \mathbb{B}^n : \text{IN\_MTS}_{f/P}(x) \Rightarrow x \models M$
1: $\phi_u = \exists P \in \mathbb{M}^{\leq k}, \exists w \in \mathbb{B}^n : TS_{f/P}(w) \models M$
2: $\vec{P} = \text{solve}(\phi_u)$
3: **while** $\vec{P}$ exists **do**
4: $\phi_{ce}(\vec{P}) = \exists x \in \mathbb{B}^n : x \not\models M \wedge \text{IN\_MTS}_{\vec{P}}(x)$
5: $\vec{x} = \text{solve}(\phi_{ce}(\vec{P}))$
6: **if** $\vec{x}$ exists **then**
7: $\phi_r(\vec{x}) = \exists y \in \mathbb{B}^n : TS_{f/P}(y) \subsetneq TS_{f/P}(\vec{x}) \wedge TS_{f/P}(y) \models M$
8: $\phi_u = \phi_u \wedge \phi_r(\vec{x})$
9: $\vec{P} = \text{solve}(\phi_u)$
10: **else**
11: **return** $\vec{P}$
12: **end if**
13: **end while**
14: **return** UNSAT

$\vec{x} = 0000$. Improving the approximation with $\phi_u \wedge \phi_r(1101) \wedge \phi_r(0000)$, we can identify now $\vec{P} = \{2 \to 1, 4 \to 0\}$ and no counter-example can be found. Then, $P = \{2 \to 1, 4 \to 0\}$ is a solution of the reprogramming problem.

3.3 Generalization to the Synthesis Problem

Recall that the synthesis problem consists in identifying a BN f in a given domain $\mathbb{F}$ so that all its minimal trap spaces match with the given marker M. This problem can be seen as a generalization of the marker reprogramming problem by considering $\mathbb{F} = \{f/P \mid P \in \mathbb{M}^{\leq k}\}$.

The CEGAR developed in the previous section can be straightforwardly applied to the synthesis problem:

$$\phi'_u = \exists f \in \mathbb{F}, \exists w \in \mathbb{B}^n, TS_f(w) \models M \tag{6}$$

$$\phi'_{ce}(\vec{f}) = \exists x \in \mathbb{B}^n \text{ s.t. } x \not\models M \text{ and } \forall y \in TS_{\vec{f}}(x), TS_{\vec{f}}(x) = TS_{\vec{f}}(y) \tag{7}$$

$$\phi'_r(\vec{x}) = \exists y \in \mathbb{B}^n : TS_f(y) \subsetneq TS_f(\vec{x}) \wedge TS_f(y) \models M \ . \tag{8}$$

BN candidates are models of ϕ'_u where, as for the reprogramming case, we already enforce the existence of a minimal trap space matching with the marker. If $\vec{f}$ is a model of ϕ'_u, $\phi'_{ce}(\vec{f})$ characterizes the counter-examples configurations, and $\phi'_r(\vec{x})$ provides the refinement from such a given counter-example $\vec{x}$.

Example 5. Consider a complete graph $\mathcal{G}$ such that all edges are positive. It is known that any BN f having $G(f) = \mathcal{G}$ has only two fixed points (0^n and 1^n) and no cyclic asynchronous attractor [1,2]. Given any M, the synthesis problem results in expression (2). Let us start by considering the under-approximation

$$\phi'_u = \exists f \in \mathbb{F}, \exists w \in \mathbb{B}^n : TS_f(w) \models M.$$

Let us assume $M = \{1 \to 0\}$ and $n = 3$. Any f, such that $G(f) = \mathcal{G}$, is a valid candidate solution $\vec{f}$ since $x = 000$ is a fixed point. However, searching for a counter-example $\vec{x}$, we will find $\vec{x} = 111$ since it is a fixed point and $\vec{x} \not\models M$. With the refinement $\phi_r(\vec{x})$, the problem turns out to be

$$\exists f \in \mathbb{F}, \exists w \in \mathbb{B}^n : TS_f(w) \models M \wedge (\exists y \in \mathbb{B}^n : TS_f(y) \subsetneq TS_f(111) \wedge TS_f(y) \models M) \quad (9)$$

which is unsatisfiable. Thus, the CEGAR approach allows us to verify the property without needing to solve the original problem (2). Remark that the known theoretical property is not implemented in the approach.

3.4 Complexity

It is important to notice that the formulas introduced in this section involve a predicate $TS_f(x)$ which returns the smallest trap space containing the given configuration x in the scope of BN f. The encoding of this predicate can affect the complexity of the formulas, by adding variables, but more importantly by potentially adding quantifiers.

As shown in [25], $TS_f(x)$ can be computed by progressively saturating a cube h, starting from $h = x$: for each component i which is fixed in h, one check whether there exists a vertex of h, $y \in c(h)$, such that $f_i(y) \neq h_i$. In that case, the i-th component is freed in the cube. This iteration can be performed up to n times. Importantly, remark that the test $\exists y \in c(h) : f_i(y) \neq h_i$ boils down to the SAT and UNSAT problems of f_i. However, it is known that the SAT/UNSAT decision can be deterministically computed in polynomial time whenever f_i is monotone (*i.e.*, the BN f is locally monotone), and, more generally, whenever f_i is given as a *propagation complete* representation [7], which includes BDDs and Petri nets. Therefore, in these cases, $TS_f(x)$ can be represented efficiently as a propositional formula (see Appendix A[1]) or using ASP, similarly to [9]. In other cases, the encoding $TS_f(x)$ involves the SAT problem.

We can then conclude that ϕ_r and the under-approximation ϕ_u are NP ($\exists$) expressions, while the check for counter-example ϕ_{ce} is 3-QBF in the general case and 2-QBF with locally monotone BNs or with propagation complete local functions (which are widely used by BN analysis tools).

4 Implementation and Performance Evaluation

We implemented the CEGAR resolution of marker reprogramming and synthesis problems in a prototype[2] relying on the Python library BoNesis[3] and Answer-Set Programming multi-shot solver CLINGO [16]. The prototype exploits a DNF representation for locally monotone local functions and a BDD representation for non-monotone local functions (see Sect. 3.4).

[1] https://doi.org/10.48550/arXiv.2305.02442.

[2] Code and dataset available at https://github.com/bnediction/cegar-bonesis (archived at https://doi.org/10.5281/zenodo.8269316).

[3] https://github.com/bnediction/bonesis (archived at https://doi.org/10.5281/zenodo.7984628).

4.1 Datasets

We considered 2 sets of BNs and markers.

The *Moon dataset* [22] consists of 10 locally monotone BNs and 1 non-monotone BN taken from biological modeling literature. Network sizes range from 13 to 75, classified as small (S1-S4), medium (M1-M3), and large (L1-L4). Importantly, each BN comes with a marker and uncontrollable components from the related biological application.

The *Trappist dataset* [30] consists of 27 locally monotone BNs and 6 non-monotones BNs ranging from 47 to 4691 components taken from biological modeling literature, either designed directly as BNs, or resulting from a conversion. Contrary to the Moon dataset, no biologically-relevant markers are specified. Therefore, for each network, we randomly generated two sets of markers from vertexes of the bottom strongly connected components of the influence graphs (the "output" components), and associating a Boolean value to them. For networks with at most 5 output components, all combinations of markers were considered. Otherwise, we generated 100 markers, where each associates 3 random output components to a random Boolean value. Then, for each network, a second set of markers of size 1 were generated with the same process. Duplicates markers were removed.

4.2 Protocol

Besides assessing the scalability of our CEGAR implementation, we aimed at benchmarking it with different variants of ϕ_r and ϕ_u, and in the case of reprogramming, with already existing approaches. Moreover, we aimed at evaluating generic QBF solvers, such as CAQE [28] or DEPQBF [20], for tackling our 3-QBF problems, for which we devised a standard quantified conjunctive normal form encoding (Appendix B[1]). However, it turned out that they failed to scale for the vast majority of our instances.

Marker Reprogramming Problem. The inputs were the BNs and their associated markers, together with the maximum size of perturbations (parameter k). Only subset-minimal perturbations were considered, using the adequate solving mode of CLINGO. Moreover, we denied perturbations involving components of the markers, and those declared as uncontrollable in the Moon dataset. In the case the instance is satisfiable, we analyzed both the time for computing the first solution, and the time to exhaust the full set of subset-minimal solutions. For the reprogramming scenario, we compared several methods:

- Enumeration and filtering: the enumeration is performed by increasing size in order to obtain only subset-minimal solutions. The filtering is performed using ϕ_{ce}. This approach somehow corresponds to the most basic CEGAR implementation without any counter-example generalization (ϕ_r is $P \neq \vec{P}$).
- Complementary: the method of [24] based on enumerating perturbations that fail the reprogramming and subtract them from $\mathbb{M}^{\leq k}$ (Sect. 2.2).

- CEGAR-2: our implementation of Algorithm 1.
- CEGAR-1: is the approach presented in Algorithm 1 with the difference that a weaker refinement, imposing only that counter-examples are not part of minimal trap spaces, is used, *i.e.*, $\phi_r(\vec{x}) = \exists y \in \mathbb{B}^n : TS_{f/P}(y) \subsetneq TS_{f/P}(\vec{x})$ (here it is not imposed that the trap space matches the marker).

The aim was therefore to investigate the scalability of the approach and the importance of the various constraints added in CEGAR-1 and CEGAR-2, and compare them to the current state of the art.

Synthesis Problem. The inputs were the influence graph of the BNs and their associated markers. The domain $\mathbb{F}$ was defined as the set of locally monotone BNs whose local functions are represented in disjunctive normal form (DNF). With these settings, the problem can be unsatisfiable as local functions of components cannot be assigned to constant functions, unless already a constant function in the original BN. Thus, the trivial solution where marker components are assigned to their corresponding value is not part of the search domain $\mathbb{F}$. For the Moon dataset, we imposed that their influence graph match exactly with the input one; while for the Trappist dataset, we relaxed this constrained by imposing only that the DNF of each local function involves all the regulators of the corresponding component with the adequate sign, and we limited the size of DNFs to 32 clauses. Here, we reported the time for deciding the existence of solution.

Contrary to the reprogramming, no other tools enable addressing this problem directly. In addition to the CEGAR-2 and CEGAR-1, we also implemented CEGAR-0 which follows the same algorithm but use no counter-example generalization, *i.e.*, $\phi_r(\vec{x})$ is $f \neq \vec{f}$. This somehow corresponds to the enumeration and filtering used for marker reprogramming. The aim is therefore to understand whether a CEGAR-based approach can be an effective first approach to attack the synthesis problem in accordance with universal marker properties.

Instances of Moon dataset were run on a desktop computer with Intel(R) Xeon(R) E-2124 CPU at 3.30 GHz and 64 GB of RAM; instances of Trappist were run on cluster nodes with AMD(R) Zen2 EPYC 7452 CPU at 2.35 GHz and 256 GB of RAM, all on a Linux operating system.

4.3 Results

Moon Instances. Table 1 gives a comparison of execution times for the implementations of the marker reprogramming, depending on the maximum size k of perturbations to identify. CEGAR-2 is the only one able to always determine the satisfiability of the instances in 10 min. Moreover, it largely outperforms other methods for the enumeration as soon as k or n is large. Through the results obtained with CEGAR-1 and CEGAR-2 on the datasets, we can conclude that imposing the trap spaces of counter-example configurations to contain a trap spaces matching with the marker helps to drastically reduce the number of iterations to exhaust the solution space. Table 2 provides a similar picture for the synthesis problem. Note that CEGAR-0 only solved instances where the first generated BN verified the universal property.

Table 1. Execution times (in sec.) for the reprogramming on the Moon dataset where n is the number of components and u the number of uncontrollable ones. Column "First" indicates the time for identifying the first solution, or the unsatisfiability of the instance (0 solutions); "Enum" the time for exhaustive solution enumeration; "Solution" the higher number of solution identified. When the problem turns out to have no solution, the time required is indicated in the column "First". For CEGAR methods, the number of identified counter-examples is in parentheses. Then, TO(-) means that no counter-examples were found in 10 min.

	k	Enum & Filter		Complementary		CEGAR-1		CEGAR-2		Solutions
		First	Enum	First	Enum	First	Enum	First	Enum	
S1, Sahin et al. (2009) n=20, u=1	2	0.03	0.9	0.03	**0.1**	0.07(4)	3.4(37)	**0.02**(1)	**0.1**(2)	9
	4	0.03	10.9	0.03	1	0.8(21)	197.3(181)	**0.02**(1)	**0.1**(2)	12
	6	0.04	TO	**0.03**	1	1(22)	TO(265)	**0.03**(1)	**0.1**(2)	12
S2, Wynn et al. (2012) n=17, u=2	2	0.6	2.5	0.1	**0.1**	0.6(21)	1.9(32)	**0.04**(3)	**0.1**(4)	9
	4	0.5	150.7	**0.1**	4.4	0.6(22)	TO(275)	**0.1**(3)	**0.2**(5)	19
	6	0.7	TO	**0.1**	20.7	0.5(20)	TO(305)	**0.1**(3)	**0.3**(4)	31
S3, Kasemeier-Kulesa et al. (2018) n=18, u=4	2	2.3	-	0.1	-	1.1(26)	-	**0.03**(2)	-	0
	4	27.4	129.1	3.6	3.6	104.6(158)	TO(266)	**0.1**(3)	**0.1**(3)	12
	6	49.7	TO	3.6	55.4	TO(-)	TO(300)	**0.1**(3)	**0.2**(3)	18
S4, Biane et al. (2019) n=13, u=2	2	**0.01**	0.6	0.02	**0.1**	0.03(2)	0.2(10)	0.02(1)	**0.1**(1)	9
	4	0.04	0.5	**0.02**	**0.1**	0.03(2)	0.3(13)	**0.02**(1)	**0.1**(1)	9
	6	0.03	0.5	**0.02**	**0.1**	0.03(2)	0.3(13)	**0.02**(1)	**0.1**(1)	9
M1, Calzone et al. (2010) n=28, u=3	2	0.3	8.4	**0.04**	**0.3**	3.1(34)	32.5(71)	0.1(3)	0.5(4)	36
	4	0.3	TO	**0.04**	49.3	19.5(73)	TO(221)	0.1(3)	**4**(9)	213
	6	0.2	TO	**0.04**	TO	37.4(95)	TO(245)	0.1(3)	**7.9**(9)	370
M2 - Cohen et al. (2015) n=32, u=6	2	14.6	-	0.6	-	**0.03**(1)	-	**0.03**(1)	-	0
	4	258	TO	6.7	78.1	1(14)	43.2(61)	**0.1**(2)	**0.6**(7)	14
	6	246.6	TO	6.7	TO	**0.02**(0)	TO(182)	**0.02**(0)	**2.3**(8)	78
M3 - Remy et al. (2015) n=35, u=4	2	4.6	21.1	0.8	0.8	255.2(116)	334.4(124)	**0.2**(5)	**0.3**(5)	3
	4	4	TO	0.9	189.6	8.6(38)	TO(173)	**0.1**(3)	**8.5**(11)	261
	6	6.4	TO	0.9	TO	8.1(37)	TO(180)	**0.1**(3)	**235.9**(17)	2015
L1, Saadatpour et al. (2011) n=59, u=3	2	98.2	-	3.7	-	0.8(11)	-	**0.04**(1)	-	0
	4	TO	TO	119.9	TO	TO(-)	TO(153)	**0.1**(2)	**2.8**(5)	83
	6	TO	TO	120.6	TO	TO(-)	TO(170)	**0.1**(3)	TO(21)	≥ 2227
L2, Singh et al. (2012) n=66, u=1	2	**0.1**	75	0.2	3.4	0.3(6)	TO(184)	**0.1**(1)	**1.2**(1)	60
	4	**0.1**	TO	0.2	TO	0.3(6)	TO(187)	**0.1**(1)	**1.2**(1)	60
	6	**0.1**	TO	0.2	TO	0.3(6)	TO(189)	**0.1**(1)	**1.2**(1)	60
L3, Grieco et al. (2013) n=53, u=3	2	24.3	59.8	2.4	2.4	**0.1**(1)	TO(159)	**0.1**(1)	**0.2**(2)	8
	4	8.9	TO	2.4	TO	**0.1**(1)	TO(171)	**0.1**(1)	**77**(23)	722
	6	18.3	TO	2.5	TO	**0.1**(2)	TO(177)	**0.1**(1)	TO(33)	≥ 2171
L4, Flobak et al. (2015) n=75, u=2	2	153.1	-	6.2	-	**0.02**(0)	-	**0.02**(0)	-	0
	4	TO	TO	228.1	TO	**0.03**(0)	TO(112)	**0.03**(0)	**78.2**(1)	1302
	6	TO	TO	227.8	TO	**0.04**(0)	TO(179)	**0.04**(0)	**518**(5)	3435

Table 2. Execution times (in sec.) for determining the satisfiability of the synthesis on the Moon dataset with a 10 min timeout (TO); † indicates unsatisfiable instances. Number of identified counter-examples is in parentheses.

	S1 (20)	S2 (17)	S3 (18)	S4 (13)	M1 (28)	M2 (32)	M3 (35)	L1 (59)	L2 (66)	L3 (53)	L4 (75)
CEGAR-0	0.2(0)	TO(59,584)	TO(53,300)	TO(63,914)	TO(44,585)	TO(4,451)	TO(9,987)	TO(6,596)	TO(24,686)	0.2(0)	1.2(0)
CEGAR-1	**0.1**(0)	1(20)†	12.4(49)†	0.04(1)	**0.2**(4)	TO(5)	249.3(5)	TO(36)	**0.1**(2)†	**0.2**(0)	1.3(0)
CEGAR-2	**0.1**(0)	**0.1**(3)†	**0.1**(3)†	**0.03**(1)	**0.2**(6)	**97.8**(1)	**84.2**(3)	**2.4**(4)	**0.1**(1)†	**0.2**(0)	**1.2**(0)

Instances size	k	# instances	Solved (UNSAT)	Enumeration
< 100	1	534	100% (75%)	100%
	2	534	100% (61%)	100%
	4	534	100% (29%)	92%
	6	534	100% (15%)	85%
≥ 100, < 400	1	2446	100% (70%)	100%
	2	2446	100% (55%)	100%
	4	2446	100% (20%)	92%
	6	2446	98% (10%)	63%
≥ 1,000	1	800	81% (52%)	80%
	2	800	77% (42%)	71%
	4	800	75% (23%)	58%
	6	800	75% (18%)	49%

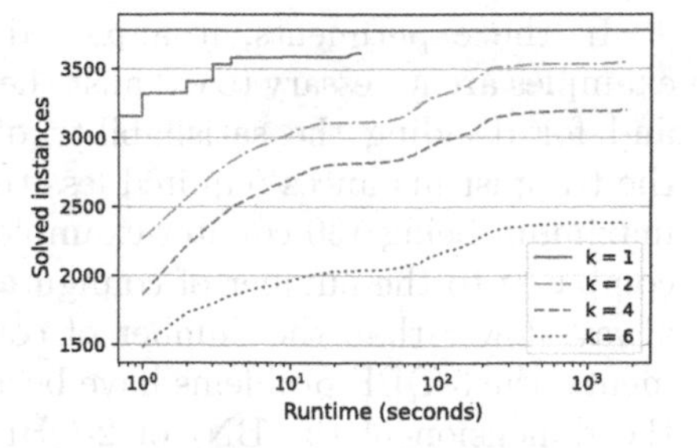

Fig. 2. Summary of results for the reprogramming of Trappist instances. (left) ratio of instances for which the satisfiability has been determined with relative ratio of unsatisfiable, and for which the exhaustive enumeration has been completed within 30min. (right) number of instances with exhaustive enumeration completed within given time. The red line indicates the total number of instances.

Instances size	$\|M\|$	# instances	Solved (UNSAT)
< 100	1	110	95% (5%)
	> 1	424	79% (22%)
≥ 100, < 400	1	886	84% (33%)
	> 1	1560	75% (42%)
≥ 1,000	1	400	50% (33%)
	> 1	400	64% (61%)

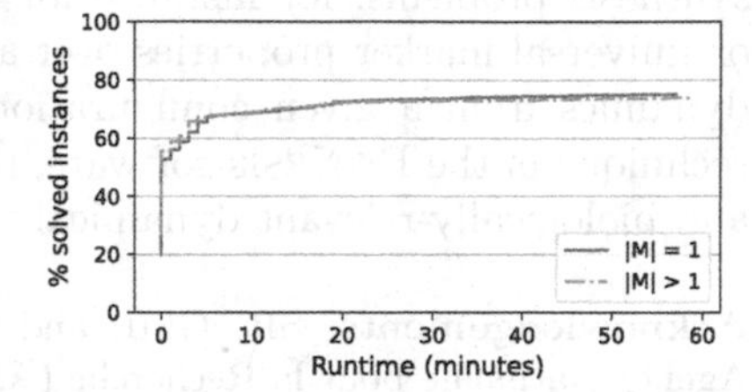

Fig. 3. Summary of results for the synthesis of Trappist instances within a 1h limit.

Trappist Instances. Figures 2 and 3 provide summary statistics of CEGAR-2 performance for the reprogramming and synthesis, grouped by network size. The execution time is limited to 30 min for reprogramming and 60 min for synthesis. Full results are provided in Appendix C[1]. The dataset consists of much larger networks than in Moon, although the employed markers may not be biologically meaningful. For the reprogramming, our prototype always decides whether the instance admits a solution for all networks up to 400 components. Failed exhaustive enumerations were then likely caused by a large combinatorics of solutions with large k. For BNs above 1,000 components, our prototype managed to determine the satisfiability of 75% of the instances within the given time limit. In the case of the synthesis, our prototype was able to solve more than 80% of the instances of the BNs below 400 components, and 50% of networks above 1,000.

5 Discussion

We demonstrated a new approach to efficiently reason on universal properties over minimal trap spaces of BNs, by iterative refinements of a logical satisfiability problem guided by counter-examples. Our prototype scaled to BNs with thousands of components for solving reprogramming and synthesis problems.

In the experiments, it appeared that only a limited number of counter-examples are necessary to exhaust the solution space of reprogramming instances and for deciding the satisfiability of the synthesis instances. Indeed, 90% of the trappist instances required less than 20 counter-example configurations, the maximum being 130 counter-example configurations for one case. It is very low compared to the number of configurations of considered networks, and actually always lower than the number of components. Thus, in the conducted experiments, the 3-QBF problems have been resolved with a (sub)linear number (with the dimension of the BN) of 2-QBF problems. Theoretical and experimental work may further investigate the connection between the number of counter-examples and the diversity of minimal trap spaces that can be generated by the candidate perturbations and Boolean functions. This could offer a finer insight on the practical complexity of tackled problems.

Our method can be employed to solve more general reprogramming and synthesis problems, for instance for enforcing the absence of cyclic attractors, or universal marker properties over attractors reachable in the most permissive dynamics from a given configuration. We plan to embed this generic solving technique in the BoNesis software, promising a scalable BN synthesis from rich and biologically-relevant dynamical properties.

Acknowledgements. SR, GML and LP We acknowledge support of the French Agence Nationale pour la Recherche (ANR) in the scope of the project "BNeDiction" (ANR-20-CE45-0001) and of the project "PING/ACK" (ANR-18-CE40-0011). Part of the experiments presented in this paper were carried out using the PlaFRIM experimental testbed, supported by Inria, CNRS (LABRI and IMB), Université de Bordeaux, Bordeaux INP and Conseil Régional d'Aquitaine (see https://www.plafrim.fr).

References

1. Aracena, J.: Maximum number of fixed points in regulatory Boolean networks. Bull. Math. Biol. **70**, 1398–1409 (2008). https://doi.org/10.1007/s11538-008-9304-7
2. Aracena, J., Demongeot, J., Goles, E.: Positive and negative circuits in discrete neural networks. IEEE Trans. Neural Netw. **15**(1), 77–83 (2004). https://doi.org/10.1109/TNN.2003.821555
3. Baral, C.: Knowledge representation, reasoning and declarative problem solving. Cambridge University Press (2003). https://doi.org/10.1017/CBO9780511543357
4. Beneš, N., Brim, L., Kadlecaj, J., Pastva, S., Šafránek, D.: AEON: attractor bifurcation analysis of parametrised Boolean networks. In: Lahiri, S.K., Wang, C. (eds.) CAV 2020. LNCS, vol. 12224, pp. 569–581. Springer, Cham (2020). https://doi.org/10.1007/978-3-030-53288-8_28
5. Biane, C., Delaplace, F.: Causal reasoning on Boolean control networks based on abduction: theory and application to cancer drug discovery. IEEE/ACM Trans. Comput. Biol. Bioinf. **16**(5), 1574–1585 (2018). https://doi.org/10.1109/tcbb.2018.2889102
6. Biere, A., Heule, M., van Maaren, H., Walsh, T. (eds.) Handbook of Satisfiability - Second Edition, vol. 336 of Frontiers in Artificial Intelligence and Applications. IOS Press (2021). https://doi.org/10.3233/FAIA336

7. Bordeaux, L., Marques-Silva, J.: Knowledge compilation with empowerment. In: Bieliková, M., Friedrich, G., Gottlob, G., Katzenbeisser, S., Turán, G. (eds.) SOFSEM 2012. LNCS, vol. 7147, pp. 612–624. Springer, Heidelberg (2012). https://doi.org/10.1007/978-3-642-27660-6_50
8. Büning, H.K., Bubeck, U.: Theory of quantified Boolean formulas. In: Handbook of Satisfiability (2021). https://doi.org/10.3233/978-1-58603-929-5-735
9. Chevalier, S., Froidevaux, C., Paulevé, L., Zinovyev, A.: Synthesis of Boolean networks from biological dynamical constraints using answer-set programming. In: 2019 IEEE 31st International Conference on Tools with Artificial Intelligence (ICTAI), pp. 34–41. IEEE (2019). https://doi.org/10.1109/ICTAI.2019.00014
10. Chevalier, S., Noël, V., Calzone, L., Zinovyev, A., Paulevé, L.: Synthesis and simulation of ensembles of Boolean networks for cell fate decision. In: Abate, A., Petrov, T., Wolf, V. (eds.) CMSB 2020. LNCS, vol. 12314, pp. 193–209. Springer, Cham (2020). https://doi.org/10.1007/978-3-030-60327-4_11
11. Cifuentes-Fontanals, L., Tonello, E., Siebert, H.: Control in Boolean networks with model checking. Front. Appl. Math. Statist. **8**, 838546 (2022). https://doi.org/10.3389/fams.2022.838546
12. Clarke, E., Grumberg, O., Jha, S., Lu, Y., Veith, H.: Counterexample-guided abstraction refinement for symbolic model checking. J. ACM (JACM) **50**(5), 752–794 (2003). https://doi.org/10.1145/876638.876643
13. Dorier, J., Crespo, I., Niknejad, A., Liechti, R., Ebeling, M., Xenarios, I.: Boolean regulatory network reconstruction using literature based knowledge with a genetic algorithm optimization method. BMC Bioinform. **17**, 1–19 (2016). https://doi.org/10.1186/s12859-016-1287-z
14. Drechsler, R., Becker, B.: Binary decision diagrams: theory and implementation. Springer Science & Business Media (2013). https://doi.org/10.1007/978-1-4757-2892-7
15. Gebser, M., Kaminski, R., Kaufmann, B., Schaub, T.: Answer set solving in practice. Synth. Lect. Artif. Intell. Mach. Learn. **6**(3), 1–238 (2012). https://doi.org/10.1007/978-3-031-01561-8
16. Gebser, M., Kaminski, R., Kaufmann, B., Schaub, T.: Multi-shot ASP solving with Clingo. Theory Pract. Logic Program. **19**(1), 27–82 (2018). https://doi.org/10.1017/s1471068418000054
17. Goldfeder, J., Kugler, H.: BRE:IN - backend for reasoning about interaction networks with temporal logic. In: Bortolussi, L., Sanguinetti, G. (eds.) CMSB 2019. LNCS, vol. 11773, pp. 289–295. Springer, Cham (2019). https://doi.org/10.1007/978-3-030-31304-3_15
18. Klarner, H., Heinitz, F., Nee, S., Siebert, H.: Basins of attraction, commitment sets, and phenotypes of Boolean networks. IEEE/ACM Trans. Comput. Biol. Bioinf. **17**(4), 1115–1124 (2018). https://doi.org/10.1109/TCBB.2018.2879097
19. Klarner, H., Siebert, H.: Approximating attractors of Boolean networks by iterative CTL model checking. Front. Bioeng. Biotechnol. **3**, 130 (2015). https://doi.org/10.3389/fbioe.2015.00130
20. Lonsing, F., Egly, U.: DepQBF 6.0: a search-based QBF solver beyond traditional QCDCL. In: de Moura, L. (ed.) CADE 2017. LNCS (LNAI), vol. 10395, pp. 371–384. Springer, Cham (2017). https://doi.org/10.1007/978-3-319-63046-5_23
21. Montagud, A., et al.: Patient-specific Boolean models of signalling networks guide personalised treatments. Elife **11**, e72626 (2022). https://doi.org/10.7554/eLife.72626

22. Moon, K., Lee, K., Chopra, S., Kwon, S.: Bilevel integer programming on a Boolean network for discovering critical genetic alterations in cancer development and therapy. Eur. J. Oper. Res. **300**(2), 743–754 (2022). https://doi.org/10.1016/j.ejor.2021.10.019
23. Moon, K., Lee, K., Paulevé, L.: Computational complexity of minimal trap spaces in Boolean networks. ArXiv e-prints (2022). https://doi.org/10.48550/arXiv.2212.12756
24. Paulevé, L.: Marker and source-marker reprogramming of most permissive Boolean networks and ensembles with BoNesis. Peer Commun. J. **3**, e30 (2023). https://doi.org/10.24072/pcjournal.255
25. Paulevé, L., Kolčák, J., Chatain, T., Haar, S.: Reconciling qualitative, abstract, and scalable modeling of biological networks. Nat. Commun. **11**(1), 4256 (2020). https://doi.org/10.1038/s41467-020-18112-5
26. Paulevé, L., Sené, S.: Non-deterministic updates of Boolean networks. In: 27th IFIP WG 1.5 International Workshop on Cellular Automata and Discrete Complex Systems (AUTOMATA: volume 90 of Open Access Series in Informatics (OASIcs), pp. 1–16. Schloss Dagstuhl - Leibniz-Zentrum für Informatik 2021 (2021). https://doi.org/10.4230/OASIcs.AUTOMATA.2021.10
27. Paulevé, L., Sené, S.: Boolean networks and their dynamics: the impact of updates. In: Systems Biology Modelling and Analysis: Formal Bioinformatics Methods and Tools. Wiley (2022). https://doi.org/10.1002/9781119716600.ch6
28. Rabe, M.N., Tentrup, L.: CAQE: a certifying QBF solver. In: 2015 Formal Methods in Computer-Aided Design (FMCAD), pp. 136–143. IEEE (2015). https://doi.org/10.1109/FMCAD.2015.7542263
29. Réda, C., Delahaye-Duriez, A.: Prioritization of candidate genes through Boolean networks. In: Petre, I., Paun, A. (eds.) Computational Methods in Systems Biology. CMSB 2022. LNCS, vol. 13447, pp. 89–121. Springer, Cham (2022). https://doi.org/10.1007/978-3-031-15034-0_5
30. Trinh, V.-G., Benhamou, B., Hiraishi, K., Soliman, S.: Minimal trap spaces of logical models are maximal siphons of their petri net encoding. In: Petre, I., Paun, A. (eds.) Computational Methods in Systems Biology. CMSB 2022. LNCS, vol. 13447, pp. 158–176. Springer, Cham (2022). https://doi.org/10.1007/978-3-031-15034-0_8
31. Yordanov, B., Dunn, S.-J., Kugler, H., Smith, A., Martello, G., Emmott, S.: A method to identify and analyze biological programs through automated reasoning. NPJ Syst. Biol. Appl. **2**(1), 1–16 (2016). https://doi.org/10.1038/npjsba.2016.10
32. Zañudo, J.G.T., et al.: Cell line-specific network models of ER+ breast cancer identify potential PI3ka inhibitor resistance mechanisms and drug combinations. Can. Res. **81**(17), 4603–4617 (2021). https://doi.org/10.1158/0008-5472.can-21-1208

SAF: SAT-Based Attractor Finder in Asynchronous Automata Networks

Takehide Soh[1(✉)], Morgan Magnin[2], Daniel Le Berre[3], Mutsunori Banbara[4], and Naoyuki Tamura[5]

[1] Kobe University, Information Infrastructure and Digital Transformation Initiatives Headquarters, 1-1, Rokko-dai, Nada, Kobe, Hyogo 657-8501, Japan
soh@lion.kobe-u.ac.jp

[2] Nantes Université, École Centrale Nantes, CNRS, LS2N, UMR 6004, 44000 Nantes, France
morgan.magnin@ec-nantes.fr

[3] Univ. Artois, CNRS, Centre de Recherche en Informatique de Lens (CRIL), 62300 Lens, France
leberre@cril-lab.fr

[4] Nagoya University, Graduate School of Informatics, Furo-cho, Chikusa-ku, Nagoya 464-8601, Japan
banbara@nagoya-u.jp

[5] Kobe University, 1-1, Rokko-dai, Nada, Kobe, Hyogo 657-8501, Japan
tamura@kobe-u.ac.jp

Abstract. In this paper, we present a SAT-based Attractor Finder (SAF) which computes attractors in biological regulatory networks modelled as asynchronous automata networks. SAF is based on translating the problem of finding attractors of a bounded size into a satisfiability problem to take advantage of state-of-the-art SAT encodings and solvers. SAF accepts an automata network and outputs attractors in ascending size order until the bound is reached. SAF's main contribution is providing an alternative to existing attractor finders. There are cases where it is able to find some attractors while other techniques fail to do so. We observed such capability on both automata networks and Boolean networks. SAF is simple to use: it is available as a command line tool as well as a web application. Finally, SAF being written in Scala, it can run on any operating system with a Java virtual machine when combined with the SAT solver Sat4j.

Keywords: Automata Networks · Attractor · Boolean Networks · SAT Solvers

1 Introduction

Asynchronous automata networks (hereafter abridged as automata networks or AN) [12,16] are a multi-valued mathematical model widely studied for the qualitative analysis of the dynamical behavior of biological regulatory networks. Figure 1 (i) is an example of automata network which contains three automata

J. Pang and J. Niehren (Eds.): CMSB 2023, LNBI 14137, pp. 175–183, 2023.
https://doi.org/10.1007/978-3-031-42697-1_12

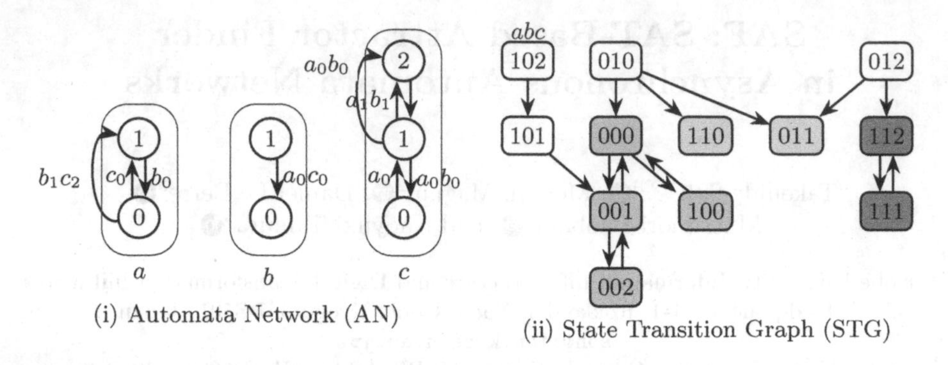

Fig. 1. Example of AN and its corresponding STG in asynchronous semantics.

a, b and c whose possible state values are $a, b \in \{0, 1\}$, $c \in \{0, 1, 2\}$. The arcs and their labels denote transitions that change the state value of an automaton with its labeled condition, e.g., a transition will change the state value of the automaton a from 0 to 1 with the condition $b = 1$ and $c = 2$. Among dynamical properties, *Attractors* raise a specific interest by their capacity to help validate the design of a biological model and predict possible asymptotic behaviors (e.g., capturing differentiation processes). *Attractors* can be defined from the state transition graph built from the automata network. Formally, a *trap domain* is a set of states that do not have outgoing arcs, meaning they cannot be escaped and thus loop indefinitely. An attractor is a subset minimal trap domain. Figure 1 (ii) represents the state transition graph of the automata network (i). It contains four attractors of sizes 1, 1, 2, and 4, corresponding to the colored states. However, few software tools are available to the general public despite the importance of attractors in automata networks. Pint [17] is a popular tool, but it cannot compute complex attractors. A method based on Answer Set Programming (ASP) presented in the paper [6, 7] can compute complex attractors, but sometimes fails to do so for large models.

To fill these gaps, we propose the software SAF using a SAT-based method for finding attractors of bounded size in asynchronous automata networks. SAF has been developed based on a recently proposed SAT-based method [19]. The existing ASP-based method [7] encodes attractors as a not-simple cycle between states, while our SAT encoding models attractors as a set of states. The complexity of our encoding is the attractor's size, while the ASP encoding's complexity lies in the cycle's length. In our running example, the attractors of size four will be computed as $\{000, 001, 002, 100\}$ by the SAT encoding while it will be computed as $(100, 000, 001, 002, 001, 000, 100)$ by the ASP encoding. SAF benefits from the following characteristics: **Efficiency.** SAF can take advantage of the wide availability of state-of-the-art SAT solvers and SAT enumerators, have received considerable attention and saw impressive progress during the last two decades [10]. The efficiency of such an approach is comparable and sometimes better than existing ASP-based methods [19]. **Portability.** The current version

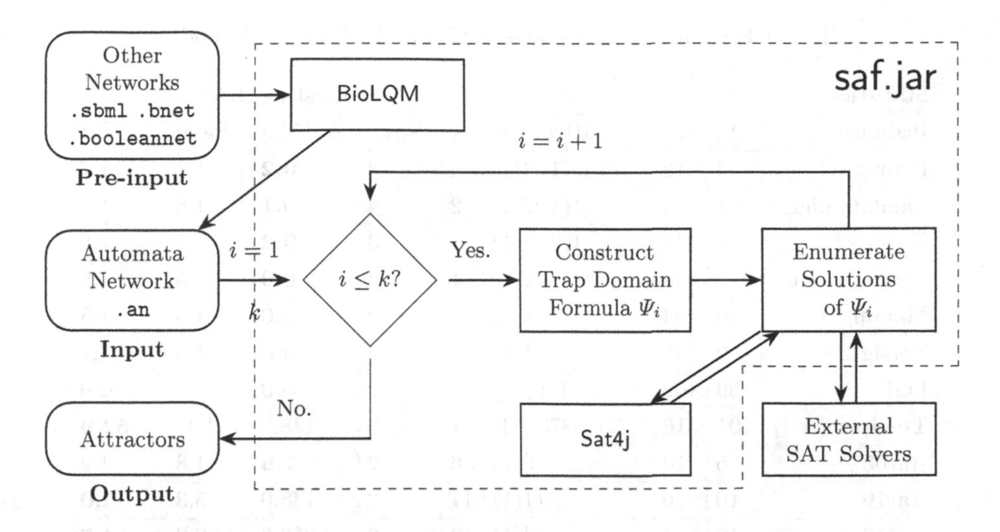

Fig. 2. SAF Architecture. The parts enclosed in dashed lines are included in saf.jar.

of SAF adopts Sat4j [15] as default back-end SAT solver. The combination of SAF and Sat4j provides a portable software tool that runs on any platform supporting Java. **Easy-to-use.** SAF is available as a command line tool as well as a web application https://saf-app.herokuapp.com/. Those tools using Sat4j can be used without installing any other dependency.

2 Architecture and Implementation

The architecture of SAF is summarized in Fig. 2. This section explains in detail its input, output, and the design of the core-engine.

I/O. SAF takes as input an automata network, but also supports other formats, e.g., `.sbml`, `.bnet`, `.booleannet`, which will be translated into automata networks using BioLQM [11]. In addition, a bound k can be given, limiting the computation to attractors of size at most or equal to k. The output is a set of attractors of increasing size, printed to the standard output as soon as the calculation is completed.

Core-Engine. The computation of attractors of a bounded size is delegated to a SAT solver, i.e., we build a formula Ψ_k such that each model of Ψ_k is a trap domain of size less than or equal to k. After finding a trap domain, we modify the formula Ψ_k to prevent finding it again. The procedure enumerates that way all trap domains of size k. Applying the procedure with an increasing value of k, it is possible to guarantee that all trap domains found are minimal and thus are also attractors of size bounded by k. The main challenges in SAT-based approaches are minimizing encoded SAT problems and enumerating numerous attractors. We address the former by employing the set-based encoding discussed above and tackle the latter by utilizing low-level integration with SAT solvers through the incremental IPASIR interface [1] and dedicated model enumerators for large attractor counts.

Table 1. Comparisons between SAF and ASP-based Method.

Statistics						Existing	SAF	
Instance	$\vert\Sigma\vert$	$\vert T\vert$	Attractors	k	$\vert S_{\max}\vert$	ASP [4]	Sat4j	Ext
Example	4	12	1(3):2(1):4(1)	4	3	**0.2**	1.7	1.7
Lambda phage	4	46	1(1):2(1)	2	4	**0.1**	1.8	1.9
Trp-reg	4	14	1(2):4(1)	4	3	**0.2**	1.7	1.8
Fission-yeast	9	43	1(1)	1	3	**0.0**	1.5	1.5
Mamm	10	34	1(1)	1	2	**0.0**	1.3	1.3
Tcrsig	40	85	1(8)	1	2	**0.0**	1.8	1.7
FGF	59	102	1(1536)	1	3	**0.0**	4.2	2.0
T-helper (AN)	101	316	1(5875504)	1	3	148.3	T.O	**32.9**
star05	5	10	6(1)	6	2	**1.6**	1.8	1.9
star10	10	20	11(1)	11	2	148.0	5.3	**4.0**
star12	12	24	13(1)	13	2	9959.3	7.8	**4.7**
star15	15	30	16(1)	16	2	T.O	26.2	**8.8**
star20	20	40	21(1)	21	2	T.O	117.0	**41.0**
star30	30	60	31(1)	31	2	T.O	8252.9	**878.8**
star40	40	80	41(1)	41	2	T.O	T.O	**9329.9**
star50	50	100	51(1)	51	2	T.O	T.O	T.O

Implementation. SAF is implemented in Scala. It can work on any system with a Java Virtual Machine (JVM) version 8 or greater using the Java SAT solver Sat4j [15]. For better performances, we also provide one state-of-the-art SAT solver, CaDiCaL [9] and SAT enumerator, BDD_Minisat_all [21], which need to be compiled on the host system. In the current implementation, by explicitly adding command line options, BDD_Minisat_all is called when $k = 1$ and CaDiCaL or Sat4j are called when $k \geq 2$.

3 Performance

We carried out experimental comparisons against the state-of-the-art tools to check the performance of SAF. We executed all experiments on a computer with a 3 GHz CPU and 16 GB of RAM. All benchmark instances and execution logs are available in https://doi.org/10.5281/zenodo.8062836.

Automata Network. We compared our system with the ASP-based tool presented in the literature [7]. Both use bounded search and are systems requiring an upper bound k. Benchmark instances are all instances from the literature [7,19], which consists in 8 biologically inspired networks, and artificial star networks that have increasing size attractors proposed in [19]. Those attractors have cycle sizes that are twice as large as their number of states (for detail, see Sect. 6.1

Table 2. Comparisons between SAF and Boolean network tools.

Instance	#Var	Boolsim [13]	Cabean [20]	Pyboolnet [14]	Pystable. [18]	Aeon [8]	SAF Ext
T-Cell 2006	40	**7/1**	0/0	**7/1**	**7/1**	**7/1**	7/0
ABA	44	**16/12**	**16/12**	**16/12**	**16/12**	**16/12**	16/3
T-LGL	61	0/0	0/0	0/0	172/88[6]	**172/142**	172/92
EMT	69	**13/0**	**13/0**	0/0	**13/0**	**13/0**	**13/0**
T-Cell 2007	101	0/0	0/0	0/0	**104/24**	**104/24**	104/0
T-helper (BN)	103	0/0	0/0	0/0	0/0	0/0	**5875504/0**

of [19]). The time limit is 3 h. Table 1 shows the result. From left to right, columns denote the name of the automata networks, the number of automata included, the number of transitions included, the sizes of the attractors, the bound k given to both solvers, the maximum number of states of automata denoted as $S_{\max}$, the CPU time of the compared ASP-based method, SAF using Sat4j and external SAT solvers denoted as Ext. In the biologically inspired benchmarks, all methods solved them within a few seconds. The main reason is that they contain small-sized attractors and thus k is also small. One other difference between the ASP-based method and SAF comes from the implementation language: C++ and Scala. SAF is implemented on Scala which is running on JVM for the modeling part and thus there is a small disadvantage on this point. One exception is `T-helper` which contains a large number of attractors. On this problem, our approach takes advantage of the SAT solver `BDD_MINISAT_ALL` dedicated to the fast enumeration of solutions. On `star` benchmarks, the difference between each encoding is more obvious. Our set-based encoding outperforms ASP cycle-based encoding. The CPU time of the ASP-based method increases exponentially, and it cannot solve `star15` within 3 h. On the other hand, although SAF+Sat4j cannot solve `star30`, SAF+external SAT solvers successfully solved them until `star40`. In summary, while the ASP-based method is faster at finding small-sized attractors, it may fail to detect some; on the other hand, SAF is more robust at finding attractors, but it faces difficulties when their size exceeds 50.

Boolean Network. We also compared our system with state-of-the-art Boolean network tools: Boolsim [13], Cabean [20], Pyboolnet [14], Pystablemotifs [18] and Aeon [8]. All compared tools execute comprehensive searches, i.e., those tools run without specifying k. So, we run SAF with large enough k and compare the number of attractors computed within the time limit 2,400 s. We employ existing Boolean network instances to facilitate the comparison since no tool is currently available for converting automata networks into Boolean networks. Benchmark instances are all real cell model instances from the literature [18]. The specific instance T-LGL may vary from 58 to 61 variables depending on the source. Our instance is available from our dataset [3]. In addition, we compare one more instance, a Boolean Network version of `T-helper` [5] downloaded from URL [2], which has a large number of attractors. Table 2 shows the result[1]. From left to right, columns denote the name of the Boolean networks, the number of atoms included, the number of attractors computed by compared tools and SAF in the form of "n/m" where n denotes the number of singleton attractors and m denotes the number of complex attractors. Unlike automata-network results, SAF did not yield conclusive results on those benchmarks. The reasons for this could be twofold: i) SAF converts Boolean networks into richer automata networks instead of directly solving them, and ii) it struggles with calculating large attractors exceeding 50 due to its process of exploring attractors of increasing size. However, Boolean network version of `T-helper`, which contains many attractors, could not be enumerated by any other tool within the time limit.

4 Availability

Command Line Tool. The command line tool, sources, and instructions are available on https://github.com/TakehideSoh/SAF. The command line tool can be executed from users' terminal as follows.

```
$ java -jar saf.jar [options] [inputFile]
```

The minimum example using Sat4j and $k = 4$ on the automata network file `runningexample.an` is as follows.

```
$ java -jar saf.jar -k 4 runningexample.an
```

To run the above command, users do not need any installation on any platform supporting Java 8 or greater. Improved performance is available by adding options of BDD_Minisat_all and CaDiCaL (or any IPASIR compatible software). In this case, the installation of those SAT solvers in the system is necessary.

Web Application. The web application version of SAF (see Fig. 3 in Appendix) is available at https://saf-app.herokuapp.com/. It currently accepts multivalued networks (`.an .sbml`), and Boolean networks (`.bnet .booleannet`). To use this application, users initially edit their network in a textbox or upload local files. Next, they specify a bound k and push "find attractors". The result is printed into another textbox. This service is hosted on HEROKU, one of the PaaS providers. Due to service limitations, each execution is limited to 20 s, but it has sufficient performance for simple use cases.

5 Conclusion

This paper presents a SAT-based Attractor Finder (SAF) for identifying attractors of biological regulatory networks modeled as dynamical multi-level discrete

[1] In our environment, Pystablemotifs version 3.3 throws an error to T-LGL on our computer. Thus, we provide the number of attractors found in their paper [18].

models. SAF was developed with the goal of being an efficient, portable, and easy-to-use tool. The core of the approach targets automata networks but, thanks to existing translators between modeling frameworks, SAF handles models specified in various Boolean network formats (Boolnet, Booleannet) and in SBML-qual. Source code, executable as a Jar file and a web application are available to make it easily usable by a wide range of users, from computer scientists to modelers. Future work includes making the output of SAF compatible with JSON format and graphical representations to make it more understandable for beginners. It is also important to apply it to challenging instances of Boolean Networks. In addition, extending SAF to other semantics and considering other attractor-related problems like bifurcation is interesting.

Acknowledgements. This work was financially supported by the "PHC Sakura" program (43009SC, JPJSBP120193213), implemented by MESRI and JSPS. This work was also supported by JSPS KAKENHI (JP21K11828, JP22K11973, JP23K11047), and by ROIS NII Open Collaborative Research 2023 (23FP04).

A Input File and Web Interface of SAF

The automata network from Fig. 1 can be described by the following automata network `runningexample.an` input file as in Fig. 3. The web application version of SAF is demonstrated in Fig. 3 as well.

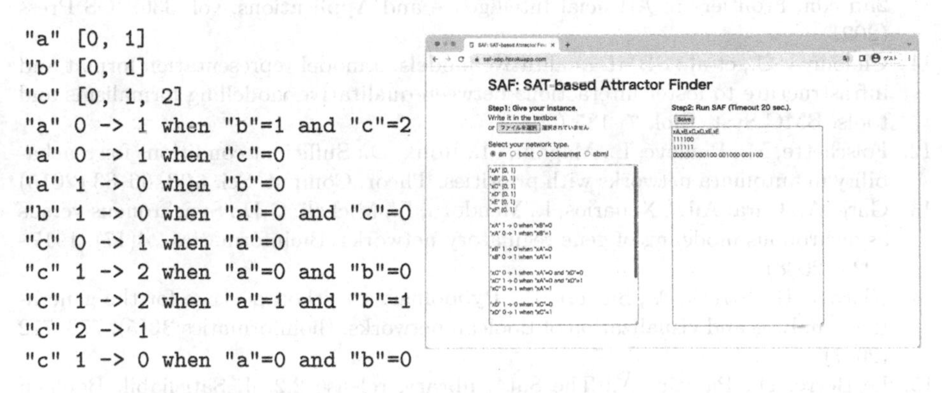
"a" [0, 1]
"b" [0, 1]
"c" [0, 1, 2]
"a" 0 -> 1 when "b"=1 and "c"=2
"a" 0 -> 1 when "c"=0
"a" 1 -> 0 when "b"=0
"b" 1 -> 0 when "a"=0 and "c"=0
"c" 0 -> 1 when "a"=0
"c" 1 -> 2 when "a"=0 and "b"=0
"c" 1 -> 2 when "a"=1 and "b"=1
"c" 2 -> 1
"c" 1 -> 0 when "a"=0 and "b"=0
runningexample.an
SAF: SAT-based Attractor Finder
Step1: Give your instance.
Step2: Run SAF (Timeout 20 sec.).
https://saf-app.herokuapp.com/

Fig. 3. Input File and Web Interface of SAF.

References

1. IPASIR. https://www.cs.utexas.edu/users/moore/acl2/manuals/current/manual/index-seo.php/IPASIR____IPASIR

2. Model of Boolean network version of T-helper. https://github.com/hklarner/pyboolnet/tree/master/pyboolnet/repository/jaoude_thdiff
3. Model of T-LGL used in Boolean netrwork experiments. https://github.com/TakehideSoh/SAF-Evaluation/blob/main/exp-boolean-network/benchmark/2176_T-LGL_Survival_Network_2008.booleannet.bnet
4. Abdallah, E.B., Folschette, M., Roux, O.F., Magnin, M.: ASP-based method for the enumeration of attractors in non-deterministic synchronous and asynchronous multi-valued networks. Algorithms Mol. Biol. **12**(1), 20:1–20:23 (2017)
5. Abou-Jaoudé, W., et al.: Model checking to assess t-helper cell plasticity. Front. Bioeng. Biotechnol. **2**, 86 (2015)
6. Ben Abdallah, E., Folschette, M., Magnin, M.: Analyzing long-term dynamics of biological networks with answer set programming. In: Systems Biology Modelling and Analysis: Formal Bioinformatics Methods and Tools, pp. 251–303 (2022)
7. Ben Abdallah, E., Folschette, M., Roux, O., Magnin, M.: Asp-based method for the enumeration of attractors in non-deterministic synchronous and asynchronous multi-valued networks. Algorithms Mol. Biol. **12**(1), 1–23 (2017)
8. Benes, N., Brim, L., Huvar, O., Pastva, S., Safránek, D., Smijáková, E.: Aeon.py: Python library for attractor analysis in asynchronous Boolean networks. Bioinformatics **38**(21), 4978–4980 (2022)
9. Biere, A., Fazekas, K., Fleury, M., Heisinger, M.: CaDiCaL, Kissat, Paracooba, Plingeling and Treengeling entering the SAT Competition 2020. In: Balyo, T., Froleyks, N., Heule, M., Iser, M., Järvisalo, M., Suda, M. (eds.) Proceedings of SAT Competition 2020 - Solver and Benchmark Descriptions. Department of Computer Science Report Series B, vol. B-2020-1, pp. 51–53. University of Helsinki (2020)
10. Biere, A., Heule, M., van Maaren, H., Walsh, T. (eds.): Handbook of Satisfiability, 2nd edn, Frontiers in Artificial Intelligence and Applications, vol. 336. IOS Press (2021)
11. Chaouiya, C., et al.: SBML qualitative models: a model representation format and infrastructure to foster interactions between qualitative modelling formalisms and tools. BMC Syst. Biol. **7**, 135 (2013)
12. Folschette, M., Paulevé, L., Magnin, M., Roux, O.: Sufficient conditions for reachability in automata networks with priorities. Theor. Comput. Sci. **608**, 66–83 (2015)
13. Garg, A., Cara, A.D., Xenarios, I., Mendoza, L., Micheli, G.D.: Synchronous versus asynchronous modeling of gene regulatory networks. Bioinformatics **24**(17), 1917–1925 (2008)
14. Klarner, H., Streck, A., Siebert, H.: Pyboolnet: a Python package for the generation, analysis and visualization of Boolean networks. Bioinformatics **33**(5), 770–772 (2017)
15. Le Berre, D., Parrain, A.: The Sat4j library, release 2.2. J. Satisfiabil. Boolean Model. Comput. **7**(2–3), 59–64 (2010)
16. Paulevé, L.: Goal-oriented reduction of automata networks. In: Bartocci, E., Lio, P., Paoletti, N. (eds.) CMSB 2016. LNCS, vol. 9859, pp. 252–272. Springer, Cham (2016). https://doi.org/10.1007/978-3-319-45177-0_16
17. Paulevé, L.: PINT: a static analyzer for transient dynamics of qualitative networks with IPython interface. In: Feret, J., Koeppl, H. (eds.) CMSB 2017. LNCS, vol. 10545, pp. 309–316. Springer, Cham (2017). https://doi.org/10.1007/978-3-319-67471-1_20
18. Rozum, J.C., Deritei, D., Park, K.H., Zañudo, J.G.T., Albert, R.: pystablemotifs: Python library for attractor identification and control in Boolean networks. Bioinformatics **38**(5), 1465–1466 (2022)

19. Soh, T., Magnin, M., Berre, D.L., Banbara, M., Tamura, N.: Sat-based method for finding attractors in asynchronous multi-valued networks. In: Ali, H., Deng, N., Fred, A.L.N., Gamboa, H. (eds.) Proceedings of the 16th International Joint Conference on Biomedical Engineering Systems and Technologies, BIOSTEC 2023, Volume 3: BIOINFORMATICS, pp. 163–174. SCITEPRESS (2023)
20. Su, C., Pang, J.: Cabean: a software for the control of asynchronous Boolean networks. Bioinformaics **37**(6), 879–881 (2021)
21. Toda, T., Soh, T.: Implementing efficient all solutions SAT solvers. ACM J. Exp. Algorithmics **21**(1), 1.12:1–1.12:44 (2016)

Condition for Periodic Attractor in 4-Dimensional Repressilators

Honglu Sun[1(✉)], Maxime Folschette[2], and Morgan Magnin[1]

[1] Nantes Université, École Centrale Nantes, CNRS, LS2N, UMR 6004, 44000 Nantes, France
honglu.sun@ls2n.fr

[2] Univ. Lille, CNRS, Centrale Lille, UMR 9189 CRIStAL, 59000 Lille, France

Abstract. One of the key questions about gene regulatory networks is how to predict complex dynamical properties based on the influence graph's topology. Earlier theoretical studies have identified conditions for complex dynamical properties, like multistability or oscillations, based on topological features, like the presence of a positive (negative) feedback loop. This work follows this path and aims to find a sufficient and necessary condition for the existence of a periodic attractor in 4-dimensional (4 genes) repressilators based on a discrete modeling framework under some dynamical assumptions. These networks are extensions of the widely studied 3-dimensional repressilator, which has been used in synthetic biology to produce synthetic oscillations. While other researchers have explored specific extensions of the 3-dimensional repressilator to improve synthetic oscillation control, our work investigates all 4-dimensional networks with only inhibitions. By uncovering new insights about periodic attractors in these small networks, our findings could aid the design of new synthetic oscillations. We search for condition for period attractor in an exhaustive manner with the guide of a decision tree model. Our major contributions include: 1) discovering that, with one exception, the relations between gene regulation thresholds do not impact the existence of periodic attractors in any of the influence graphs considered in this study; 2) identifying a sufficient and necessary condition of simple form for the existence of a periodic attractor when the exception is ignored; 3) identifying new topological features of influence graphs that are necessary for predicting the existence of periodic attractor in 4-dimensional repressilators.

Keywords: Periodic attractor · Discrete dynamical system · Repressilator · Decision tree · Gene regulatory networks

1 Introduction

Gene expression is not an isolated biological process as the expression of a single gene could activate or inhibit the expression of one or multiple other genes. These

Supported by China Scholarship Council.

J. Pang and J. Niehren (Eds.): CMSB 2023, LNBI 14137, pp. 184–201, 2023.
https://doi.org/10.1007/978-3-031-42697-1_13

complex interdependencies among genes constitute the gene regulatory network, which is always represented mathematically as a directed graph known as the influence graph. The vertices in the influence graph correspond to individual genes, while the arcs denote the relations between genes.

Our work focuses on a fundamental question about gene regulatory networks: how can we predict a system's dynamical properties from its influence graph? In fact, the dynamical properties do not only depend on the topology of the influence graph, but also depend on the dynamical model used to model the gene regulatory network. In the literature, different modeling frameworks have been applied to model gene regulatory networks, mainly continuous models [3,4,13], discrete models [6,7,14,15,32,33] and hybrid models [5,9,10,31]. This work is based on Thomas' discrete modeling framework of gene regulatory networks [32,33], where the continuous expression values of all genes in the system are abstracted by a vector of integers, called discrete state, describing the discrete expression levels of all genes, and the system's dynamics are then captured by the transitions between discrete states. Using this discrete modeling framework, we can describe the dynamical properties of the system in terms of the existence of different attractors. There are two types of attractors: fixed-point attractors, which correspond to stable states, and periodic attractors, which correspond to oscillations. One major advantage of using discrete modeling is its simplicity in implementation and analysis.

Our study investigates all 4-dimensional (with 4 genes) gene regulatory networks with only inhibition relations, called 4-dimensional repressilators, as they can be considered extensions of the 3-dimensional canonical repressilator. The 3-dimensional canonical repressilator is a network of three genes having a unique feedback loop with only inhibitions between genes [8,11,23,30]. It is widely studied in synthetic biology due to its ability to generate synthetic oscillations. However, controlling these synthetic oscillations remains an open problem. To address this issue, there are works studying extensions of 3-dimensional canonical repressilator by adding more genes into the network [12,19,22,34,35]. Most of these works focus on some specific networks and are mostly based on ordinary differential equations. In our work, we use discrete dynamical models to study all possible combinations of 4-dimensional networks with only inhibitions, under some dynamical assumptions. Our goal is to identify a sufficient and necessary condition for the existence of a periodic attractor. The existence of a periodic attractor in discrete models is linked to the presence of oscillations in certain hybrid models, which is significant from the perspective of synthetic biology as sometimes the goal is to have synthetic oscillations.

The method to find such conditions for a periodic attractor has two major steps. Firstly, we exhaustively generate all discrete models of 4-dimensional repressilators and compute various topological features of their influence graphs. Using these features we construct a highly accurate machine learning model which predicts the existence of periodic attractor with an accuracy close to 1.0. Secondly, we manually analyze this machine learning model to develop a sufficient and necessary condition for periodic attractor. In fact, the idea of studying certain classes of networks in an exhaustive manner has been used in the literature [16,17,24] but, as far as we know, it has not yet been applied specifically to 4-dimensional repressilators.

It is worth noting that some theoretical works have investigated the relationship between the dynamical properties of discrete models and the topological structures of influence graphs [21,25,27–29]. Specifically, these studies have employed topological characteristics such as the presence of positive or negative feedback loops in the influence graph to describe conditions that lead to the emergence of complex dynamical properties, such as multistability, oscillations or attractors. In line with these studies, we also adopt topological features to develop the desired condition. Furthermore, we have identified new features that are essential for predicting the existence of a periodic attractor in 4-dimensional repressilators.

We also want to highlight that the machine learning-based approach used in this work for exhaustively searching for conditions could be potentially extended for the search of conditions that determine various properties in other systems.

We summarize the main contributions of this work as follows:

- We show that with the exception of one particular influence graph, the relations between the thresholds of gene regulations have no impact on the existence of a periodic attractor; in other words, the existence of a periodic attractor is solely dependent on the topology of the influence graph.
- We introduce new features, namely *total out-degree of cycles of length n*, which characterize the influence graph's topology and are essential to describe the condition for the existence of a periodic attractor in 4-dimensional repressilators.
- Based on these new features, we find a new sufficient and necessary condition for the existence of a periodic attractor in 4-dimensional repressilators, except for the aforementioned particular influence graph.

The paper is organized as follows. In Sect. 2, we introduce the discrete modeling framework used in this work. In Sect. 3, we present the 4-dimensional repressilators and some dynamical assumptions used in this work. In Sect. 4, we represent step by step how we develop a sufficient and necessary condition for periodic attractor. Finally, in Sect. 5, we make a conclusion and discuss our future works.

2 Discrete Modeling of Gene Regulatory Networks

This section introduces the pre-existing discrete modeling of *gene regulatory networks* used in this work. A gene regulatory network can be described by a directed graph $IG = (V, A)$ called *influence graph*, where V is the set of vertices describing the genes in the system and A is the set of arcs describing the regulations (activation or inhibition) between genes. For example, Fig. 1-Left represents an influence graph of a 3-dimensional gene regulatory network. In this gene regulatory network, the expression of gene G_0 (resp. G_1) activates the expression of gene G_1 (resp. G_2) (the sign "+" on the arc denotes an activation) and the expression of G_0 inhibits the expression of G_2 (the sign "−" on the arc denotes an inhibition).

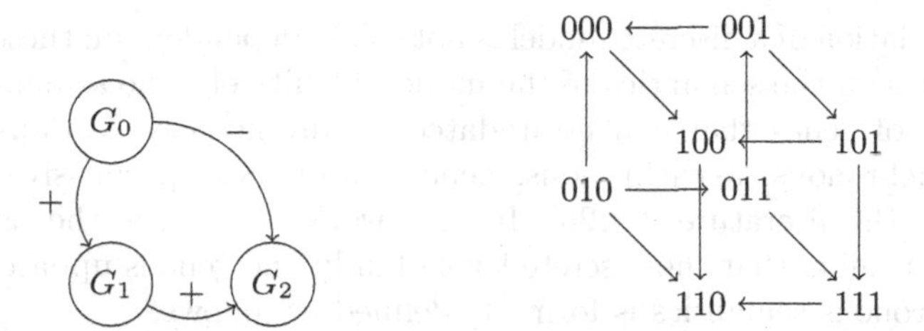

Fig. 1. Left: a 3-dimensional influence graph. Right: a transition graph of discrete states, based on the influence graph on the left and the logic program of Eq. 3.

An influence graph can only describe partially a gene regulatory network, as it lacks the description of some dynamical properties. In this work, we use a discrete modeling framework of gene regulatory networks. In this discrete modeling framework, the continuous expression of a gene is abstracted by an integer (e.g. 0, 1, 2, ...), called discrete level, which describes the discrete expression level of a gene. More formally, for any gene $G \in V$, there exists a set of integers $a(G)$ which gives all possible discrete levels of G; for instance, if G has only two discrete levels, then $a(G) = \{0, 1\}$. A discrete model is a logic program which is a set of logic rules. The form of a logic rule is shown as follows:

$$G_i = k \leftarrow \phi_i \tag{1}$$

where $G_i \in V$ is a gene in the system, $k \in a(G_i)$ is a possible discrete level of G_i, and ϕ_i is a logic formula. The form of a logic formula ϕ is given as follows:

$$\phi :== \emptyset \mid G \sim k \mid \phi_1 \wedge \phi_2 \mid \phi_1 \vee \phi_2 \tag{2}$$

where $k \in a(G)$ is a possible discrete level of gene G, $\sim$ is one of the relations $\{>, <, =, \geq, \leq\}$, ϕ_1 and ϕ_2 are also logic formulas.

A logic rule (see Eq. 1) indicates that if ϕ_i at discrete time t (t is an integer) is satisfied, then at time $t+1$ the value of G_i can be updated to k. For example, a possible discrete model of the influence graph in Fig. 1-Left is given as follows:

$$\begin{aligned} G_0 &= 1 \leftarrow \emptyset \\ G_1 &= 1 \leftarrow (G_0 = 1) \\ G_1 &= 0 \leftarrow (G_0 = 0) \\ G_2 &= 1 \leftarrow (G_0 = 0) \wedge (G_1 = 1) \\ G_2 &= 0 \leftarrow (G_0 = 1) \vee (G_1 = 0) \end{aligned} \tag{3}$$

In this example, the second line indicates that, if at time t the discrete level of G_0 is 1, then at time $t+1$ the discrete level of G_1 can be updated to 1; this logic rule corresponds to the activation from G_0 to G_1 in the influence graph. The first line indicates that for any moment t, the discrete level of G_0 can be updated to 1 at time $t+1$ (in fact, once the discrete level of G_0 reaches 1, it will remain at 1).

The simulation of a discrete model is not solely dependent on these logic rules, but also relies on the semantics of the model. Intuitively, the semantics dictates the number of genes that can be updated simultaneously. Various semantics, such as synchronous, asynchronous, general and most permissive, have been proposed in the literature [20,26]. In this work, we adopt the asynchronous semantics, meaning that the discrete level of only one gene is updated at a time. The asynchronous semantics is formally defined as follows.

We consider a system with N genes noted as $G_1, G_2, ..., G_N$. We define that a discrete state d_s of a system is an integer vector of length N, which assigns the discrete level d_s^i to gene G_i, where $i \in \{1, 2, 3, ..., N\}$ and d_s^i is the ith component of d_s. For any discrete state d_s at time t, if there exists a logic rule $G_{i_0} = k \leftarrow \phi$ where $i_0 \in \{1, 2, 3, ..., N\}$, such that d_s satisfies ϕ (meaning that the assignment $G_i = d_s^i$ for $i \in \{1, 2, 3, ..., N\}$ satisfies ϕ) and $d_s^{i_0} \neq k$, then at time $t + 1$, the system can reach the new discrete state d'_s such that $d'^i_s = d_s^i$ for $i \in \{1, 2, 3, ..., N\} \setminus \{i_0\}$ and $d'^{i_0}_s = k$.

Based on the choice of semantics, we can get the transition graph of discrete states of a discrete model, which is a directed graph containing all possible transitions between discrete states from t to t+1. The transition graph of discrete states describes the dynamics of a discrete model. Note that depending on the semantics, the dynamics can be non-deterministic, as is the case for the asynchronous semantics. For example, Fig. 1-Right presents the transition graph of discrete states, derived by asynchronous semantics, of the discrete model described by the logic program in Eq. 3. Consider the discrete state 101 (representing the assignment $G_0 = 1, G_1 = 0, G_2 = 1$), according to the logic program in Eq. 3, G_1 can be updated from 0 to 1 and G_2 can be updated from 1 to 0. Since we use asynchronous semantics, only one gene can be updated, so it can reach 100 or 111, but it cannot make two updates at the same time to reach 110.

3 4-Dimensional Repressilator

The scope of this work is limited to 4-dimensional gene regulatory networks where genes are linked only through inhibition, and where every gene has an impact on at least one other gene. In fact, the 3-dimensional network with a unique negative feedback loop with only inhibitions, called canonical repressilator, has been proved in the literature to be able to generate oscillations, while the understanding of oscillations in its 4-dimensional extensions is still limited. We call these networks 4-dimensional repressilators. An influence graph of such networks is shown in Fig. 2.

Our analysis of these 4-dimensional repressilators is based on two underlying assumptions about their dynamics.

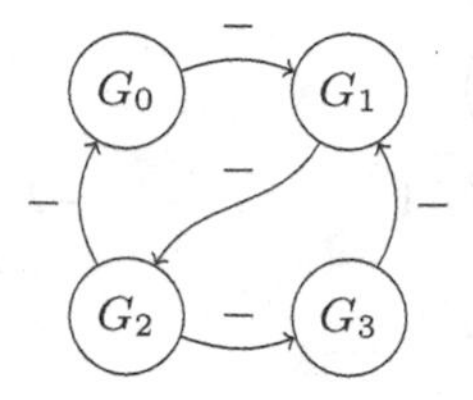

Fig. 2. An influence graph of a 4-dimensional repressilator.

Assumption 1. *If one gene influences m different genes, then it has m distinct thresholds that correspond to each of these m genes.*

Consider the influence graph of Fig. 2, G_2 inhibits G_0 and G_3, so G_2 has two distinct thresholds because of Assumption 1, meaning that it has three discrete levels: $0, 1, 2$. Since other genes only influence one other gene, they only have two discrete levels: $0, 1$. A similar assumption can be found in [1] for example.

The relations between the two thresholds of G_2 can have impact on the dynamical properties of the system. To show the relations between thresholds on an influence graph, we introduce the notion of influence graph with thresholds which is defined as $IGS = (V, A, s)$ where V and A are the sets of genes and regulations between genes respectively, as in the definition of an influence graph, and the function s assigns an integer to each regulation that represents the minimum discrete level of the source gene necessary to inhibit the target gene. Thus, the function s also characterizes the relationship between thresholds.

From the influence graph of Fig. 2, by considering all different relations between thresholds, we can get two different influence graphs with thresholds as illustrated in Fig. 3. For the regulation $G_2 \to G_3$ in the left influence graph with thresholds, $s(G_2 \to G_3) = 2$ (which is the number on the arc) means that G_3 is inhibited by G_2 if the discrete level of G_2 is bigger or equal to 2. For the regulation $G_2 \to G_0$ in the same graph, $s(G_2 \to G_0) = 1$ means that G_0 is inhibited by G_2 if the discrete level of G_2 is bigger or equal to 1. We can see that, in the left influence graph with thresholds, the threshold of G_2 triggering the inhibition of G_0 is smaller than the threshold triggering the inhibition of G_3, while in the right influence graph with thresholds, the situation is reversed. We can also see that the function s gives all possible discrete levels of the system.

A priori, different discrete models (logic programs) can be associated to the same influence graph with thresholds, particularly when one gene is inhibited by several genes. Moreover, different choices of logic programs lead to different transition graphs, in other words different dynamical properties. In this work, we make an assumption about the dynamics when one gene is inhibited by several genes.

Assumption 2. *In an influence graph with threshold $IGS = (V, A, s)$, for any gene G, its discrete level can decrease by 1 if there exists a regulation from G' to G and the current discrete level of G' is bigger or equal to $s(G' \to G)$, otherwise its discrete level can increase by 1.*

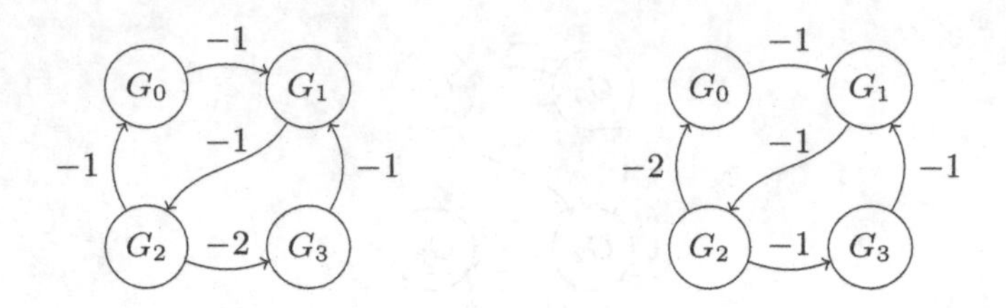

Fig. 3. Different influence graphs with thresholds corresponding to the influence graph of Fig. 2.

In fact, Assumption 2 is equivalent to assume that the inhibitions are disjunctive, meaning that only one inhibitor is enough to decrease the target gene. Similar assumptions about the disjunction or conjunction of gene regulation can be found in [2,18]. Consider the influence graph with thresholds on the left of Fig. 3; Assumption 2 leads to the following logic program:

$$\begin{aligned}
G_0 &= 0 \leftarrow (G_2 \geq 1)\\
G_0 &= 1 \leftarrow (G_2 < 1)\\
G_1 &= 0 \leftarrow (G_0 \geq 1) \vee (G_3 \geq 1)\\
G_1 &= 1 \leftarrow (G_0 < 1) \wedge (G_3 < 1)\\
G_2 &= 0 \leftarrow (G_1 \geq 1) \wedge (G_2 \leq 1)\\
G_2 &= 1 \leftarrow (G_1 \geq 1) \wedge (G_2 = 2)\\
G_2 &= 1 \leftarrow (G_1 < 1) \wedge (G_2 = 0)\\
G_2 &= 2 \leftarrow (G_1 < 1) \wedge (G_2 \geq 1)\\
G_3 &= 0 \leftarrow (G_2 \geq 2)\\
G_3 &= 1 \leftarrow (G_2 < 2)
\end{aligned} \tag{4}$$

Note that, for instance, there needs to be two rules in order to make G_2 increase to the expression level 2: one to update it from level 0 to level 1 (line 7) and one to make it increase from 1 to 2 (line 8); this is because we didn't constraint the dynamics to be unitary and we thus need to encode this property inside the rules. Using Assumption 2, we get a unique discrete model from any influence graph with thresholds, which simplifies the analysis. The transition graph of discrete states corresponding to the discrete model of Eq. 4 is illustrated in Fig. 4.

The general logic rules for an arbitrary *IGS* are given in Eq. 5.

$$\begin{aligned}
G_i = k+1 &\leftarrow (G_i = k) \wedge (k < Max(a(G_i))) \wedge (\forall G \in reg(G_i), G < s(G \to G_i))\\
G_i = k-1 &\leftarrow (G_i = k) \wedge (k > 0) \wedge (\exists G \in reg(G_i), G \geq s(G \to G_i))
\end{aligned} \tag{5}$$

where $Max(a(G_i))$ is the maximum discrete level of G_i and $reg(G_i)$ is the set of all genes that inhibit G_i. Obviously, the rules of Eq. 4 can be derived from the ones of Eq. 5 by simplification. Some simplifications also involve the knowledge of the dynamics given in Sect. 2.

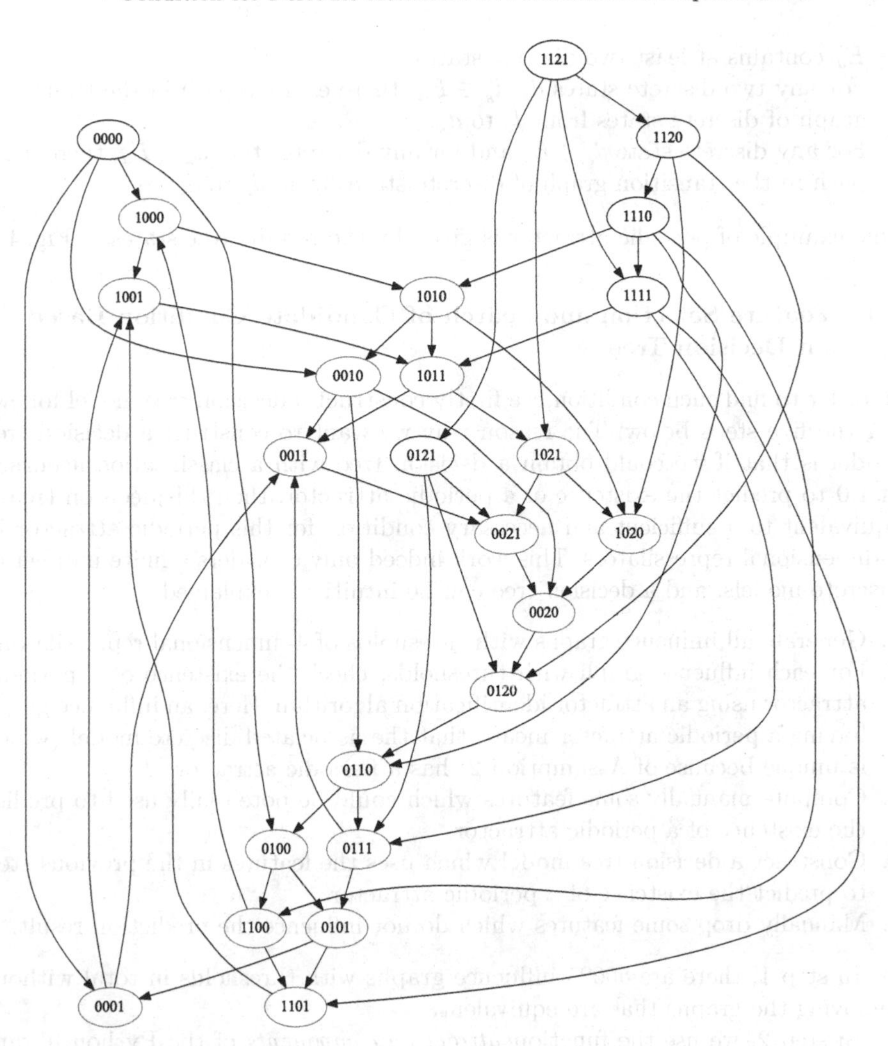

Fig. 4. Transition graph of discrete states of a model of 4-dimensional repressilator (corresponding to Eq. 4). Red discrete states represent a periodic attractor. (Color figure online)

4 Condition for a Periodic Attractor

In this section, we present our contribution: a step by step method to find a sufficient and necessary condition for the existence of a periodic attractor in 4-dimensional repressilators. Additionally, we also discuss the number of oscillatory dimensions in these periodic attractors. The definition of a periodic attractor is given as follows.

Definition 1. A periodic attractor is a set of discrete states E_a such that:

- E_a contains at least two discrete states.
- For any two discrete states $d_s, d'_s \in E_a$, there exists a path in the transition graph of discrete states from d_s to d'_s.
- For any discrete state $d_s \in E_a$ and for any discrete state $d'_s \notin E_a$, there is no path in the transition graph of discrete states from d_s to d'_s.

One example of periodic attractor is given by the red discrete states in Fig. 4.

4.1 Feature Selection and Search of Candidate Condition Based on Decision Tree

In order to find such condition, we firstly construct a decision tree model following the five steps below. The reason why we want to construct a decision tree model is that if we could obtain a decision tree with a classification accuracy of 1.0 to predict the existence of a periodic attractor, then this decision tree is equivalent to a sufficient and necessary condition for this periodic attractor in 4-dimensional repressilators. This work indeed only considers a finite number of discrete models, and a decision tree can be intuitively explained.

1. Generate all influence graphs with thresholds of 4-dimensional repressilators.
2. For each influence graph with thresholds, check the existence of a periodic attractor using an attractor identification algorithm. Here, an influence graph having a periodic attractor means that the associated discrete model (which is unique because of Assumption 2) has a periodic attractor.
3. Compute manually some features which could be potentially used to predict the existence of a periodic attractor.
4. Construct a decision tree model which uses the features in the previous step to predict the existence of a periodic attractor.
5. Manually drop some features which do not influence the prediction result.

In step 1, there are 50625 influence graphs with thresholds in total without removing the graphs that are equivalent.

In step 2, we use the function *attracting_components* of the Python library NetworkX to verify the existence of a periodic attractor. We find that any influence graph with thresholds which has a periodic attractor has only one periodic attractor.

In step 3, we compute two classes of features on the influence graph to describe the topology of the influence graph: the *number of cycles of length* n and the *total out-degree of cycles of length* n.

For the first class, since the system has 4 genes, there are only cycles of length 2, 3 and 4. We use $C2$, $C3$ and $C4$ to represent the numbers of cycles of length 2, 3 and 4, respectively. For example, for the influence graph in Fig. 2, there is no cycle of length 2 or 4 ($C2 = 0, C4 = 0$) and two cycles of length 3 ($C3 = 2$). It is logical to use these features to predict the existence of a periodic attractor because, in this class of repressilators, the length of a cycle determines whether it is a negative feedback loop or a positive feedback loop as there is only inhibition regulations and the presence of loops is related to the existence of attractor(s).

For example, it is already known that the presence of a negative feedback loop is a necessary condition for sustained oscillations [27] and the presence of positive feedback loop is a necessary condition for multistability [28].

The second class of features is a new class of features introduced in this work which is defined formally as follows.

Definition 2 (Total out-degree of cycles of length n). The total out-degree of cycles of length n is the total number of arcs which go from a vertex which belongs to a cycle of length n to a vertex which does not belong to this cycle.

For example, for the influence graph in Fig. 2, the arc $G_2 \to G_3$ goes from the cycle of length 3: $G_0 \to G_1 \to G_2$, to G_3, which does not belong to this cycle. There are two arcs like this in this influence graph: $G_2 \to G_3$ (for the cycle $G_0 \to G_1 \to G_2$) and $G_2 \to G_0$ (for the cycle $G_1 \to G_2 \to G_3$). So the total out-degree of cycles of length 3 is 2. Since, in this influence graph, there is no cycle of length 2, the total out-degree of cycles of length 2 is 0. Since the graph considered in this work has only 4 genes, the total out-degree of cycles of length 4 is always 0. We use $OD2$ and $OD3$ to denote the total out-degree of cycles of length 2 and 3, respectively.

To explain the motivation about these features describing the total out-degree of cycles, let's consider the two influence graphs in Fig. 5. The dynamical properties of these two influence graphs are different: any influence graph with thresholds associated to the left influence graph has a periodic attractor while any influence graph with threshold associated to the right influence graph does not have a periodic attractor. However, the topologies of these two influence graphs are similar: the numbers of cycles of length 2, 3 and 4 of these two graphs are identical and they both have 6 arcs. In order to find a condition for periodic attractor, we need to find a way to exhibit the topological difference between these two graphs, and these new features are effective: for the left graph, $OD2 = 1, OD3 = 3$; for the right one, $OD2 = 2, OD3 = 2$. Note that these features do not depend on the relations between thresholds.

In step 4, we construct a decision tree to predict the existence of a periodic attractor based on the 5 features $C2$, $C3$, $C4$, $OD2$ and $OD3$ using all influence graphs with thresholds considered in this work. This decision tree is constructed automatically using the decision tree model of the Python library Scikit-learn.

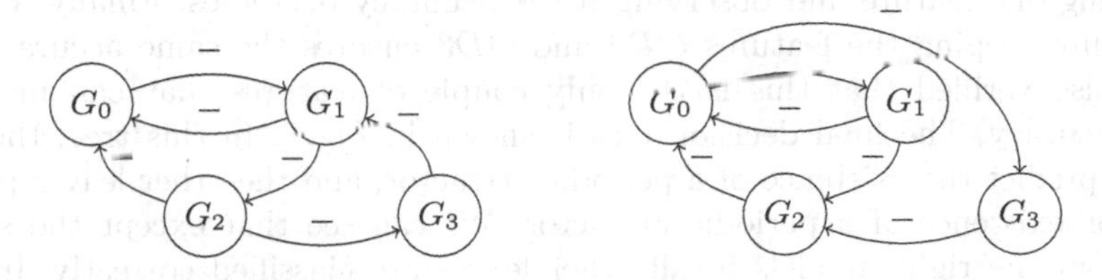

Fig. 5. Left: influence graph always having a periodic attractor. Right: influence graph never having a periodic attractor.

The accuracy of prediction of this decision tree is nearly 0.9990. Initially, we wished this accuracy to be 1 because in that case, the decision tree would provide a sufficient and necessary condition for the existence of a periodic attractor. This small lack of accuracy is actually caused by a few influence graphs with thresholds all related to the same influence graph. By analyzing this influence graph, a very interesting result arises:

There exists one particular influence graph such that, for any influence graph with thresholds that is not associated to this influence graph (or any isomorphism), the existence of a periodic attractor does not depend on the relations between thresholds and can be predicted by this decision tree with an accuracy of 1.

This particular influence graph is shown in Fig. 6. This figure also presents all influence graphs with thresholds, associated to this influence graph, having a periodic attractor. In fact, with the exception of the relation between the thresholds presented in this figure (the only arcs assigned with numbers), the relations between the thresholds of G_2 do not influence the existence of a periodic attractor, meaning that for any order of thresholds of G_2 added to this figure, it always has a periodic attractor.

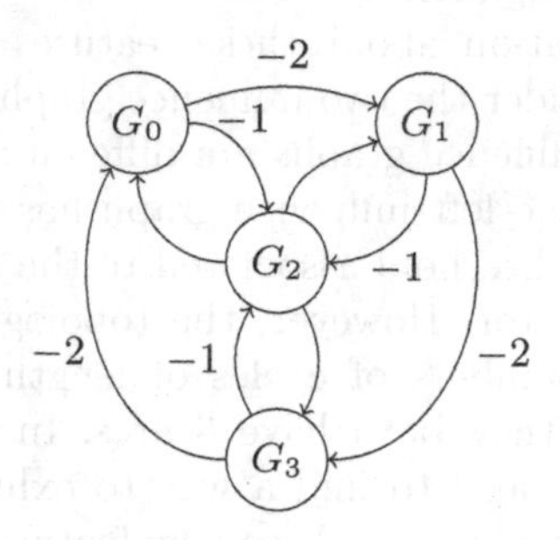

Fig. 6. The particular influence graph whose the relations between thresholds influence the existence of a periodic attractor. Amongst all influence graphs with thresholds associated to this influence graph, only a subset has a periodic attractor; this subset is characterized by the thresholds depicted in the figure.

In step 5, we manually drop features that do not influence the accuracy of the decision tree model. To do so, we re-train the decision tree each time after dropping one feature and observing if the accuracy decreases. Finally, we find that only keeping the features $OD2$ and $OD3$ ensures the same accuracy. We have also verified that this is the only couple of features that can maintain this accuracy. The final decision tree is shown in Fig. 7. In this tree, the blue leaves predict the existence of a periodic attractor, and the other leaves predict the non-existence of a periodic attractor. We can see that except the second leaf from the right, models in all other leaves are classified correctly. In fact, this second leaf from the right contains all models associated to the particular influence graph of Fig. 6. This means that apart from this particular influence graph, this tree describes a sufficient and necessary condition for the existence of a periodic attractor in 4-dimensional repressilators.

4.2 Condition Simplification

In this subsection, we compute a simplified sufficient and necessary condition for the existence of a periodic attractor based on the decision tree of Fig. 7. The four paths which end at blue leaves, which are the leaves related to the existence of a periodic attractor, are equivalent to the following logic rules:

$$\begin{aligned}(OD2 \leq 4 \wedge OD3 \leq 4 \wedge OD2 \leq 1 \wedge OD3 \leq 2 \wedge OD3 > 1 \wedge OD2 \leq 0) \vee \\ (OD2 \leq 4 \wedge OD3 \leq 4 \wedge OD2 \leq 1 \wedge OD3 > 2) \vee \\ (OD2 \leq 4 \wedge OD3 \leq 4 \wedge OD2 > 1 \wedge OD2 \leq 2 \wedge OD3 > 3) \vee \\ (OD2 \leq 4 \wedge OD3 > 4)\end{aligned} \tag{6}$$

Since $OD2$ and $OD3$ are integers, these logic rules can be simplified as follows.

$$\begin{aligned}(OD2 = 0 \wedge OD3 = 2) \vee \\ (OD2 \in \{0, 1\} \wedge OD3 \in \{3, 4\}) \vee \\ (OD2 = 2 \wedge OD3 = 4) \vee \\ (OD2 \in \{0, 1, 2, 3, 4\} \wedge OD3 \in \{5, 6, 7, ...\})\end{aligned} \tag{7}$$

Moreover, for all influence graphs, $OD2$ and $OD3$ are not independent and they are linked by the following constraints. Since there is a finite number of models, these constraints can be easily obtained by enumerating all models and comparing the values of $OD2$ and $OD3$.

$$\begin{aligned}\text{If } OD2 = 0 \text{ then } OD3 \in \{0, 1, 2, 3\} \\ \text{If } OD2 = 1 \text{ then } OD3 \in \{0, 1, 2, 3, 4\} \\ \text{If } OD2 = 2 \text{ then } OD3 \in \{0, 1, 2, 3, 4, 5\} \\ \text{If } OD2 = 3 \text{ then } OD3 \in \{0, 1, 2, 3\} \\ \text{If } OD2 = 4 \text{ then } OD3 \in \{0, 1, 2, 3, 4, 6, 8\}\end{aligned} \tag{8}$$

By combining Eq. 7 and Eq. 8, we get the following result, which is a sufficient and necessary condition for the existence of a periodic attractor in case that the influence graph is not equivalent to the one in Fig. 6:

$$\begin{aligned}(OD2 = 0 \wedge OD3 \in \{2, 3\}) \vee \\ (OD2 = 1 \wedge OD3 \in \{3, 4\}) \vee \\ (OD2 = 2 \wedge OD3 \in \{4, 5\}) \vee \\ (OD2 = 4 \wedge OD3 \in \{6, 8\})\end{aligned} \tag{9}$$

This result is of simple form and we can also find some patterns in it: the values of $OD2$ are powers of 2 $(2^0, 2^1, 2^2)$ except 0, and $OD3$ increases as $OD2$ increases. These patterns might lead to some general theoretical results for N-dimensional repressilators.

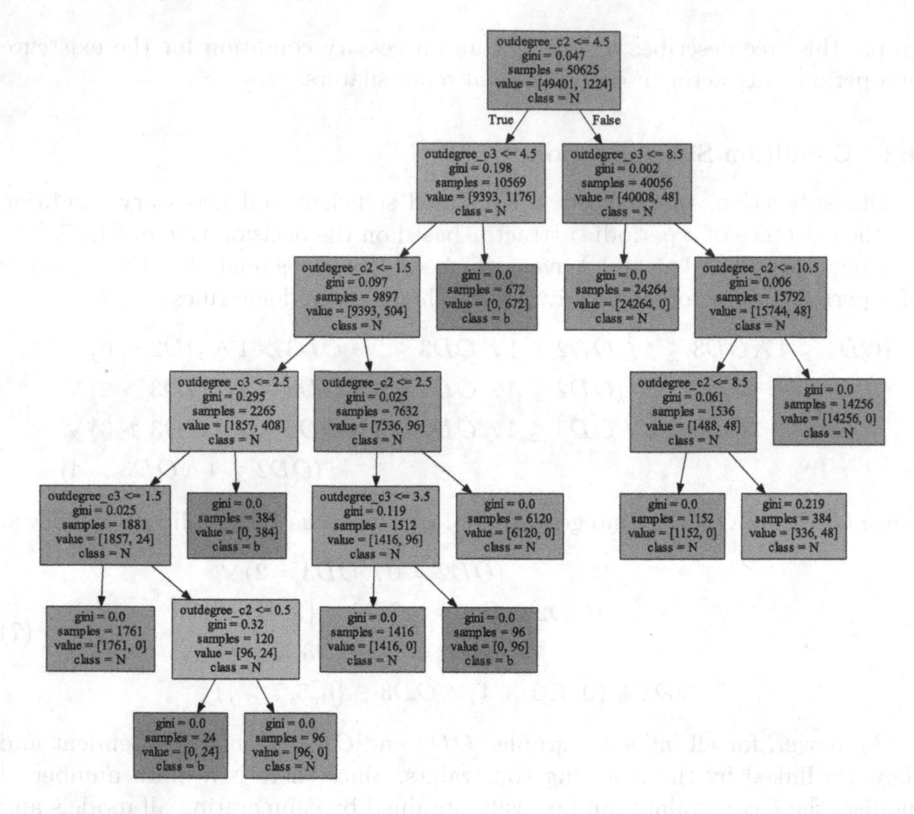

Fig. 7. A decision tree model to predict the existence of a periodic attractor. Blue leaves represent the models classified as having a periodic attractor and orange leaves represent the models classified as not having a periodic attractor. "gini" describes the purity of models in a node regarding the two classes considered here: models having a periodic attractor and models not having a periodic attractor; if all models in a node belong to the same class then gini = 0, otherwise gini > 0 ($gini = 1-(\frac{number_{class1}}{number_{total}})^2 - (\frac{number_{class2}}{number_{total}})^2$). "sample" represents the number of models in a node. The first value of "value" represents the number of models not having a periodic attractor and the second value of "value" represents the number of models having a periodic attractor. (Color figure online)

4.3 Number of Oscillatory Dimensions in a Periodic Attractor

In this subsection, we also investigate the number of oscillatory dimensions in the periodic attractors. For a periodic attractor, oscillating in 3 dimensions means that there exists one dimension i_0 and an integer a such that for any discrete state d_s in this periodic attractor, $d_s^{i_0} = a$, and for any dimension i which differs from i_0, we can find two discrete states d_{s1}, d_{s2} in this periodic attractor, such that $d_{s1}^i \neq d_{s2}^i$. An example of a periodic attractor that oscillates in 3 dimensions is given in Fig. 8 where there is no oscillation in the first dimension. Oscillating in 4 dimensions means that for any dimension i, we can find two discrete states

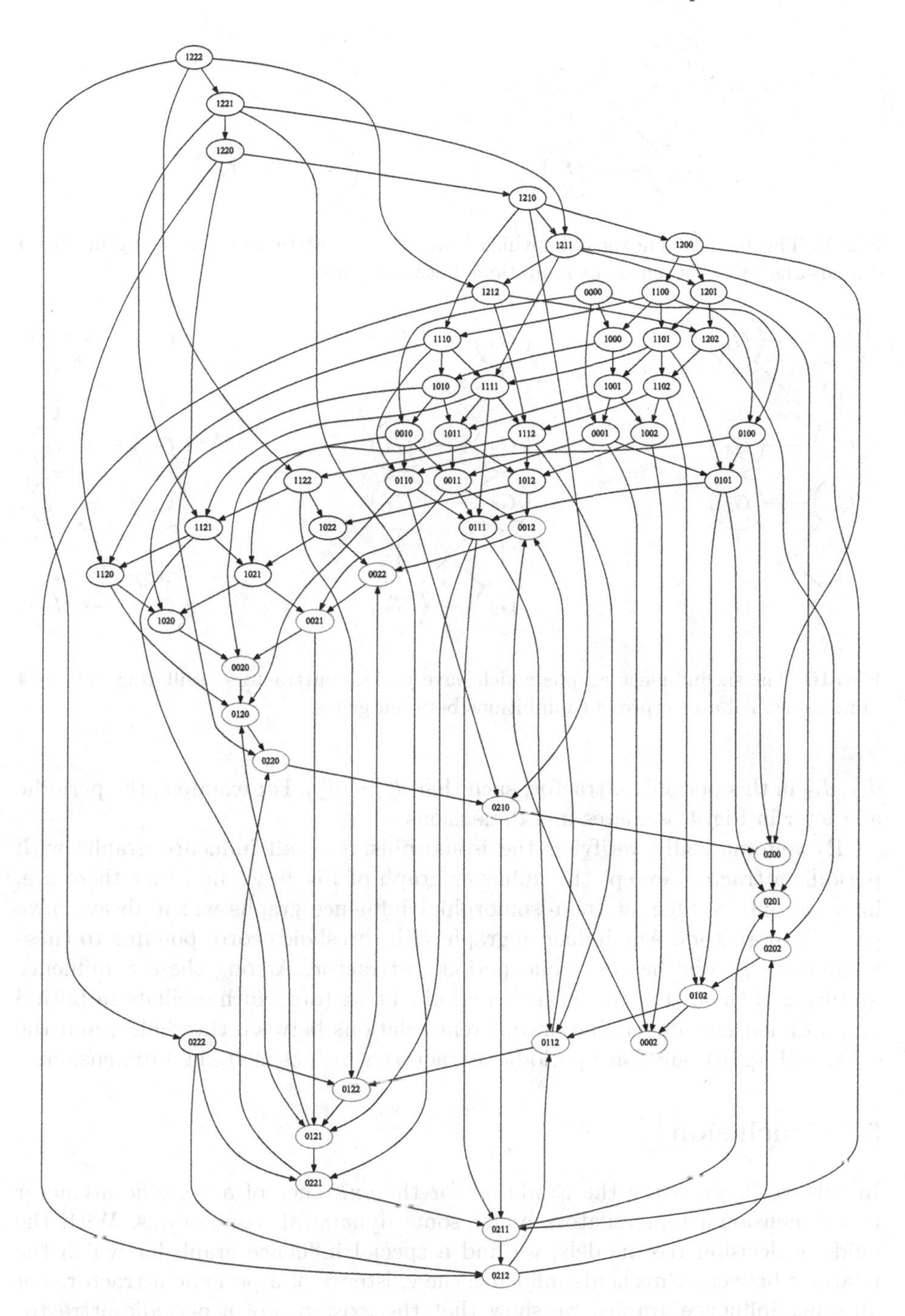

Fig. 8. Example of a periodic attractor (red discrete states) that oscillates in 3 dimensions. The discrete model corresponds to the left influence graph in Fig. 9, with $s(G_1 \to G_0) = s(G_2 \to G_0) = s(G_3 \to G_0) = 1$. (Color figure online)

Fig. 9. The two influence graphs which have periodic attractors oscillating in 3 or 4 dimensions. All arcs represent inhibitions between genes.

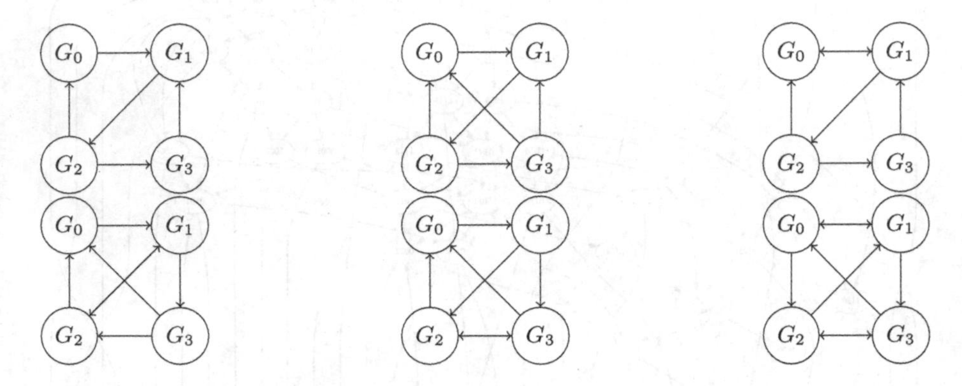

Fig. 10. The six influence graphs which have periodic attractors oscillating only in 4 dimensions. All arcs represent inhibitions between genes.

d_{s1}, d_{s2} in this periodic attractor, such that $d^i_{s1} \neq d^i_{s2}$. For example, the periodic attractor in Fig. 4 oscillates in 4 dimensions.

By automatically verifying the isomorphisms of all influence graphs with periodic attractors except the influence graph of Fig. 6, we find that there are, in total, only 8 different (non-isomorphic) influence graphs which always have periodic attractors. Any influence graph with thresholds corresponding to these 8 influence graphs has only one periodic attractor. Among these 8 influence graphs, 2 of them (Fig. 9) can have periodic attractors which oscillate in both 3 and 4 dimensions depending on different relations between thresholds, and the other 6 (Fig. 10) only have periodic attractors which oscillate in 4 dimensions.

5 Conclusion

In this work, we study the condition for the existence of a periodic attractor in 4-dimensional repressilators under some dynamical assumptions. With the guide of decision tree models, we find a special influence graph for which the relations between thresholds influence the existence of a periodic attractor. For all other influence graphs, we show that the existence of a periodic attractor does not depend on the relations between thresholds and we find a sufficient and necessary condition with a simple form, describing the topology of the influence graph, for the existence of a periodic attractor.

In this work, we use an exhaustive and computational approach to find this condition and we find some patterns in this condition. In our next step, we would like to prove this result in a more mathematical way, and try to extend this result for repressilators in N dimensions.

We also want to apply this method for other models of interest in biology, attempting to generalize this work to show that certain dynamic patterns are caused by particular topological features.

The topological feature used in this work, that is, the total out-degree of cycles of length n, could be potentially simplified based on other more common features. If it is possible, the simplified result could be more easily used for the design of new synthetic circuits.

Meanwhile, only the topology of influence graphs is considered in this paper. For future works, we will also investigate how the relations between thresholds influence some complex dynamical properties.

Finally, in this work, we only use a decision tree to guide the search of the condition. For future works, we will also try other learning methods based on logic programming which could be more adapted for this kind of problem.

Acknowledgements. We would like to thank Gilles Bernot and Jean-Paul Comet for their fruitful discussions. We would also like to thank Coraly Soto for her internship report which helped us in the early stage of this work to find features to predict the existence of a periodic attractor.

References

1. Abou-Jaoudé, W., Ouattara, D.A., Kaufman, M.: From structure to dynamics: frequency tuning in the p53-mdm2 network: I. logical approach. J. Theor. Biol. **258**(4), 561–577 (2009)
2. Akutsu, T., Kosub, S., Melkman, A.A., Tamura, T.: Finding a periodic attractor of a boolean network. IEEE/ACM Trans. Comput. Biol. Bioinf. **9**(5), 1410–1421 (2012)
3. Almeida, S., Chaves, M., Delaunay, F.: Control of synchronization ratios in clock/cell cycle coupling by growth factors and glucocorticoids. Royal Soc. Open Sci. **7**(2), 192054 (2020)
4. Barik, D., Baumann, W.T., Paul, M.R., Novak, B., Tyson, J.J.: A model of yeast cell-cycle regulation based on multisite phosphorylation. Molec. Syst. Biol. **6**(1), 405 (2010)
5. Behaegel, J., Comet, J.P., Bernot, G., Cornillon, E., Delaunay, F.: A hybrid model of cell cycle in mammals. J. Bioinf. Comput. Biol. **14**(01), 1640001 (2016)
6. Bernot, G., Comet, J.P., Khalis, Z.: Gene regulatory networks with multiplexes. In: European Simulation and Modelling Conference Proceedings, pp. 423–432 (2008)
7. Boyenval, D., Bernot, G., Collavizza, H., Comet, J.-P.: What is a cell cycle checkpoint? the `TotemBioNet` answer. In: Abate, A., Petrov, T., Wolf, V. (eds.) CMSB 2020. LNCS, vol. 12314, pp. 362–372. Springer, Cham (2020). https://doi.org/10.1007/978-3-030-60327-4_21
8. Buşe, O., Pérez, R., Kuznetsov, A.: Dynamical properties of the repressilator model. Phys. Rev. E **81**(6), 066206 (2010)

9. Comet, J.-P., Fromentin, J., Bernot, G., Roux, O.: A formal model for gene regulatory networks with time delays. In: Chan, J.H., Ong, Y.-S., Cho, S.-B. (eds.) CSBio 2010. CCIS, vol. 115, pp. 1–13. Springer, Heidelberg (2010). https://doi.org/10.1007/978-3-642-16750-8_1
10. Cornillon, E., Comet, J.P., Bernot, G., Enée, G.: Hybrid gene networks: a new framework and a software environment. Adv. Syst. Synth. Biol. (2016)
11. Elowitz, M.B., Leibler, S.: A synthetic oscillatory network of transcriptional regulators. Nature **403**(6767), 335–338 (2000)
12. Goh, K.I., Kahng, B., Cho, K.H.: Sustained oscillations in extended genetic oscillatory systems. Biophys. J. **94**(11), 4270–4276 (2008)
13. Karlebach, G., Shamir, R.: Modelling and analysis of gene regulatory networks. Nat. Rev. Molec. Cell Biol. **9**(10), 770–780 (2008)
14. Kauffman, S.A.: Metabolic stability and epigenesis in randomly constructed genetic nets. J. Theor. Biol. **22**(3), 437–467 (1969)
15. Khalis, Z., Comet, J.P., Richard, A., Bernot, G.: The smbionet method for discovering models of gene regulatory networks. Genes Genom. Genomics **3**(1), 15–22 (2009)
16. Li, Z., Liu, S., Yang, Q.: Incoherent inputs enhance the robustness of biological oscillators. Cell Syst. **5**(1), 72–81 (2017)
17. Ma, W., Lai, L., Ouyang, Q., Tang, C.: Robustness and modular design of the drosophila segment polarity network. Molec. Syst. Biol. **2**(1), 70 (2006)
18. Melkman, A.A., Tamura, T., Akutsu, T.: Determining a singleton attractor of an and/or boolean network in o (1.587 n) time. Inf. Process. Lett. **110**(14–15), 565–569 (2010)
19. Page, K.M.: Oscillations in well-mixed, deterministic feedback systems: beyond ring oscillators. J. Theor. Biol. **481**, 44–53 (2019)
20. Paulevé, L., Kolčák, J., Chatain, T., Haar, S.: Reconciling qualitative, abstract, and scalable modeling of biological networks. Nat. Commun. **11**(1), 4256 (2020)
21. Paulevé, L., Richard, A.: Static analysis of Boolean networks based on interaction graphs: a survey. Electron. Notes Theor. Comput. Sci. **284**, 93–104 (2012)
22. Perez-Carrasco, R., Barnes, C.P., Schaerli, Y., Isalan, M., Briscoe, J., Page, K.M.: Combining a toggle switch and a repressilator within the ac-dc circuit generates distinct dynamical behaviors. Cell Syst. **6**(4), 521–530 (2018)
23. Potvin-Trottier, L., Lord, N.D., Vinnicombe, G., Paulsson, J.: Synchronous long-term oscillations in a synthetic gene circuit. Nature **538**(7626), 514–517 (2016)
24. Qiao, L., Zhao, W., Tang, C., Nie, Q., Zhang, L.: Network topologies that can achieve dual function of adaptation and noise attenuation. Cell Syst. **9**(3), 271–285 (2019)
25. Remy, É., Ruet, P., Thieffry, D.: Graphic requirements for multistability and attractive cycles in a boolean dynamical framework. Adv. Appl. Math. **41**(3), 335–350 (2008)
26. Ribeiro, T., Folschette, M., Magnin, M., Inoue, K.: Learning any semantics for dynamical systems represented by logic programs (2020)
27. Richard, A.: Negative circuits and sustained oscillations in asynchronous automata networks. Adv. Appl. Math. **44**(4), 378–392 (2010)
28. Richard, A., Comet, J.P.: Necessary conditions for multistationarity in discrete dynamical systems. Disc. Appl. Math. **155**(18), 2403–2413 (2007)
29. Richard, A., Tonello, E.: Attractor separation and signed cycles in asynchronous boolean networks. Theor. Comput. Sci., 113706 (2023)

30. Sun., H., Comet., J., Folschette., M., Magnin., M.: Condition for sustained oscillations in repressilator based on a hybrid modeling of gene regulatory networks. In: Proceedings of the 16th International Joint Conference on Biomedical Engineering Systems and Technologies - BIOINFORMATICS, pp. 29–40. INSTICC, SciTePress (2023). https://doi.org/10.5220/0011614300003414
31. Sun, H., Folschette, M., Magnin, M.: Limit cycle analysis of a class of hybrid gene regulatory networks. In: Computational Methods in Systems Biology: 20th International Conference, CMSB 2022, Bucharest, Romania, 14–16 September 2022, Proceedings, pp. 217–236. Springer, Heidelberg (2022). DOI: September
32. Thomas, R.: Boolean formalization of genetic control circuits. J. Theor. Biol. **42**(3), 563–585 (1973)
33. Thomas, R.: Regulatory networks seen as asynchronous automata: a logical description. J. Theor. Biol. **153**(1), 1–23 (1991)
34. Tomazou, M., Barahona, M., Polizzi, K.M., Stan, G.B.: Computational re-design of synthetic genetic oscillators for independent amplitude and frequency modulation. Cell Syst. **6**(4), 508–520 (2018)
35. Zhang, F., et al.: Independent control of amplitude and period in a synthetic oscillator circuit with modified repressilator. Commun. Biol. **5**(1), 23 (2022)

Attractor Identification in Asynchronous Boolean Dynamics with Network Reduction

Elisa Tonello[1(✉)] and Loïc Paulevé[2]

[1] Freie Universität Berlin, Berlin, Germany
elisa.tonello@fu-berlin.de

[2] University of Bordeaux, CNRS, Bordeaux INP, LaBRI, UMR 5800, 33400 Talence, France
loic.pauleve@labri.fr

Abstract. Identification of attractors, that is, stable states and sustained oscillations, is an important step in the analysis of Boolean models and exploration of potential variants. We describe an approach to the search for asynchronous cyclic attractors of Boolean networks that exploits, in a novel way, the established technique of elimination of components. Computation of attractors of simplified networks allows the identification of a limited number of candidate attractor states, which are then screened with techniques of reachability analysis combined with trap space computation. An implementation that brings together recently developed Boolean network analysis tools, tested on biological models and random benchmark networks, shows the potential to significantly reduce running times.

Keywords: Boolean networks · Attractors · Reduction · Trap spaces

1 Introduction

Boolean networks are adopted as modelling tools to organise knowledge and explore possible behaviours emerging in biological processes [5,20,27]. From the logic describing the influence between species, dynamics are defined to express the evolution of variables at play. Update schemes that implement asynchrony are of particular interest, as they express one form of inherent stochasticity [22]. Attractors are fundamental structures that ought to capture the fate or stable behaviours of modelled systems. Recently, the topic of identification of attractors of asynchronous dynamics has seen renewed interest and several developments. To look at the last two years alone, [17] suggested an algorithm that combines different techniques such as motif detection and time reversal; [1,2] developed a symbolic approach that can handle large transition systems, adopting binary decision diagrams to represent Boolean networks, and supporting partially defined update functions; [24,25] described approaches based on feedback vertex set identification and model checking reachability analysis. The new

J. Pang and J. Niehren (Eds.): CMSB 2023, LNBI 14137, pp. 202–219, 2023.
https://doi.org/10.1007/978-3-031-42697-1_14

techniques have enabled the handling of models of increasing complexity, as summarized in [24]. Additional references and an overview of approaches to attractor identification for some classes of synchronous and asynchronous Boolean networks can be found in the mini review [7].

With this work we want to investigate the usefulness of network reduction in the investigation of attractors of asynchronous state transition graphs. We refer specifically to the popular reduction method that consists in the iterative elimination of non-autoregulated components, as described in [10,11,26]. When a variable is selected for elimination, the regulatory functions of its targets are changed to remove the selected variable and replace it with its regulatory function. The asynchronous dynamics and regulatory structures of the original network and the network obtained with this reduction process are related in a useful way; for instance, the networks have the same number of steady states, and the number of attractors of the reduced network cannot be smaller than the original one. Properties of network reduction have already been exploited for attractor identification. [18,19] suggested the removal of special "mediator" nodes, nodes with only one incoming and one outgoing edge; even when limited to this simple case, however, variable elimination can change the number of attractors [21] and therefore require further analysis for their correct identification. A different type of reduction that has been proposed for attractor detection is based on the iterative calculation of stable motifs [28], and is not guarantee to identify all the attractors. The approach has been extended in [17] and combined with other techniques, including variable elimination to find upper bounds on the number of attractors. In this work, we show that a systematic use of variable elimination, adopted as the first step of the network analysis, can significantly reduce running times for attractor identification when compared to state-of-the-art tools, while providing information on the nature of attractors and the relationship between the attractors of the full model and of simplified versions. With the approach we suggest, elimination can be applied to any component that is not autoregulated, without restrictions on the number of targets or regulators. From the attractors of reduced versions, we investigate the identification of attractor states of the original network using information on trap spaces and reachability analysis.

More in detail, the standing of variable elimination as a useful tool for identification of asynchronous attractors comes from the following property. For each attractor $\mathcal{A}$ of a Boolean network f, if $\tilde{f}$ is obtained from f by iteratively eliminating some non-autoregulated components, there exists at least one attractor of $\tilde{f}$ from which states in $\mathcal{A}$ can be reconstructed, by tracing the reduction process backwards. This property allows us to identify, from the attractors of $\tilde{f}$, a limited set of "candidate" states that cover all the attractors of f (Sect. 2.2). Then, a screening of those candidate states is required for filtering out those outside the attractors of f. By calculating the minimal trap spaces [6,23] we can first check if a candidate state is part of an attractor which is contained in a minimal trap space, and is the sole attractor contained in that minimal trap space. If there is more than one candidate attractor in a minimal trap space, or if there are candidate states outside of minimal trap spaces, other checks are required to determine whether they are actually part of an attractor. More computationally demanding techniques, for example from model checking, are useful for this step [14,25] to check if a previously identified attractor, or a trap space not containing the candidate state, can be reached from the candidate state.

We tested these ideas on both biological and random networks by implementing variable elimination using `colomoto`'s `minibn` [9]. The reduced network were then studied using AEON [1] and mtsNFVS [24], to identify candidate attractor states. We used `trappist` to find the minimal trap spaces [23] and enable the first screening of the candidate states. For the check of the remaining candidate states we used mtsNFVS's reachability analysis software [24]. We found, in general, high potential for reduction of computational times (Sect. 3.2). In addition, we observed that the reachability check, which is the most computationally expensive task in the pipeline, needs to be invoked only occasionally. In fact, although variable elimination can lead to an increased number of attractors, this appears to happen in a quite limited number of cases, meaning that the number of candidate states is often very close, if not equal, to the number of attractors. As a result, the amount of work necessary to handle such situations is also contained. In the last section (Sect. 4) we comment on some avenues that could be explored for possible further improvements of the techniques discussed.

2 Background

2.1 Boolean Networks

We call V the set of variables or species of interest, and set $n = |V|$. A Boolean network is defined by a map $f = (f_1, \ldots, f_n)\colon 2^n \to 2^n$. The set 2^n is called the set of states or configurations of the Boolean network. For $i = 1, \ldots, n$ and $x \in 2^n$, we denote by $\bar{x}^i$ the state that coincides with x on all components except i. A *subspace* is a subset of the state space consisting of all the states that share the same values for a set of components. More formally, $S \subseteq 2^n$ is a subspace or subcube if there exists a state $x \in 2^n$ and a subset of components I such that $S = \{y \in 2^n \mid y_i = x_i \text{ for all } i \in I\}$. In this case, the set I is called the set of *fixed* variables of the subspace. For instance, for $n = 2$, the set $\{00, 01\}$ is a subspace, consisting of all the states with first component equal to zero. The set of fixed variables of this subspace is $\{1\}$. The set $\{00, 01, 10\}$, for example, is not a subspace.

The Boolean function f_i, $i = 1, \ldots, n$, is sometimes called the *update function* of variable i. The *influence graph* $G(f)$ of f is the signed multi-digraph with set of nodes $\{1, \ldots, n\}$ and edges capturing the dependence of update functions on each variable: there exist an edge (or *regulation*) (i, j) in $G(f)$ with sign s if and only if, for some $x \in 2^n$, $f_j(\bar{x}^i) \neq f_j(x)$ and $s = (f_j(\bar{x}^i) - f_j(x))(\bar{x}^i_i - x_i)$. The influence graph can admit parallel edges of different signs. If (i, j) is an edge in $G(f)$, j is called a target of i and i is called a regulator of j.

A *state transition graph* or *dynamics* associated to f is a graph with set of vertices 2^n and set of edges (also called *transitions*) that depends on the chosen semantics. This work deals with the *asynchronous dynamics*, a form of non-deterministic dynamics which includes only transitions between states differing by one component. Specifically, an edge exists from a state x to a state $\bar{x}^i$ if and only if $f_i(x) \neq x_i$. We will write $\Gamma(f)$ to denote the asynchronous state transition graph of f. Examples of asynchronous dynamics are depicted in Fig. 1. The following are central definitions in this work.

Definition 1. A subset $A \subseteq 2^n$ is a *trap set* for the asynchronous dynamics $\Gamma(f)$ of f if, for all transitions (x, y) in $\Gamma(f)$, $x \in A$ implies $y \in A$. A trap set A is called an *attractor* of f if, for all trap sets $A' \subseteq 2^n$, $A' \subseteq A$ implies $A' = A$.

Attractors are therefore subsets that do not admit any outgoing transition, and are minimal with respect to inclusion. They can consist of singleton states, which are called *steady states* or *fixed points*, or can involve more than one state, in which case they are referred to as *cyclic* or *complex attractors*.

Definition 2. A subspace S is *trap space* for the asynchronous dynamics $\Gamma(f)$ of f if it is a trap set. S is a *minimal* trap space, if, for all trap spaces T, $T \subseteq S$ implies $T = S$.

Consider for instance the 2-dimensional asynchronous dynamics $\Gamma(f)$ in Fig. 1, left. The subspace $\{00, 01\}$ is not a trap space, since an outgoing transition exists (from 00 to 10). On the other hand, the set $\{00, 01, 10\}$ is a trap set, albeit not minimal. It contains the set $\{00, 10\}$, which is a subspace and a trap space, as well as the unique attractor.

Determining whether a given state belongs to an attractor is a PSPACE-complete problem with asynchronous or synchronous dynamics [15]. In practice, computations usually rely on computing (partly and symbolically) the state transition graph, which can be significantly more complex in the asynchronous than in the synchronous case. Compared to attractor identification, computation of trap spaces and related properties is a much more tractable problem, due to a largely reduced theoretical complexity: determining whether a subspace is a minimal trap space is Π_2^{P}-complete or coNP-complete depending on the representation of the update functions f_i [6]. Minimal trap spaces are particularly interesting, since each of them must contain at least one attractor. Often minimal trap spaces of Boolean networks that serve as biological models contain only one attractor [3], meaning, for instance, that they can replace the attractor they contain as reachability targets in reachability analysis. Although some properties of the regulatory structure that can guarantee a "good" attractor landscape (attractors only in minimal trap spaces, uniqueness of attractors in minimal trap spaces) have been identified [12,16], these have limited application. In general, establishing whether a Boolean network admits multiple attractors in a minimal trap space, or attractors outside of minimal trap spaces (the "motif-avoidant" attractors of [17]), remains a difficult task.

Based on these considerations, we can lay out the following classification of attractors of Boolean asynchronous dynamics into four categories, which will be useful in our discussion later:

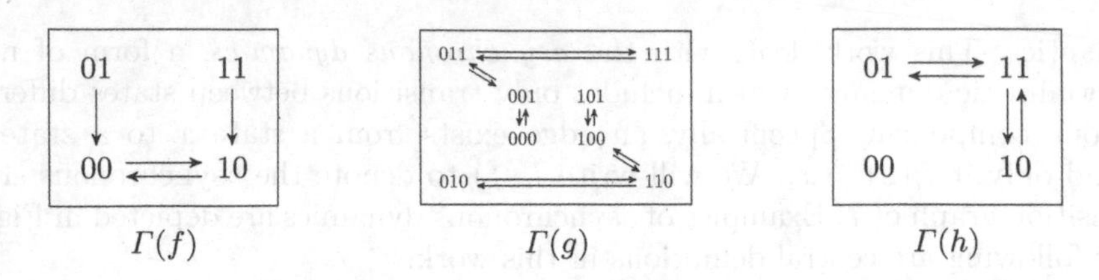

Fig. 1. Asynchronous state transition graphs of the Boolean networks f, g and h of Example 1. $\Gamma(f)$ has one minimal univocal attractor, $\Gamma(g)$ has two minimal nonunivocal attractors, and $\Gamma(h)$ has a minimal univocal and a nonminimal attractor.

(A) *steady states*: these are "easy" to find. They coincide with the minimal trap spaces where all variables are fixed variables.
(B) *minimal univocal*: by definition, each minimal trap space must contain at least one attractor. If a minimal trap space contains only one attractor, we say that the trap space and the attractor are *univocal*, using the term introduced in [3].
(C) *minimal nonunivocal*: if a minimal trap space contains more than one attractor, we call these *nonunivocal.*
(D) *nonminimal* or *motif-avoidant* [17]: these are found outside of minimal trap spaces and are the most difficult to detect or exclude the existence of.

In the examples, to lighten the notation, the symbol for **and** takes precedence and is omitted.

Example 1. The asynchronous state transition graphs of the maps

$$\begin{aligned} f(x_1, x_2) &= (x_1x_2 \vee \bar{x}_1\bar{x}_2, 0), \\ g(x_1, x_2, x_3) &= (x_2\bar{x}_1 \vee x_1\bar{x}_2, x_1(x_2x_3 \vee \bar{x}_2\bar{x}_3) \vee \bar{x}_1(x_2\bar{x}_3 \vee x_3\bar{x}_2), x_2x_3 \vee \bar{x}_2\bar{x}_3), \\ h(x_1, x_2) &= (x_1\bar{x}_2 \vee \bar{x}_1x_2, x_1\bar{x}_2 \vee \bar{x}_1x_2) \end{aligned}$$

are represented in Fig. 1.

$\Gamma(f)$ has a cyclic attractor and minimal trap space $\{00, 10\}$.

$\Gamma(g)$ has two attractors that are nonunivocal, since they are found in the same minimal trap space (the full space).

$\Gamma(h)$ has two attractors, one steady state 00 and one nonminimal attractor $\{01, 10, 11\}$. The steady state is the unique minimal trap space.

In the next section we review how the removal of components works in Boolean networks and discuss how it can help with identification of attractors.

2.2 Reduction

A popular reduction method for asynchronous dynamics of Boolean networks iteratively eliminates variables that are not autoregulated [10,11,26]. The approach has been recently extended to negatively autoregulated components [21].

Although all our observations here can be extended, *mutatis mutandis*, to negatively autoregulated components, we only discuss the standard case for sake of simplicity.

Suppose that $G(f)$ has a node that does not have a loop. Without loss of generality, we can assume that the node is n. We write $\pi\colon 2^n \to 2^{n-1}$ for the projection on the first $n-1$ variables.

By definition of $G(f)$, we have that $f_n(x,0) = f_n(x,1)$ for all $x \in 2^{n-1}$. In particular, the state transition graph of f admits exactly one of the transitions from $(x,0)$ to $(x,1)$ or from $(x,1)$ to $(x,0)$. We can therefore define a map $\mathcal{S}^n$ that associates to a "reduced" state a state in the larger space, as follows:

$$\begin{aligned} \mathcal{S}^n\colon 2^{n-1} &\to 2^n \\ x &\mapsto (x, f_n(x,0)) = (x, f_n(x,1)). \end{aligned}$$

The reduction $\tilde{f}$ of f obtained by elimination of component n is then defined as

$$\tilde{f} = (f_1 \circ \mathcal{S}^n, \ldots, f_{n-1} \circ \mathcal{S}^n)\colon 2^{n-1} \to 2^{n-1}.$$

Each update function f_i is therefore changed by replacing every occurrence of the last variable with its update function f_n. This simple operation is not without consequences on the reachability properties of the state transition graph. Given $y \in 2^n$, the state $\mathcal{S}^n(\pi(y))$ can be thought of as the "representative state" of the pair $(\pi(y), 0)$, $(\pi(y), 1)$, or the state that "survives the reduction", since all transitions leaving the state $\mathcal{S}^n(\pi(y))$ have a corresponding transition in $\Gamma(\tilde{f})$, whereas the transitions leaving the state $\overline{\mathcal{S}^n(\pi(y))}^n$ are not guaranteed to be preserved [11]. The following example shows a very simple occurrence of this phenomenon (other instances, as well as possible consequences on the attractor landscape, can be observed in other examples later in this section).

Example 2. Consider the Boolean network $f(x_1, x_2) = (x_1 \vee \bar{x}_2, 1)$, and the elimination of the second variable, which is not autoregulated. The state 01 (resp. 11) is the "representative state" for the pair 00, 01 (resp. 10, 11). The definition of $\tilde{f}$ above reads $\tilde{f}(x_1) = f_1(x_1, f_2(x_1, 1)) = f_1(x_1, 1) = x_1$. The state transition graph of f has a transition involving a change in the first variable, from the non-representative state 00 to 10, whereas $\Gamma(\tilde{f})$ has no transition.

It was shown in [10,11] that f and $\tilde{f}$ admit the same number of steady states, and the number of attractors of $\tilde{f}$ is greater or equal to the number of attractors of f. The latter property does not hold for the synchronous dynamics, as can be seen by considering for example the network $f(x_1, x_2) = (x_2, x_1)$. Hence, the techniques we describe here are not transferable to the synchronous case. The following result, which forms the basis for our method, is a simple consequence of properties proved in [11].

Theorem 1. *If $\mathcal{A}$ is an attractor of f, then there exists at least one attractor for $\tilde{f}$ in $\pi(\mathcal{A})$, and for each $x \in \pi(\mathcal{A})$ contained in an attractor of $\tilde{f}$, $\mathcal{S}^n(x) \in \mathcal{A}$.*

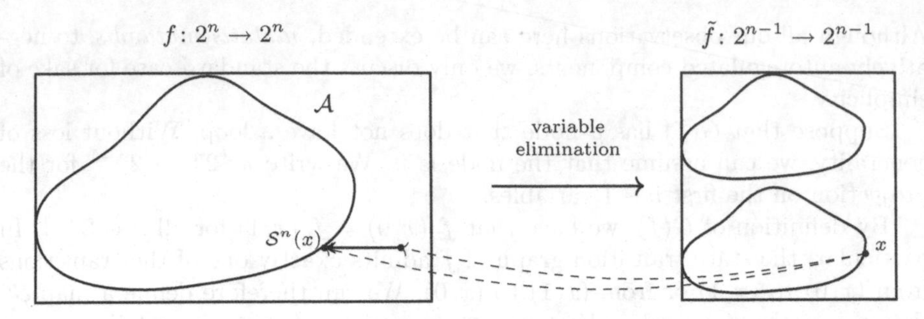

Fig. 2. Idea behind the approach: states in attractors of reduced networks $\tilde{f}$ can be used to find candidate states in attractors of f.

Proof. The first part is a consequence of the observation that, if B is a trap set for $\Gamma(f)$, then $\pi(B)$ is a trap set for $\Gamma(\tilde{f})$.

Given a state x in an attractor for $\tilde{f}$ contained in $\pi(\mathcal{A})$, by definition of π either $\mathcal{S}^n(x)$ or $\overline{\mathcal{S}^n(x)}^n$ is in $\mathcal{A}$, and since $\mathcal{S}^n(x)$ is reachable from $\overline{\mathcal{S}^n(x)}^n$, $\mathcal{S}^n(x)$ must be in $\mathcal{A}$. □

The theorem gives us the following property: for each attractor for f, we are able to identify one state of this attractor by finding the attractors of the reduced version $\tilde{f}$, sampling a state from each of these attractors, and "lifting" the states to the original space using $\mathcal{S}^n$ (Fig. 2).

Before we discuss more in detail how we can use the reduced network to identify attractors of the original network, let us look at some examples, which can give an idea of what can happen to attractors with variable elimination.

Example 3. Consider the map f from Example 1. Since the second variable is not autoregulated, it can be removed using the elimination method described in this section. The constant value 0 replaces x_2 in the update function of x_1, giving the reduced network $\tilde{f}(x_1) = \bar{x}_1$. We have $\mathcal{S}^2(0) = 00$ and $\mathcal{S}^2(1) = 10$. This simple example is sufficient to illustrate how, given a state x in the attractor of a reduced network (say $x = 0$ here), to retrieve a state in an attractor of the original network we cannot pick any state in the preimage $\pi^{-1}(x)$ (the state 01 does not work), but we need to apply the function $\mathcal{S}^2$.

Example 4. If $\mathcal{A}$ is an attractor for a Boolean network and x is a state in $\mathcal{A}$, does $\pi(x)$ necessarily belong to an attractor of the reduced network? The answer is negative. Take for instance

$$\hat{f}(x_1, x_2, x_3) = (\bar{x}_1\bar{x}_2 \vee \bar{x}_1\bar{x}_3 \vee x_2\bar{x}_3, x_1\bar{x}_2\bar{x}_3 \vee \bar{x}_1\bar{x}_2x_3, x_1\bar{x}_2 \vee \bar{x}_1x_2).$$

By removing the third component, we obtain the function f of Example 1. The state 011 is in the unique attractor of $\hat{f}$, while $\pi(011) = 01$ does not belong to any attractor of f.

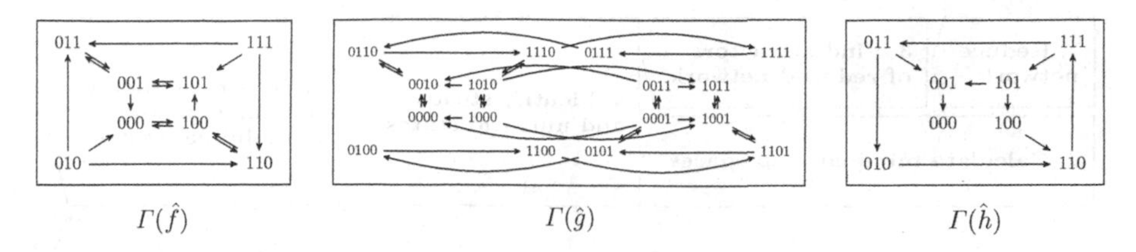

Fig. 3. Asynchronous state transition graphs of the Boolean networks $\hat{f}$, $\hat{g}$, $\hat{h}$ of Examples 5 to 4.

In general, therefore, we are not able to retrieve all states of an attractor of a Boolean network by lifting states in attractors of its reduction. We will only find some states, from which we can visit the attractor if required.

Example 5. The Boolean network

$$\hat{g}(x_1,x_2,x_3,x_4) = (x_2\bar{x}_4 \vee \bar{x}_2 x_4, x_4(x_2x_3 \vee \bar{x}_2\bar{x}_3) \vee \bar{x}_4(x_2\bar{x}_3 \vee x_3\bar{x}_2), x_2x_3 \vee \bar{x}_2\bar{x}_3, x_1)$$

has one cyclic attractor, that fills the whole state space (see Fig. 3). By removing variable x_4 we obtain the network g in Example 1, which has two attractors.

Example 6. The Boolean network

$$\hat{h}(x_1,x_2,x_3) = (x_1\bar{x}_3 \vee x_2\bar{x}_3, x_1\bar{x}_3 \vee x_2\bar{x}_3, x_1x_2)$$

has a unique attractor, the fixed point 000. By eliminating variable x_3 we obtain the Boolean network h in Example 1, which has two attractors.

Suppose that $\tilde{f}\colon 2^m \to 2^m$ is obtained from f by iteratively eliminating variables $n, n-1, \ldots, m+1$, and that $\mathcal{A}_1, \ldots, \mathcal{A}_M$ are attractors of $\tilde{f}$. Take one state $x^1, \ldots, x^M \in 2^m$ in each attractor. We can reconstruct the corresponding states in 2^n by applying the map $\mathcal{S} = \mathcal{S}^n \circ \mathcal{S}^{n-1} \circ \cdots \circ \mathcal{S}^{m+1}$, obtaining a set of candidate states $C = \{\mathcal{S}(x^1), \ldots, \mathcal{S}(x^M)\}$. How can we establish whether each of these states is in an attractor of f, and how many attractors f has?

We first make the following observation.

Corollary 1. *Let T be a trap space for f that contains k states of C. Then T contains at most k attractors for f.*

Proof. By considering the restriction of the dynamics of f to T, that is, the subgraph of $\Gamma(f)$ induced by the nodes in T, we can assume that T is the full state space. Then k is the number of attractors of $\tilde{f}$, and the conclusion follows from the fact that the number of attractors of $\tilde{f}$ is greater or equal to the number of attractors of f [11]. □

Consider $x \in \{x^1, \ldots, x^M\}$.

(a) If x is a steady state of $\tilde{f}$, then $\mathcal{S}(x)$ is a steady state of f (and all steady states of f can be calculated in this fashion).

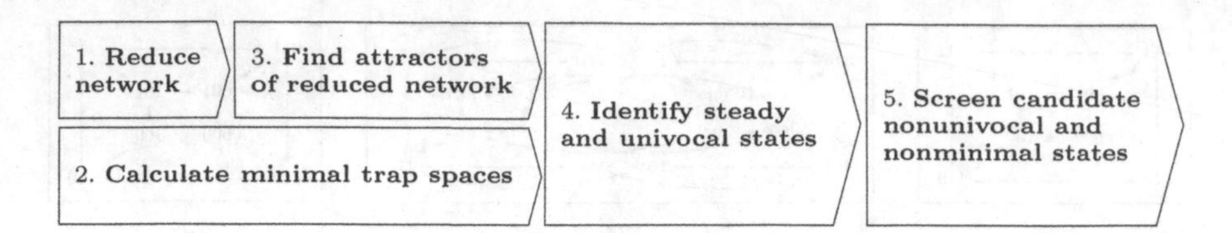

Fig. 4. Main steps of the algorithm.

(b) If $\mathcal{S}(x)$ belongs to a minimal trap space T of f, and is the only state in C that belongs to T, then, by Corollary 1, T contains only one attractor, meaning that $\mathcal{S}(x)$ belongs to an attractor of f that is minimal univocal. In this case we call $\mathcal{S}(x)$ a *univocal* state.

(c) If $\mathcal{S}(x)$ is contained in a minimal trap space T of f, and is not the only state in C that is contained in this minimal trap space, then the number of states in C contained in T gives an upper bound on the number of attractors contained in T (Corollary 1). We need to study the dynamics in T to clarify whether each candidate state contained in T belongs to an attractor, and whether the candidate states in T belong to different attractors. We can call these states *candidate nonunivocal.*

(d) If $\mathcal{S}(x)$ is not contained in any minimal trap space, then $\mathcal{S}(x)$ *might* belong to a nonminimal attractor of f. To establish whether this is the case, and to find the number of nonminimal attractors, we need to do additional work. We refer to these states as *candidate nonminimal.*

Note that these observations would have to be slightly changed were one to consider the elimination of negatively autoregulated components [21]. We can now give the description of the approach to the identification of attractors based on elimination of components.

3 Method

Based on the analysis of the relationship between attractors of reduced versions of a Boolean network and attractors of the original network, we propose the following pipeline (see Fig. 4).

First reduce the network f by eliminating variables (Step 1), and, possibly in parallel, find the minimal trap spaces of the Boolean network f (Step 2). Then, identify one state for each attractor of the reduced network $\tilde{f}$ that is not a steady state (Step 3), obtaining a set of candidate attractor states for f.

Steps 4 and 5 deal with the screening of these candidate attractor states. Step 4 is the easy part: we check whether each candidate state is contained in a minimal trap space. If a minimal trap space contains only one candidate state, then this is a univocal state and we identified a univocal attractor for f (point (b) of the last section).

Step 5 takes care of the remaining candidate states. These can be either multiple states contained in the same minimal trap space (point (c) of the last section), or states that do not belong to any minimal trap space (point (d) of the last section). In both scenarios, we have to study in some way the state transition graph to understand whether the candidate states belong to an attractor, and how many attractors exist (some candidate states might be part of the same attractor, which has been "split" during the reduction process, like in Example 5, or part of an attractor that was created by the reduction process as in Example 6). For this step one can use model checking approaches (see the next section), possibly combined with other techniques (see the discussion in the last section).

3.1 Implementation

Software. For our implementation, we use the Python library `colomoto`'s `minibn` [9] to compute the network reduction (Step 1 in Fig. 4). In `minibn`, the update functions of the Boolean network are represented in propositional logic formula, with usual Boolean algebra. Given a component i to reduce, our implementation substitutes x_i with the expression of f_i in all its targets. Then, basic Boolean expression simplifications are applied, which may result in variable elimination.

For Step 2, we use `trappist` to calculate the minimal trap spaces [23]. Because `trappist` relies on a Petri net representation of the Boolean network, this step involves a transformation of each local Boolean function f_i in two expressions in DNF form (one for $f_i(x) = 1$ and one for $f_i(x) = 0$), from which the Petri net transitions are derived.

For Step 3, the identification of attractors of the reduced network, we consider two recently developed methods, AEON [1] and mtsNFVS [24]. In the analysis of [24], they have been shown to be the two fastest methods available, while implementing two very different approaches. The outputs of the two methods are also quite different: AEON, for instance, can give information on the attractors (cardinality, list of states) that is not directly available with mtsNFVS, and can deal with partially specified networks, do control tasks, etc. Here we are only looking at the performance of the methods in regard to the identification of the number of attractors and their nature (steady or cyclic), and we are mostly concerned with understanding whether reduction might be useful in this respect.

The pieces are put together and Step 4 performed in `python`. For Step 5, we run the `java` tool `mtsNFVS.jar` available as part of mtsNFVS. The tool uses `Pint` [14] for a first check via static analysis followed by bounded and exact model checking [25]. We noticed however that mtsNFVS (without reduction) produces wrong results in several instances (some examples are provided in our project repository). Hence, the results obtained with mtsNFVS should be taken with caution. For the biological networks we considered, we checked that all methods found the same number of steady states and cyclic attractors. As we will discuss in the Results section, the reachability analysis step has limited impact on the overall process, since nonminimal candidates appear only rarely, and no nonunivocal candidates are found in any of the biological or random tests.

Elimination Order and Other Considerations. The choice of order for the variables to be eliminated has an influence on the reduced network as well as on the running times. Suppose that variable i is removed, and that j is a target of i. By substituting x_i with f_i in the update function f_j, the regulator i of j is replaced by the r regulators of j, unless simplifications lead to the elimination of some variables. This means that each pair of regulations (k, i) and (i, j) might be replaced in the reduced network by a regulation (k, j). If i has t targets and r regulators, the regulations involving i could be replaced, in the worst case scenario, by $r \cdot t$ regulations. We therefore pick for elimination, at each step, one of the variables for which this product is minimum. A more systematic study of how the elimination order influences the running time could lead to identification of better elimination orders. For instance, one might try to favour elimination orders that have the least impact on the number of trap spaces.

Another crucial point when using variable elimination is deciding where to stop the iterative process. Unfortunately, there does not seem to be a simple answer to this question. In some cases, at some point in the elimination process reduction can introduce complicated update functions that can slow the processing steps that follow. In addition, although reduction seems to be beneficial as a preliminary step for the methods we tested, we will see that different levels of reduction might favour different methods variably. In our implementation we considered two possible parameters to stop the reduction process. One specifies the minimum size for the reduced network in terms of number of nodes (`stop_at`). The other looks at the minimum product of number of regulators and number of targets for each removable node (that is, the parameter used to choose the variable to eliminate), and sets a maximum for this value (`max_product`).

We also investigated the impact of the simplification of Boolean expressions on running times. We found that, although simplification is expensive, it leads to faster elimination. In the next section we will discuss the impact of the elimination step on running times for both biological and random networks.

3.2 Results

The experiments on the biological network models were run on a desktop computer with an Intel(R) Core(TM) i7-8700 processor, 32GB of RAM, operating system Debian GNU/Linux 11. The experiments on the random networks were run on an HPC cluster nodes with an AMD(R) Zen2 EPYC 7702 @ 2 GHz, 1TB of RAM, operating system, operating system CentOS Linux. The code is available at https://github.com/etonello/attractors-with-reduction.

We run our implementation using AEON [1] and mtsNFVS [24] to find the attractors of the reduced networks, and compared the running times to those of AEON and mtsNFVS applied on the networks without reduction. We consider the seven models examined in [24] as available in the tool repository, as well as additional models extracted from the PyBoolNet repository [4] and the

Table 1. Information on the sourcing of the bnet models considered, their size and the number of steady states and cyclic attractors.

Model	file name	file source	n. nodes	n. edges	n. steady states	n. cyclic attractors
MAPK	grieco_mapk	PyBoolNet	53	108	12	6
IL-6	IL_6_Signalling	mtsNFVS	55	95	28672	4096
EMT	selvaggio_emt	PyBoolNet	56	159	1452	0
T-LGL	TLGLSurvival	mtsNFVS	58	193	172	146
CACC	Colitis_associated_...	mtsNFVS	66	144	2	8
AD	A_model	mtsNFVS	74	198	0	2
AGS	id-148-AGS-CELL-FATE-...	biodivine	83	193	1	0
CC	cell_cycle_2019	mtsNFVS	87	467	8	0
SP	id-192-SEGMENT-POLA...	biodivine	102	432	65	0
SIPC	SIGNALING-IN-PROST...	mtsNFVS	116	428	2460	300
DSP	id-210-DRUG-SYNERGY-...	biodivine	144	367	0	1
C3.0	CASCADE3	mtsNFVS	176	449	0	1
EP	id-211-EPITHELIAL-DER...	biodivine	183	602	0	1

Table 2. Reduction scenarios considered for the biological models listed in Table 1. `max_product=` k indicates that the elimination ends when the product of number of regulators and number of targets is above k for all elimination candidates. The elimination is also set to stop when the number of nodes reaches 10.

Model	(1) `max_product=20`			(2) `max_product=50`			(3) `max_product=100`		
	n. nodes	n. edges	time	n. nodes	n. edges	time	n. nodes	n. edges	time
MAPK	10	41	0.1	10	41	0.0	10	41	0.1
IL-6	17	28	0.0	17	28	0.0	17	28	0.0
EMT	19	94	0.1	17	130	0.1	17	130	0.1
T-LGL	21	91	0.0	18	111	0.0	18	111	0.0
CACC	14	81	0.0	11	56	0.0	11	56	0.0
AD	11	93	0.0	10	97	0.0	10	97	0.0
AGS	2	4	0.0	2	4	0.0	2	4	0.0
CC	38	415	0.2	35	490	0.3	29	669	23.9
SP	35	234	0.0	33	273	0.1	32	357	0.1
SIPC	41	390	0.3	32	465	0.6	28	522	8.7
DSP	14	108	0.0	10	71	0.0	10	71	0.0
C3.0	23	191	0.0	14	175	0.1	13	193	0.1
EP	33	322	0.1	25	365	0.1	21	316	0.1

biodivine-boolean-models repository [13]. We use these models as they are sufficiently large and thus challenging to analyse, and have no constant input variable, and can thus be run through all tools without any preprocessing. Information on the source of each model, as well as the size of the networks is given in Table 1. We refer the reader to the respective sources for references detailing

Table 3. Running times for AEON [1] and mtsNFVS [24] without reduction and for the three reduction scenarios described in Table 2. The running times in the three reduction scenarios include the time for network reduction. Running times for mtsNFVS are averaged over five iterations, and show the standard deviation if this is higher than 1 s. "DNF" indicates that the processing did not complete within one hour.

Model	AEON running times				mtsNFVS running times			
	No red	(1)	(2)	(3)	No red	(1)	(2)	(3)
MAPK	5.7	0.3	0.3	0.3	28.9 ±5.7 (3 DNF)	0.7	0.7	0.7
IL-6	774.6	6.0	6.0	6.0	14.8 ±1.8	7.4	7.4	7.4
EMT	25.6	0.8	0.7	0.7	DNF	1.3	1.4	1.4
T-LGL	17.5	0.9	0.9	1.1	2.2	1.8	1.9	2.7
CACC	9.3	0.3	0.3	0.3	0.5	0.8	0.7	0.7
AD	361.9	0.3	0.4	0.4	0.7	1.0	0.8	0.8
AGS	1.7	0.3	0.3	0.3	0.6	0.7	0.7	0.7
CC	DNF	27.0	11.1	26.9	8.2 ±3.5	2.2	6.0	39.4
SP	DNF	0.9	0.8	0.8	DNF	1.0	1.1	1.4
SIPC	DNF	28.2	6.9	14.4	1664.7 ±506.3	47.8 ±5.6	53.7 ±7.4	138.8
DSP	DNF	0.4	0.4	0.4	2.3	0.8	0.7	0.7
C3.0	DNF	0.5	0.4	0.4	2.1	1.0	1.0	0.9
EP	DNF	1.5	0.6	0.5	62.7 ±64.2	2.5	2.4	1.9

each model, which explain if and how the models have been modified from their original sources.

As we mentioned in the previous section, different levels of reduction can have different impacts on running times of both the elimination procedure and the remainder of the attractor identification process. Here we report running times obtained in three scenarios, where we set the parameter `max_product` described in the previous section (maximum value accepted for the minimum product of in- and out-regulations over all nodes) to three levels, 20 in scenario (1), 50 in scenario (2) and 100 in scenario (3). In all three cases, we set the minimum size of the reduced network to 10. In some cases these parameters lead to networks with fewer than 10 nodes, because constants variables might be generated by the elimination process, and these are always eliminated. In Table 2 we show the number of nodes and signed regulations of the reduced network obtained in each of the three scenarios, as well as the time required by the reduction process. The time for reduction is below one second except for two networks, CC and SIPC, where it goes up to 24 and 9 s, respectively.

Table 3 shows the running times for the two methods and the different scenarios (no reduction, reduction scenario (1), (2) and (3)). The algorithm of mtsNFVS relies on some random choices, which cause its running times to vary, sometimes significantly. We report the minimum and maximum running times obtained on five iterations. We set a timeout of one hour, and "DNF" indicates that the processing did not complete within this time.

It is clear from the results that approaching the attractor identification problem by first reducing the network can speed up the processing significantly. We can also observe that a more aggressive reduction does not necessarily translate to faster attractor identification times. This can be observed for both attractor identification methods for the networks CC and SIPC. Looking at the numbers for these two networks, we can also see that the reduction levels giving the shortest processing times are not the same for AEON and mtsNFVS. In particular, mtsNFVS seems to work better with more conservative reduction levels, and in one case (network CC, scenario (3)) the time required for the reduction process was significantly higher than the running time for mtsNFVS without reduction.

Importantly, only for one of the networks (TLGL) nonminimal candidates were identified, and no network presented non-univocal candidates, meaning that reduction had overall a limited impact on the general attractor configuration. Looking at what happens on randomly-generated networks might help to clarify whether these behaviours could be specific of biological models or more general.

Random Benchmarks. As in [24], we generate random networks by invoking `generateRandomNKNetwork` from the R package BoolNet ([8]), fixing $K = 2$ for the number of regulators for each variable. We consider networks with $n = 100k$ nodes, with $k = 1, \ldots, 5$, and create 10 networks for each size. Other choices of benchmark generators and parameters are possible and will be considered in the future for a more extensive analysis. For this work, we decided to use the same settings that were adopted in [24] to give some continuity to the comparison.

We tested several stopping conditions for the reduction process, and did not identify a general rule that would give optimal times in all cases. To give an idea of the range of possible results, we report minimum and maximum running times for three reduction scenarios, where we set the parameter `max_product` to $\frac{n}{2}$, n and $2n$, while `stop_at` is set to $\frac{n}{10}$. The details of the three scenarios are shown in Table 4. In Table 5, we show the average running times for networks that were processed within the timeout of one hour, and the number of networks that were not processed in time.

We observe again that, by adopting the reduction approach, the number of networks that can be successfully analysed within the timeout given increases. AEON without reduction could process 4 out of the 10 networks of size 100, and no network of larger size. With reduction and AEON, all networks of size 200 and 8 out of the 10 networks of size 300 could be processed, as well as one network of size 400. The running times are just a few seconds for networks of size 100, and vary significantly for larger networks. The tool mtsNFVS without reduction could process all networks of size 100 and some networks of larger size. With reduction, more networks could be handled within the time limit, although with running times that can be higher than without reduction for the stopping conditions considered, possibly as a result of having to perform computationally expensive conversions to DNF. As we pointed out in Sect. 3.1,

Table 4. Statistics for three reduction scenarios on 10 random networks (generated with BoolNet [8], setting $K = 2$). `max_product=` k indicates that the elimination ends when the product of number of regulators and number of targets is above k for all nodes. The elimination is also set to stop when one-tenth of the number of nodes of the original network is reached.

Model size	(1) `max_product=n/2`			(2) `max_product=n`			(3) `max_product=2n`		
	min-max n. nodes	min-max n. edges	min-max time	min-max n. nodes	min-max n. edges	min-max time	min-max n. nodes	min-max n. edges	min-max time
100	10-18	69-313	0.1-0.4	10-15	69-305	0.1-1.8	10-14	69-265	0.1-1.8
200	20-30	375-908	0.2-1.8	18-27	375-1017	0.2-16.4	18-25	375-1005	0.2-987.2
300	30-43	1111-1748	0.7-6.0	30-39	1111-1849	0.7-38.5	28-35	1111-1934	0.7-516.2
400	43-55	1951-2555	0.8-6.0	40-48	1868-2972	5.5-122.3	40-45	1868-3118	11.3-1355.9
500	52–67	2516-4043	2.0-14.5	50-62	2679-4580	6.2-51.0	50-58	2679-5247	23.6–419.5

the results generated by mtsNFVS might require further validation, as several failures were identified in test networks. We can nevertheless, by observing the results obtained with AEON, note that the number of candidate states that need processing via reachability analysis remains very limited. Only five nonminimal candidate states were identified for the networks that were processed with reduction and AEON, and no nonunivocal candidate states. This suggests that nonminimal and nonunivocal attractors might be rare phenomena.

Reduction allowed the size of networks to be drastically reduced, but the results shown illustrate how different number of reduction steps might affect networks in different ways. Some networks could only be processed in time in scenario (3), with the highest number of eliminated variables; for others, processing times were lower when a smaller number of variables was removed.

4 Discussion

We investigated how reduction can help in the process of attractor identification for asynchronous Boolean networks. Although the influence on running times can depend on the adopted attractor detection approach and on the chosen level of reduction, we observed that variable elimination generally reduces running times for biological models significantly, and can enable the processing of larger random networks. At the same time, the pipeline we implemented provides additional information about attractors (for example, whether they are in one-to-one correspondence with minimal trap spaces). For better application of the method we suggested, we think it might be interesting to conduct deeper investigations on the impact of reduction on specific attractor detection tools (for instance, for mtsNFVS, one might want to investigate different reduction levels, or elimination orders that do not increase the size of minimal feedback vertex sets). Further improvements concern the investigation of shape and size of cyclic attractors:

Table 5. Average running times for AEON [1] and mtsNFVS [24], on random networks without reduction and in the three reduction scenarios of Table 4. In parentheses is the number of tests that did not terminate within one hour. Since running times for mtsNFVS have a high variability, we run each test five times. The mean and standard deviation shown are over all tests for the given size.

Model size	AEON running times			
	No red	(1)	(2)	(3)
100	1104.6 ±676.0 (6)	2.3 ±1.3	3.0 ±1.5	3.2 ±1.5
200	(10)	83.7 ±205.7	48.9 ±113.7	146.1 ±402.4
300	(10)	1331.3 ±1394.7 (5)	696.0 ±684.1 (3)	888.9 ±1185.0 (2)
400	(10)	(10)	2994.6 (9)	3033.2 (9)
500	(10)	(10)	(10)	(10)
Model size	mtsNFVS running times			
	No red	(1)	(2)	(3)
100	3.3	6.9 ±2.0	7.6 ±2.1	7.7 ±2.0
200	148.4 ±536.0 (10)	543.5 ±767.9	575.4 ±1015.6 (1)	379.8 ±654.0 (5)
300	191.4 ±616.9 (24)	2192.8 ±648.1 (21)	2635.3 ±602.6 (23)	2388.1 ±712.1 (41)
400	577.2 ±562.9 (31)	1974.8 ±599.1 (10)	2353.6 ±677.8 (30)	2291.0 ±565.9 (23)
500	1102.5 ±1388.6 (47)	2113.7 ±724.5 (34)	1978.3 ±749.5 (40)	1754.3 ±602.6 (44)

our current implementation singles out one state per attractor, and does not give information on the number of states or oscillating variables.

Although candidate states that require reachability analysis are only rarely encountered, this step might benefit from additional developments. One improvement concerns the screening of nonunivocal candidate states, as the technique of mtsNFVS currently can fail to detect the existence of multiple attractors contained in the same minimal trap space. In addition, other techniques could be incorporated for the exclusion of existence of nonminimal attractors. For instance, at the moment only minimal trap spaces are used. Larger trap spaces (the maximal trap spaces that do not contain the candidate state) can provide bigger targets for reachability analysis.

Finally, attractor detection tools are capable of other tasks, for instance, AEON can perform detection of bifurcations and source-target control. There is the potential that reduction could be used sensibly to speed up these activities too. Each task needs to be studied individually and carefully for the implications of variable elimination.

Acknowledgements. ET was supported by the Deutsche Forschungsgemeinschaft (DFG) under Germany's Excellence Strategy - The Berlin Mathematics Research Center MATH+ (EXC-2046/1, project ID 390685689). LP was supported by the French Agence Nationale pour la Recherche (ANR) in the scope of the project "BNeDiction" (grant number ANR-20-CE45-0001). Experiments presented in this paper were carried out using the PlaFRIM experimental testbed, supported by Inria, CNRS (LABRI and IMB), Université de Bordeaux, Bordeaux INP and Conseil Régional d'Aquitaine (see https://www.plafrim.fr).

References

1. Beneš, N., et al.: AEON.py: python library for attractor analysis in asynchronous Boolean networks. Bioinformatics **38**(21), 4978–4980 (2022). https://doi.org/10.1093/bioinformatics/btac624
2. Beneš, N., Brim, L., Pastva, S., Šafránek, D.: Computing bottom SCCs symbolically using transition guided reduction. In: Silva, A., Leino, K.R.M. (eds.) CAV 2021. LNCS, vol. 12759, pp. 505–528. Springer, Cham (2021). https://doi.org/10.1007/978-3-030-81685-8_24
3. Klarner, H., Siebert, H.: Approximating attractors of Boolean networks by iterative CTL model checking. Front. Bioeng. Biotechnol. **3**, 130 (2015). https://doi.org/10.3389/fbioe.2015.00130
4. Klarner, H., Streck, A., Siebert, H.: PyBoolNet: a python package for the generation, analysis and visualization of Boolean networks. Bioinformatics **33**(5), 770–772 (2017). https://doi.org/10.1093/bioinformatics/btw682
5. Montagud, A., et al.: Patient-specific Boolean models of signalling networks guide personalised treatments. eLife **11**, e72626 (2022). https://doi.org/10.7554/elife.72626
6. Moon, K., Lee, K., Paulevé, L.: Computational Complexity of Minimal Trap Spaces in Boolean Networks. arXiv preprint arXiv:2212.12756 (2022). https://doi.org/10.48550/arXiv.2212.12756
7. Mori, T., Akutsu, T.: Attractor detection and enumeration algorithms for Boolean networks. Comput. Struct. Biotechnol. J. (2022). https://doi.org/10.1016/j.csbj.2022.05.027
8. Müssel, C., Hopfensitz, M., Kestler, H.A.: BoolNet-an R package for generation, reconstruction and analysis of Boolean networks. Bioinformatics **26**(10), 1378–1380 (2010). https://doi.org/10.1093/bioinformatics/btq124
9. Naldi, A., et al.: The CoLoMoTo interactive notebook: accessible and reproducible computational analyses for qualitative biological networks. Front. Physiol. **9**, 680 (2018). https://doi.org/10.3389/fphys.2018.00680
10. Naldi, A., Remy, E., Thieffry, D., Chaouiya, C.: A reduction of logical regulatory graphs preserving essential dynamical properties. In: Degano, P., Gorrieri, R. (eds.) CMSB 2009. LNCS, vol. 5688, pp. 266–280. Springer, Heidelberg (2009). https://doi.org/10.1007/978-3-642-03845-7_18
11. Naldi, A., Remy, E., Thieffry, D., Chaouiya, C.: Dynamically consistent reduction of logical regulatory graphs. Theoret. Comput. Sci. **412**(21), 2207–2218 (2011). https://doi.org/10.1016/j.tcs.2010.10.021
12. Naldi, A., Richard, A., Tonello, E.: Linear cuts in Boolean networks. arXiv preprint arXiv:2203.01620 (2022). https://doi.org/10.48550/arXiv.2203.01620
13. Pastva, S., Safranek, D., Benes, N., Brim, L., Henzinger, T.: Repository of logically consistent real-world Boolean network models. bioRxiv, pp. 2023–2306 (2023). https://doi.org/10.1101/2023.06.12.544361
14. Paulevé, L.: PINT: a static analyzer for transient dynamics of qualitative networks with IPython interface. In: Feret, J., Koeppl, H. (eds.) CMSB 2017. LNCS, vol. 10545, pp. 309–316. Springer, Cham (2017). https://doi.org/10.1007/978-3-319-67471-1_20
15. Paulevé, L., Sené, S.: Boolean networks and their dynamics: the impact of updates. In: Systems Biology Modelling and Analysis: Formal Bioinformatics Methods and Tools. Wiley (2022). https://doi.org/10.1002/9781119716600.ch6

16. Richard, A., Tonello, E.: Attractor separation and signed cycles in asynchronous Boolean networks. Theoret. Comput. Sci. **947**, 113706 (2023). https://doi.org/10.1016/j.tcs.2023.113706
17. Rozum, J.C., Gómez Tejeda Zañudo, J., Gan, X., Deritei, D., Albert, R.: Parity and time reversal elucidate both decision-making in empirical models and attractor scaling in critical Boolean networks. Sci. Adv. **7**(29), eabf8124 (2021). https://doi.org/10.1126/sciadv.abf8124
18. Saadatpour, A., Albert, I., Albert, R.: Attractor analysis of asynchronous Boolean models of signal transduction networks. J. Theoret. Biol. **266**(4), 641–656 (2010). https://doi.org/10.1016/j.jtbi.2010.07.022
19. Saadatpour, A., Albert, R., Reluga, T.C.: A reduction method for Boolean network models proven to conserve attractors. SIAM J. Appl. Dyn. Syst. **12**(4), 1997–2011 (2013). https://doi.org/10.1137/13090537X
20. Schwab, J.D., Ikonomi, N., Werle, S.D., Weidner, F.M., Geiger, H., Kestler, H.A.: Reconstructing Boolean network ensembles from single-cell data for unraveling dynamics in the aging of human hematopoietic stem cells. Comput. Struct. Biotechnol. J. **19**, 5321–5332 (2021). https://doi.org/10.1016/j.csbj.2021.09.012
21. Schwieger, R., Tonello, E.: Reduction for asynchronous Boolean networks: elimination of negatively autoregulated components. arXiv preprint arXiv:2302.03108 (2023). https://doi.org/10.48550/arXiv.2302.03108
22. Stoll, G., et al.: MaBoSS: 2.0 an environment for stochastic Boolean modeling. Bioinformatics **33**(14), 2226–2228 (2017). https://doi.org/10.1093/bioinformatics/btx123
23. Trinh, V.G., Benhamou, B., Hiraishi, K., Soliman, S.: Minimal trap spaces of Logical models are maximal siphons of their Petri net encoding. In: Petre, I., Păun, A. (eds.) Computational Methods in Systems Biology, CMSB 2022. LNCS, vol. 13447, pp. 158–176. Springer, Cham (2022). https://doi.org/10.1007/978-3-031-15034-0_8
24. Trinh, V.G., Hiraishi, K., Benhamou, B.: Computing attractors of large-scale asynchronous Boolean networks using minimal trap spaces. In: Proceedings of the 13th ACM International Conference on Bioinformatics, Computational Biology and Health Informatics, pp. 1–10 (2022). https://doi.org/10.1145/3535508.3545520
25. Van Giang, T., Hiraishi, K.: An improved method for finding attractors of large-scale asynchronous Boolean networks. In: 2021 IEEE Conference on Computational Intelligence in Bioinformatics and Computational Biology (CIBCB), pp. 1–9. IEEE (2021). https://doi.org/10.1109/CIBCB49929.2021.9562947
26. Veliz-Cuba, A.: Reduction of Boolean network models. J. Theoret. Biol. **289**, 167–172 (2011). https://doi.org/10.1016/j.jtbi.2011.08.042
27. Zañudo, J.G.T., et al.: Cell line-specific network models of ER+ breast cancer identify potential PI3ka inhibitor resistance mechanisms and drug combinations. Cancer Res. **81**(17), 4603–4617 (2021). https://doi.org/10.1158/0008-5472.can-21-1208
28. Zañudo, J.G., Albert, R.: An effective network reduction approach to find the dynamical repertoire of discrete dynamic networks. Chaos Interdiscipl. J. Nonlinear Sci. **23**(2), 025111 (2013). https://doi.org/10.1063/1.4809777

3D Hybrid Cellular Automata for Cardiac Electrophysiology: A Concept Study

Lilly Maria Treml[1,2](✉)

[1] Department Life Science Engineering, UAS Technikum Wien, Vienna, Austria
[2] Institute of Computer Engineering, TU Wien, Vienna, Austria
lilly.treml@tuwien.at

Abstract. A heartbeat is the emerging collective behavior of billions of cells. However, if this synchronization fails, lethal arrhythmias can appear. Mathematical and computational models offer an ethical alternative to in-vivo analyses of cardiac electrophysiological properties. However, the inherent multiscale complexity of the underlying nonlinear dynamics still limits model applicability and predictability. In previous contributions, we implemented a unidirectional Hybrid Cellular Automata (HCA) model reproducing cardiac cell cables and introducing the concept of a statistically distributed cell-cell resistance. Here, we generalize the theoretical framework by considering a bidirectional coupling and simulate physiological and pathological conditions in three-dimensional domains. The work compares two HCA approaches reproducing critical spatiotemporal phenomena and contrasts them with well-established model formulations. We discuss the limits and applicability of discrete vs. continuum approaches in view of improved numerical performances.

Keywords: Cardiac Simulation · Computational Biology · Hybrid Cellular Automata

1 Introduction

Each heartbeat results from billions of cells simultaneously contracting. The highly complex biochemical processes responsible for cell contraction also implement natural resilience to single failures. However, various underlying conditions can overwrite this fail-safe, and cardiac arrhythmias can occur. In several cases, they can lead to Atrial Fibrillation (AFib).

According to the US Centers for Disease Control and Prevention (CDC), in 2019, AFib was the underlying cause of death in 26 535 deaths [9]. The many causes of cardiac arrhythmia and generalized treatment are still open issues, requiring extensive research.

Starting from the seminal works of Hodgkin-Huxley [20] notable contributions were introduced for safe, non-invasive, ethical alternatives to in-vitro

This work has been supported by the Doctoral College Resilient Embedded Systems, which is run jointly by the TU Wien's Faculty of Informatics and the UAS Technikum Wien.

J. Pang and J. Niehren (Eds.): CMSB 2023, LNBI 14137, pp. 220–235, 2023.
https://doi.org/10.1007/978-3-031-42697-1_15

research: using a set of (partial) differential equations to model the electrical propagation in excitable cells and tissues.

The concept was soon modified to describe pacemaker cells [30] and multi-dimensional cardiac models [31]. The models evolved into a non-invasive and simple solution study of unexpected, chaotic properties of cardiac dynamics [2,10]. Even more recently, we continue to understand better which additional parameters, like temperature [11,13] influence cardiac electrophysiology. Another factor discovered was anisotropy [36], directionally dependent signal propagation due to fiber orientation. Some models are even already used in clinically oriented applications [8].

Although the mathematical models enabled the simulation of cardiac dynamics, they have a significant disadvantage. Due to the (partial) differential equations, most models are inherently highly complex. Therefore, the simulations need many computational resources. Many aimed to simplify the models by decreasing the physiological complexity, like [10] and [5], who summarized the electrical currents and relied on reproducing phenomena instead of accurate processes. Others explored a different branch of mathematics: automata theory. The basic concept of automata is to encode complex processes in states and transitions depending on the input. Hence enabling a less complex representation of cardiac membrane potential [17]. With the advancement of high-performance hardware like Graphic Processing Units, [22,38] and FPGAs [12] research took another step toward real-time simulation.

Our work uses concepts of a particular automaton approach, namely Cellular Automata (CA). CA is well-known and widely used to reproduce natural phenomena like pattern formation and growth [28,39]. First proposed by [29], a CA generally refers to a grid of cells, each transitioning within a set of discrete states. A set of rules incorporating the neighborhood control the state change. With this simple approach, it is possible to simulate observed phenomena in excitable media like the nervous system [24] and cardiac tissue [4].

The simplicity renders CAs inaccurate, as pure state transition can only describe limited behavior. Hence, another well-established approach considers Hybrid Automata (HA). A HA uses a set of ODEs to gain fine control over the modeled system. [19] The combination of hybrid-automata methods with technology enabled the accurate reproduction of a cardiac action potential [1].

The overall concept of our Hybrid-Cellular automaton (HCA) is to model each cell's internal reaction using an arbitrary cell model and the propagation of ionic currents in the neighborhood using CA concepts. Using a free resistance variable enables us to model various degrees of inhomogeneity, even anisotropy. Thus we control the degree of accuracy regarding the representation of the biological processes. Our previous work presented such an approach for a non-linear unidirectional 1D HCA [37], using the Minimal Model by [5]. In this work, we show the progress towards actual cardiac tissue, adapting [37] to an inhomogeneous tissue slab with bidirectional signal propagation. For this work, we used a cardiac cell model with fewer differential equations for the cell reaction by [10]. First, we used discrete states to control activation. Naturally, the discrete nature of CA, when used to model continuous processes, may require additional effort to provide sufficient accuracy. Hence we also present a second approach using an extension of a CA: a continuous CA. Instead of discrete, we use continuous mapping functions. This work aims to discuss the parameter sets of each approach and highlight the behavioral differences and respective limitations. Although we use GPUs to accelerate our simulations, an analysis of the efficiency is out of the scope of this work.

This paper is organized as follows: First, in Sect. 2, we introduce the reader to the mathematical frame of the proposed HCAs and analysis techniques. In Sect. 3 , we summarize our experiments' main findings. We compare the results for each approach in Sect. 4 and conclude in Sect. 5 by discussing the future perspectives of our work.

2 Methodology

In this section, we formally define our HCA and present the underlying mathematical models. Then we give our analysis goals and the different computational approaches to simulate them.

2.1 Hybrid Cellular Automata (HCA) Concept

A healthy myocyte has a resting membrane potential of approximately -90 mV. Rapid ion exchange occurs once the myocyte receives an electrical impulse from the neighborhood, creating a brief change in the membrane potential known as the cardiac action potential. Hence we formalize the approach as tuple $(V^{\mathrm{u}}, S, N, \mathrm{u}, \theta_{\mathrm{a}}, \theta_{\mathrm{r}})$, such that V^{u} is a 3-dimensional grid of cells represented by their membrane potentials, u an element of V^{u}, S is a set of states, N is the neighborhood and θ_{a} & θ_{r} are thresholds. First, we derive the global transition function from the underlying cell model, such that each u produces an action potential in dependence of N and S. For this dependency, we consider three relevant domains for state transition: extracellular space, intracellular space, and the cell membrane. For the cell membrane, we apply the free resistance variable to a composite signal of the surrounding cells, yielding the incoming signal I_{in}. We incorporate the other two domains as activation constraints Q & E. E regulates the incoming signal depending on the cell's state of excitement, while Q ensures that propagation only occurs if the neighborhood is excited enough. Thus, we define the state transition function u : $S|N| \rightarrow S$ as the output of the HCA $I_{\mathrm{out}}\left(I_{\mathrm{in}}\right)$ to each cell, yielding Eq. (1). Each approach utilizes state-mapping functions depending on the thresholds to obtain the finite set of states S such that $V^{\mathrm{u}} \rightarrow S$. The value of the thresholds and the formulation of the global transition function depend on the model for the membrane potential.

$$I_{\mathrm{out}}\left(I_{\mathrm{in}}\right) = I_{\mathrm{in}} Q E \tag{1}$$

2.2 Global Transition Function

We designed the cell grid of the HCA such that an arbitrary cardiac cell model defines the behavior of each cell to internal and external stimuli. The membrane potential described by the model provides information about the state of the cell. In this work, we use the dimensionless variable u from the 3-Variable Fenton-Karma model [10] (FKM) for the membrane potential, given in Eq. (2a).

The FKM [10] derives from the Beeler-Reuter [2] and Luo-Rudy [25] models. They define u as the summation of three currents $J_{\mathrm{fi}}(u,v)$, $J_{\mathrm{so}}(u)$ and $J_{\mathrm{si}}(u,w)$. According to [10], $J_{\mathrm{fi}}(u,v)$, given in Eq. (2b), is the fast inward current depends on the inactivation-reactivation gate v and is responsible for the depolarization and analogous to the gates of the sodium current in the base models. The slow outward current $J_{\mathrm{so}}(u)$ of Eq. (2c), is responsible for the repolarization and analogous to the time-independent potassium current in these models [10]. Lastly, the slow inward current $J_{\mathrm{si}}(u,w)$ in Eq. (2d) is derived from the calcium current and balances $J_{\mathrm{so}}(u)$ during the plateau phase, it depends on gating variable w [10]. [10] also gives a detailed explanation of the parameters τ and constants u_{c}.

We denote the standard Heaviside function in all equations as Θ.

$$\partial_t u = \nabla \cdot \left(\tilde{D}\nabla u\right) - \left(J_{\mathrm{fi}}(u,v) + J_{\mathrm{so}}(u) + J_{\mathrm{si}}(u,w)\right) \tag{2a}$$

with the currents

$$J_{\mathrm{fi}}(u,v) = -\frac{v}{\tau_d}\Theta(u-u_{\mathrm{c}})(1-u)(u-u_{\mathrm{c}}) \tag{2b}$$

$$J_{\mathrm{so}}(u) = \frac{u}{\tau_{\mathrm{o}}}\Theta(u_{\mathrm{c}}-u) + \frac{1}{\tau_{\mathrm{r}}}\Theta(u-u_{\mathrm{c}}) \tag{2c}$$

$$J_{\mathrm{si}}(u,w) = -\frac{w}{2\tau_{\mathrm{si}}}\left(1+\tanh\left(k(u-u_{\mathrm{c}}^{\mathrm{si}})\right)\right) \tag{2d}$$

Replacing the spatial change $\nabla \cdot \left(\tilde{D}\nabla u\right)$ with the output $I_{\mathrm{out}}\left(I_{\mathrm{in}}\right)$ of the HCA transition function, yields Eq. (3). During simulation, we approximate the solution of the equation with the Euler method.

$$u_{\mathrm{t}+1} = u_{\mathrm{t}} - \left(\left(J_{\mathrm{fi}}(u_{\mathrm{t}},v) + J_{\mathrm{so}}(u_{\mathrm{t}}) + J_{\mathrm{si}}(u_{\mathrm{t}},w) - I_{\mathrm{out}}\left(I_{\mathrm{in}}\right)\right)\right)\Delta T \tag{3}$$

where ΔT refers to the time integration step of 50 µs.

2.3 Neighborhood Stencil

Cardiac cell dynamics are inherently dependent on their surroundings, as are cellular automata. To keep the HCA as simple as possible, we define a 6-connected neighborhood, which is the comon used von Neumann neighborhood. In three-dimensional space, it forms a 6-cell octahedron [3], such that all Lattice elements satisfy Definition 1.

Definition 1. $(x',y',z') \in L \iff (x',y',z') \equiv (x \pm 1, y, z) \oplus (x',y',z') \equiv (x, y \pm 1, z) \oplus (x',y',z') \equiv (x, y, z \pm 1)$

However, depending on the grid position and overall virtual geometry, a cell might have fewer neighbors. Thus, we use N in all equations to refer to the number of existing neighbors.

2.4 Incoming Signals

In our previous work [37], we defined a one-dimensional HCA with uni-directional coupling between cells. However, in three-dimensional space, the coupling needs to be bi-directional. Hence, we redefined the computation propagation as an internal impulse I_{in} reflecting signals leaking into a cell to ensure bi-directionality. We achieve this by utilizing the Laplacian operator, yielding Eq. (4). We compute the membrane potential u and the neighborhood potential u_n in each iteration according to Eq. (3). The free resistance variable enables the HCA to model various grades of inhomogeneity and even anisotropy (R^{v}) as it can differ for each neighbor (n) .

$$I_{\text{in}} = \sum_{n=1}^{N} (u_n - u)\, R^{\text{v}}_n \tag{4}$$

2.5 Activation Constraints and State Mapping

Both approaches implement functions to derive the cell's current state from the cell membrane potential. This state represents the excitability E of the cell, the first constraint we apply to our HCA. As a turning point of excitability, we define a refractory threshold (θ_{r}), the minimum potential for the refractory phase. During the refractory phase, cardiac cells contract. Therefore they can not be activated. In the discrete HCA approach (HCA^D), E is determined by a standard Heaviside function Θ. We consider two states: *off* if u is below the θ_{r} and *on* otherwise. As only *off* cells can be excited, we define the mapping as shown in Eq. (5).

$$E = \Theta(\theta_{\text{r}} - u) \tag{5}$$

For the continuous HCA approach (HCA^C), we utilize a sigmoidal activation function (abbreviated SAF) known from neural network applications for state mapping. Early results showed that a hyperbolic tangent is more feasible for our approach than other SAFs. However, we must increase the activation time in the HCA^C as we could only partially recover valid electrophysiology. We needed to strengthen the output signal via the states to achieve this. First, for the internal state E, we established that the natural range of a hyperbolic tangent from [-1,1] is problematic for the model. We do not want propagation if the cell is excited, so we concluded that a [0,1] range would be adequate. However, despite shifting the hyperbolic tangent function, the results were unsatisfactory. Thus, to solve this, we replaced the difference with the ratio between the cell's potential and θ_r, as the function converges faster.

$$E = 1 - \tanh\left(\frac{u}{\theta_r}\right) \tag{6}$$

The second transition constraint is the neighborhood excitement Q. Propagation is only allowed if enough cells exceed an activation threshold (θ_a). We summarize the neighborhood potential as average u_{avg}, defined in Eq. (7), representing the neighboring cells.

$$u_{avg} = \left(\frac{1}{N}\sum_{n=1}^{N} u_n\right) \tag{7}$$

Formalized by Eq. (8), we define Q for HCA^D, such that there must be at least one cell *on* in the neighborhood.

$$Q = \Theta\left(u_{avg} - \left(\theta_a \frac{1}{N}\right)\right) \tag{8}$$

Like E, we ran through an initial adaption phase for Q of the HCA^C to increase activation time. We concluded that a range of [1,2] is necessary as a boost. We achieve this, as seen in Eq. (9), by shifting the hyperbolic tangent function in combination with using the ratio.

$$Q = 1 + \left(\tanh\left(\frac{u_{avg}}{(\theta_a \frac{1}{N})}\right)\right) \tag{9}$$

2.6 Analysis and Benchmarks

We used a simple 3D tissue slab of cells to evaluate and compare our approaches. All cells are assumed spheres, with a diameter of 100 µm. Thus the slab dimensions are 40 × 40 x 1.6 mm.

Threshold Sensitivity. One aspect we analyze is the influence of the thresholds. By setting the free resistance variable to 1, the propagation becomes solemnly dependent on the threshold parameters.

We examined the typical FKM action potential produced by u. First, we assessed two boundaries to separate the fully resting and full refractory states, yielding the mid cross indicated in Fig. 1. We consider the mid-cross as the point where the cell transitions between excitement and rest. Hence, we selected a range of threshold sets for θ_r and θ_a around this point, presented in Table 1.

One widely-used benchmark to evaluate electrophysiological behavior is the conduction velocity (CV). We use initial conditions to induce a single stable wave parallel to an axis, such that the wave traverses in either x, y, or z-direction. Then we measure the CV in these directions and compare the results of the different

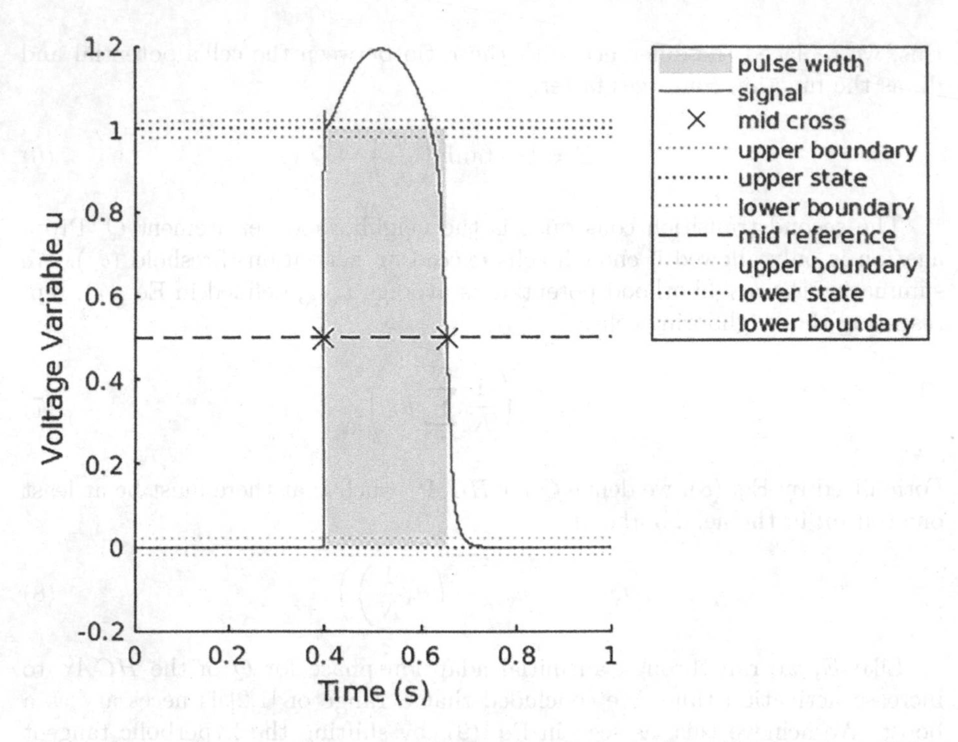

Fig. 1. Fenton-Karma action potential plotted with the 'pulsewidth' function of Matlab. The cut-off value for fully resting cells is 0.0 and for full refractory 1.0.

threshold sets. Once we establish a baseline, we compile a brief overview of the impact of the free resistance variable. Similar to our previous work, we use a random resistance. We draw a random sample from a von Mises distribution for each neighbor with mean $\mu = 2.0$ and concentration $\kappa = 8$.

Action Potential Duration (APD) Stability. We investigate the slab behavior under regular and irregular pacing using the two threshold sets. We always apply external stimuli to the first slice for 1 ms. The protocol starts with a stable pre-pace of four S1 beats with a frequency of 500 ms and a normalized amplitude of 1. Then follows the basic cycle of three S1 and one 40% lower S2 beat. These two phases let us research the behavior under a stable beat. However, after the first cycle, we decrease the interval of the S2 by 50 ms. The increased irregularity should lead to a reduction in the APD.

Table 1. Threshold sets for sensitivity analysis.

Set	I	II	III	IV	V	VI
θ_r	0.4	0.5	0.4	0.5	0.4	0.5
θ_a	0.5	0.5	0.6	0.6	0.7	0.7

Spiral Pattern Evolution. Pattern evolution with initial conditions is a common approach to investigating automata behavior. In excitable media, one prominent pattern is spiral waves. Characteristically these traveling waves rotate outward from the center of origin, forming a spiral. We can force this behavior using initial conditions. The initial conditions for such a wave require three areas with different action potential (AP) phases. These areas are rectangular, and we apply them to the slab's first (lowest) slice: All cells in the left half are resting, in the top right active, and in the bottom right full-refractory. The different excitability of the states forces the wave to form a spiral. If we reproduce this behavior, our HCA can model complex electrophysiological behavior in cardiac cells.

3 Results

In this section, we present the results of the introduced experiments. We begin with the benchmarks and show the primary differences between the discrete and continuous HCA. Then we move on to the complex behavior and spiraling pattern we could reproduce.

3.1 Basic Electrophysiology

The conduction velocity (CV) gives insights into the pathology of cardiac tissue, therefore, is a valuable benchmark for our approaches. We use histograms to visualize the differences in the CV for each threshold set. First, we induce three single waves using initial conditions, each parallel to another axis. As shown in Fig. ??, the signal propagates in all three directions, with slight discrepancies per direction. The CV in the discrete model is way slower than in the continuous. In the next step we increased the CV and force inhomogenity in the slab. We depict the results in Fig. 2.

The cardiac APD of the FKM, ranges from 200–300 ms, hence our model should reproduce this width. We used the initial conditions for a single wave traversing along each axis to compute the APD using a constant R^v shown in Figs. 3. Applying the random R^v drawn from a von Mises distribution w. mean $\mu = 2.0$ and concentration $\kappa = 8$ yields a APD of 250ms in all scenarios.

3.2 Spiraling

One characteristic of cardiac cells is the unique patterns generated by specific stimulation. We aimed to capture the capability of generating spiral wave

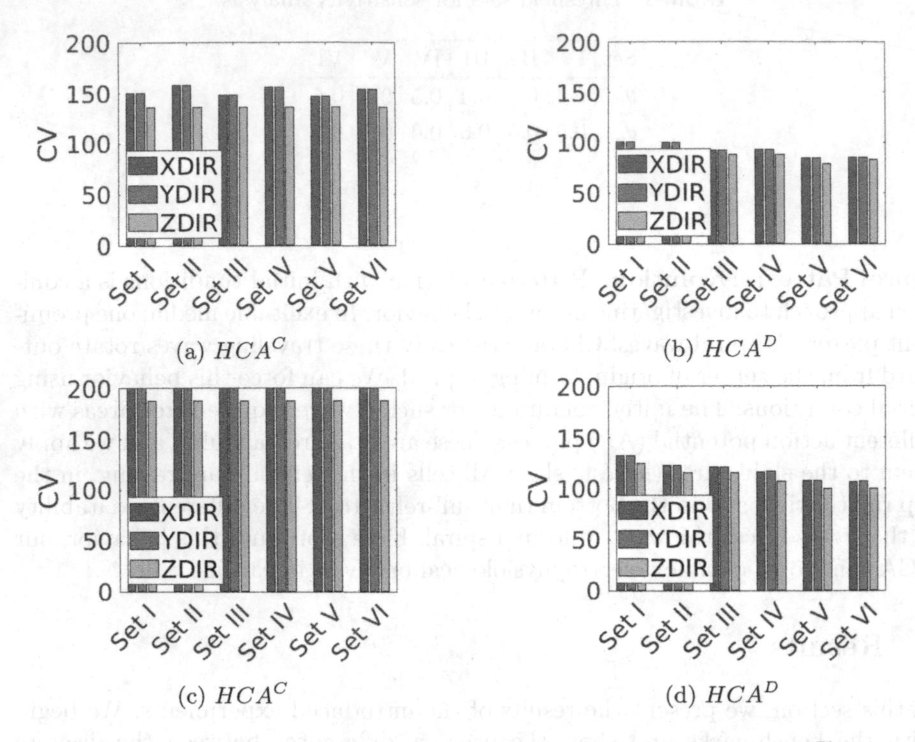

Fig. 2. Parameter set comparison over the average conduction velocity ($\frac{\mu m}{ms}$) of three single waves, each induced parallel to an axis. Where 2a and 2a show the CV with $R^v = 1$ and 2c and 2d with a random resistance R^v drawn each step for each neighbor. The parameters for the von Mises distribution where such that $\mu = 2.0$ and $\kappa = 8$.

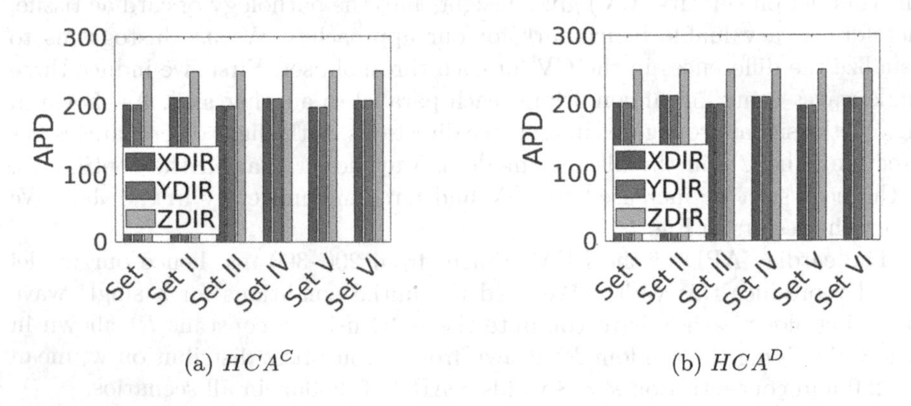

Fig. 3. Parameter set comparison over the average APD (in ms) of three single waves, each induced parallel to an axis. Covering a sample population covering 30% of the slab with a constant $R^v = 1.0$.

patterns using each approach and threshold set. However, threshold sets with $\theta_r = 0.5$ did not produce a spiral, neither did $\theta_a = 0.7$. Hence, we focused on these two remaining sets for the experiments. We present snapshots of the spiral pattern evolution in the first slice for the HCA^C in Fig. 4 and for the HCA^D in Fig. 5. The color bar indicates the value of the membrane potential u, where yellow indicates a high excitement. The pattern for both thresholds is highly similar, but Fig. 4b indicates an earlier wave breakup.

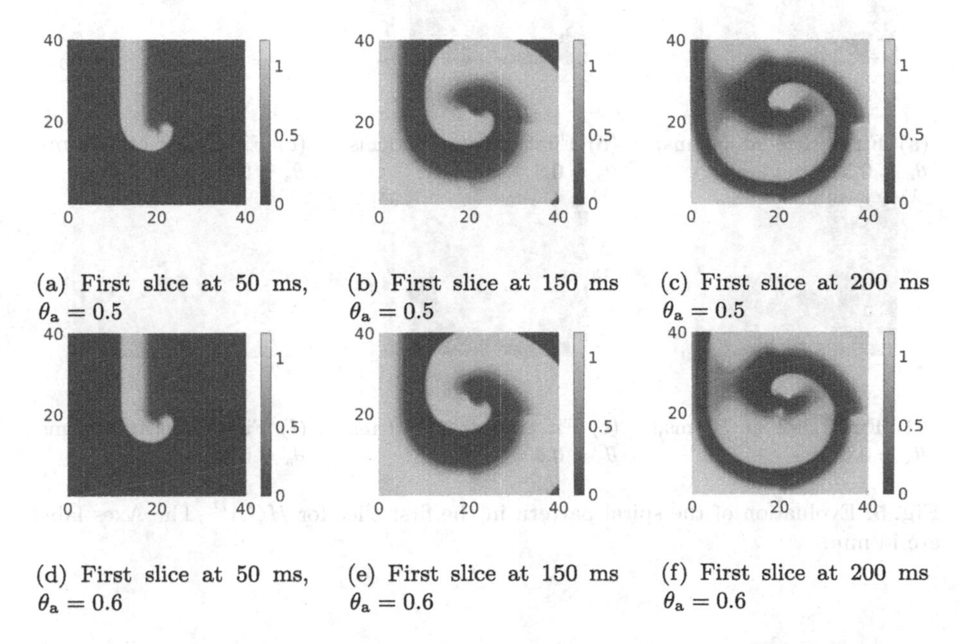

(a) First slice at 50 ms, $\theta_a = 0.5$ (b) First slice at 150 ms $\theta_a = 0.5$ (c) First slice at 200 ms $\theta_a = 0.5$

(d) First slice at 50 ms, $\theta_a = 0.6$ (e) First slice at 150 ms $\theta_a = 0.6$ (f) First slice at 200 ms $\theta_a = 0.6$

Fig. 4. Evoluation of the spiral pattern in the first slice for HCA^C. The Axes show the length in X and Y-direction in mm.

The HCA^D does not produce an actual spiral pattern. Further, the higher θ_a leads to a more hexagonal shape, evident in Fig. 5e. When we look at the whole slab, the HCA^C populates the spiral correctly. The HCA^D, on the other hand, produces a U-shaped wave in the slab. We depict both in Fig. 6. Due to the minimal discrepancies between the thresholds, we show results for $\theta_a = 0.6$

3.3 APD Stability

Previously we only considered three separate single wave traversing along each axis, but our model's physiological system calls for a stable beat, presented in Fig. 7. Thus we use the two stable phases in the S1S2 protocol to estimate the APD stability under a regular beat. We use a built-in Matlab function to measure the pulse width from the action potentials. We draw uniformly distributed random samples covering 30% of the slab to speed up the analysis.

After the stable phase, the interval between the S2 decreases in each cycle. Leading to a decreased S2 APD, as cells have not fully recovered yet. We present the results for $\theta_a = 0.5$ in Fig. 8a and for $\theta_a = 0.6$ in Fig. 8b.

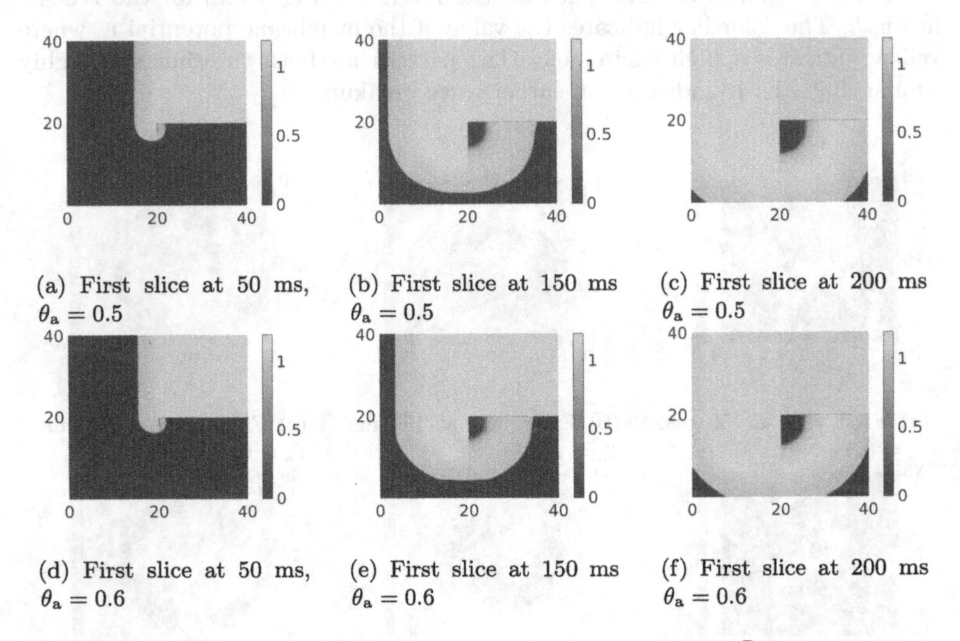

(a) First slice at 50 ms, $\theta_a = 0.5$ (b) First slice at 150 ms $\theta_a = 0.5$ (c) First slice at 200 ms $\theta_a = 0.5$

(d) First slice at 50 ms, $\theta_a = 0.6$ (e) First slice at 150 ms $\theta_a = 0.6$ (f) First slice at 200 ms $\theta_a = 0.6$

Fig. 5. Evoluation of the spiral pattern in the first slice for HCA^D. The Axes labels are in mm.

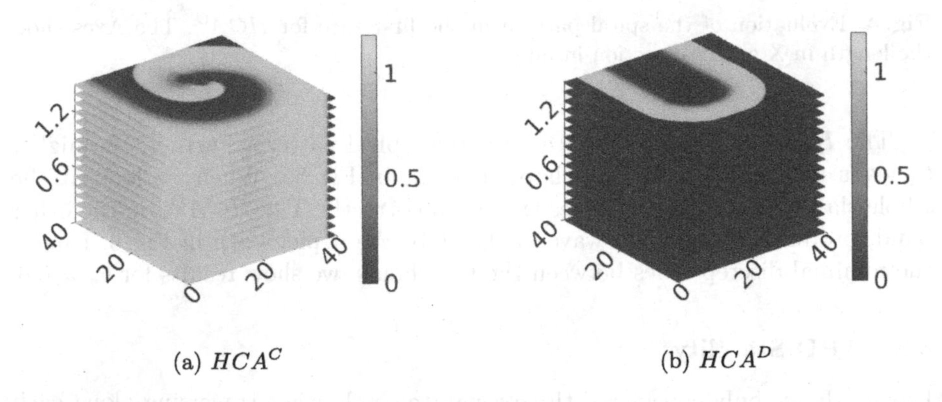

(a) HCA^C (b) HCA^D

Fig. 6. Spiral pattern in the whole slab at 150ms w. $\theta_a = 0.6$. The axes labels are in mm.

The rapid upstroke, visible in Fig. 8, results from the detection method. If we can not capture an S2, we compute the APD for the next S1.

4 Discussion

This survey aims to give insights into the electrophysiological modeling capabilities and differences between the discrete and continuous HCA approaches.

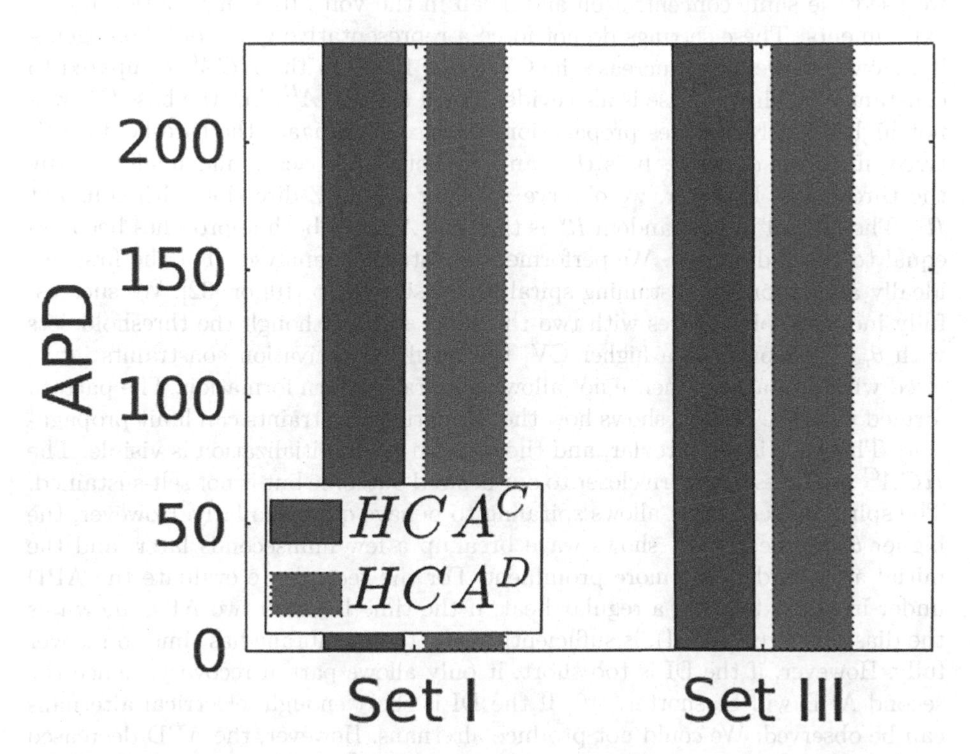

Fig. 7. Average APD (ms) of the pre-pacing phase. Computed using random samples covering 30% of the slab.

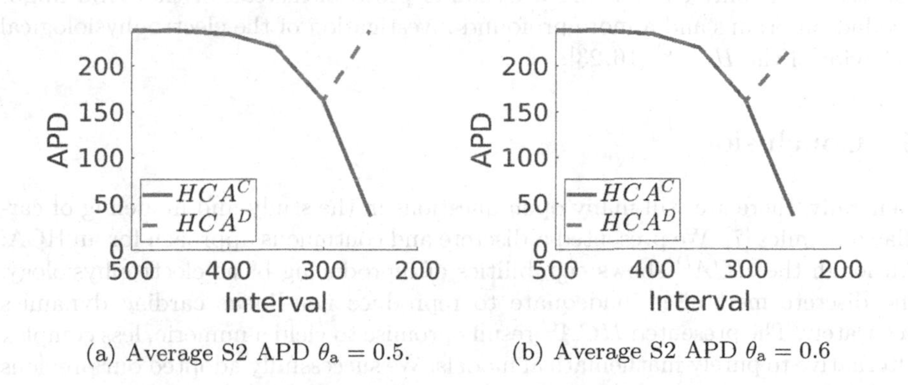

(a) Average S2 APD $\theta_a = 0.5$.

(b) Average S2 APD $\theta_a = 0.6$

Fig. 8. Decrease of the S2 APD (in ms). Computed using a random sample population covering 30% of the slab. The interval (in ms) denotes the time between the last S1 and the S2. The dashed line denotes the HCA^D.

We evaluate the propagation over the CV. With constant R^{v}, there are discrepancies between the directions and parameter sets. Along the x- and y-axis, we have higher velocities for parameter sets with $\theta_{\mathrm{r}} = 0.4$. Although we can observe this in both approaches, the HCA^{D} is slower by roughly 5- $8\frac{\mathrm{cm}}{\mathrm{s}}$. In this work, we used the same concentration and mean in the von Mises distribution for all experiments. These settings do not force a representative grade of inhomogeneity. However, we could increase the CV up to $10\frac{\mathrm{cm}}{\mathrm{s}}$ in the HCA^{C} compared to constant R^{v}. The increase is also evident with the HCA^{D}, but the base CV was not high initially. Besides propagation, we also investigate the threshold sensitivity in terms of APD. It is the same in both approaches and unaffected by the thresholds. However, we observe a longer AP in Z-direction with constant R^{v}. The impact of the random R^{v} is that the APD in both approaches becomes equal to the Z-direction. We performed two stability analyses. For the first, we ideally aimed for self-sustaining spiral waves similar to [10] or [32]. We successfully induced spiral waves with two threshold sets. Although the threshold sets with $\theta_{\mathrm{r}} = 0.5$ produce a higher CV, the resulting activation constraints interfered with propagation hence not allowing spiral-pattern formation. The pattern formed with the HCA^{D} shows how the activation constraints can limit propagation. The wave is not circular, and the pattern from initialization is visible. The HCA^{C} produces a pattern closer to our desired outcome but is not self-sustained. The spiral-pattern effect allows spiraling to behave quite similarly. However, the higher θ_{a} in the HCA^{C} shows wave breakup a few milliseconds later, and the initial area borders are more prominent. For the second, we evaluate the APD under irregular beat. In a regular beat, if the time between two APs, known as the diastolic interval (DI), is sufficiently long, the membrane has time to recover fully. However, if the DI is too short, it only allows partial recovery; hence the second APD will be shorter. [10] If the DI is short enough, electrical alternans can be observed. We could not produce alternans. However, the APD decreased under the S1S2 protocol. The last S2 for the HCA^{D} could be recovered at a DI of 350 ms, with a reduction of 33% of the width. The width of the final S2 in the HCA^{C} is only 84%. Using a dynamic protocol instead of the S1S2 might produce alternans and a more profound investigation of the electrophysiological behavior of the HCA^{C} [16,23].

5 Conclusion

Generally, there are still many open questions in the study and modeling of cardiac dynamics [7]. We presented a discrete and continuous approach for an HCA. Although the HCA^{D} shows capabilities of reproducing basic electrophysiology, the discrete method is inadequate to reproduce non-linear cardiac dynamics accurately. The presented HCA^{C} results promise to yield a numeric, less complex alternative to purely mathematical models. We successfully adopted our previous work [37] to three-dimensional propagation. The generalization of the activation allows us to exchange the underlying cell model freely. Hence, we aim to evaluate our HCA computational approach in a more physiological context [2,21].

Further, we want to investigate the advantages and disadvantages of our approach compared to more common approaches such as finite elements [15,18] and partial differential equations [26,35]. In order to develop efficient computation of ECG signals [27], we also need to adapt our slab and individual cell properties because [6] showed that boundaries and overall geometry influence spatial propagation. Further, in [14] the effects of fiber orientations are discussed. We can mimic the latter with our free resistance variable. Regarding geometry, we need to consider species-specific properties [34] and fiber architecture [33]. A simulation of this capacity is quite expansive. Hence, we consider using multi-GPU systems to accelerate the simulation for future research.

References

1. Andalam, S., Ramanna, H., Malik, A., Roop, P., Patel, N., Trew, M.L.: Hybrid automata models of cardiac ventricular electrophysiology for real-time computational applications. In: 2016 38th Annual International Conference of the IEEE Engineering in Medicine and Biology Society (EMBC), pp. 5595–5598. IEEE (2016). https://doi.org/10.1109/embc.2016.7591995
2. Beeler, G.W., Reuter, H.: Reconstruction of the action potential of ventricular myocardial fibres. J. Physiol. **268**(1), 177–210 (1977)
3. Breukelaar, R., Bäck, T.: Using a genetic algorithm to evolve behavior in multi dimensional cellular automata: emergence of behavior. In: Proceedings of the 7th Annual Conference on Genetic and Evolutionary Computation, pp. 107–114 (2005)
4. Bub, G., Shrier, A., Glass, L.: Bursting in cellular automata and cardiac arrhythmias. In: Chaos, CNN, Memristors and Beyond: A Festschrift for Leon Chua With DVD-ROM, composed by Eleonora Bilotta, pp. 135–145. World Scientific (2013)
5. Bueno-Orovio, A., Cherry, E.M., Fenton, F.H.: Minimal model for human ventricular action potentials in tissue. J. Theor. Biol. **253**(3), 544–560 (2008). https://doi.org/10.1016/j.jtbi.2008.03.029
6. Cherry, E.M., Fenton, F.H.: Effects of boundaries and geometry on the spatial distribution of action potential duration in cardiac tissue. J. Theor. Biol. **285**(1), 164–176 (2011)
7. Clayton, R., et al.: Models of cardiac tissue electrophysiology: progress, challenges and open questions. Progress Biophys. Molecular Biol. **104**(1), 22–48 (2011). https://doi.org/10.1016/j.pbiomolbio.2010.05.008, https://www.sciencedirect.com/science/article/pii/S0079610710000362, cardiac Physiome project: Mathematical and Modelling Foundations
8. Dierckx, H., Fenton, F.H., Filippi, S., Pumir, A., Sridhar, S.: Simulating normal and arrhythmic dynamics: from sub-cellular to tissue and organ level. Front. Phys. **7**, 89 (2019). https://doi.org/10.3389/978-2-88963-067-7
9. for Disease Control, C., Prevention: CDC atrial fibrilation (2022). https://www.cdc.gov/heartdisease/atrial_fibrillation.htm
10. Fenton, F., Karma, A.: Vortex dynamics in three-dimensional continuous myocardium with fiber rotation: filament instability and fibrillation. Chaos: Interdisc. J. Nonlinear Sci. **8**(1), 20–47 (1998)
11. Fenton, F.H., Gizzi, A., Cherubini, C., Pomella, N., Filippi, S.: Role of temperature on nonlinear cardiac dynamics. Phys. Rev. E **87**(4), 042717 (2013). https://doi.org/10.1103/physreve.87.042717

12. Georgis, G., Lentaris, G., Reisis, D.: Acceleration techniques and evaluation on multi-core CPU, GPU and FPGA for image processing and super-resolution. J. Real-Time Image Process. **16**(4), 1207–1234 (2019). https://doi.org/10.1007/s11554-016-0619-6
13. Gizzi, A., et al.: Nonlinear diffusion and thermo-electric coupling in a two-variable model of cardiac action potential. Chaos: Interdisc. J. Nonlinear Sci. **27**(9), 093919 (2017). https://doi.org/10.1063/1.4999610
14. Gizzi, A., Cherry, E.M., Gilmour, R.F., Jr., Luther, S., Filippi, S., Fenton, F.H.: Effects of pacing site and stimulation history on alternans dynamics and the development of complex spatiotemporal patterns in cardiac tissue. Front. Physiol. **4**, 71 (2013)
15. Göktepe, S., Kuhl, E.: Computational modeling of cardiac electrophysiology: a novel finite element approach. Int. J. Numer. Methods Eng. **79**(2), 156–178 (2009)
16. Goldhaber, J.I., Xie, L.H., Duong, T., Motter, C., Khuu, K., Weiss, J.N.: Action potential duration restitution and alternans in rabbit ventricular myocytes: the key role of intracellular calcium cycling. Circ. Res. **96**(4), 459–466 (2005)
17. Grosu, R., et al.: From cardiac cells to genetic regulatory networks. In: Gopalakrishnan, G., Qadeer, S. (eds.) CAV 2011. LNCS, vol. 6806, pp. 396–411. Springer, Heidelberg (2011). https://doi.org/10.1007/978-3-642-22110-1_31
18. Heidenreich, E.A., Ferrero, J.M., Doblaré, M., Rodríguez, J.F.: Adaptive macro finite elements for the numerical solution of monodomain equations in cardiac electrophysiology. Ann. Biomed. Eng. **38**, 2331–2345 (2010)
19. Henzinger, T.: The theory of hybrid automata. In: Proceedings 11th Annual IEEE Symposium on Logic in Computer Science, pp. 278–292 (1996). https://doi.org/10.1109/LICS.1996.561342
20. Hodgkin, A.L., Huxley, A.F.: A quantitative description of membrane current and its application to conduction and excitation in nerve. J. Physiol. **117**(4), 500 (1952)
21. Hund, T.J., Rudy, Y.: Rate dependence and regulation of action potential and calcium transient in a canine cardiac ventricular cell model. Circulation **110**(20), 3168–3174 (2004)
22. Kaboudian, A., Velasco-Perez, H.A., Iravanian, S., Shiferaw, Y., Cherry, E.M., Fenton, F.H.: A comprehensive comparison of GPU implementations of cardiac electrophysiology models. In: Bartocci, E., Cleaveland, R., Grosu, R., Sokolsky, O. (eds.) From Reactive Systems to Cyber-Physical Systems. LNCS, vol. 11500, pp. 9–34. Springer, Cham (2019). https://doi.org/10.1007/978-3-030-31514-6_2
23. Koller, M.L., Riccio, M.L., Jr., R.F.G.: Dynamic restitution of action potential duration during electrical alternans and ventricular fibrillation. Am. J. Physiol.-Heart Circ. Physiol. **275**(5), H1635–H1642 (1998)
24. Lehotzky, D., Zupanc, G.K.: Cellular automata modeling of stem-cell-driven development of tissue in the nervous system. Dev. Neurobiol. **79**(5), 497–517 (2019). https://doi.org/10.1002/dneu.22686
25. Luo, C.H., Rudy, Y.: A dynamic model of the cardiac ventricular action potential. i. simulations of ionic currents and concentration changes. Circ. Res. **74**(6), 1071–1096 (1994)
26. Marcotte, C.D., Grigoriev, R.O.: Implementation of PDE models of cardiac dynamics on GPUs using OpenCL. arXiv preprint arXiv:1309.1720 (2013)
27. McSharry, P.E., Clifford, G.D., Tarassenko, L., Smith, L.A.: A dynamical model for generating synthetic electrocardiogram signals. IEEE Trans. Biomed. Eng. **50**(3), 289–294 (2003)
28. Murray, J.D. (ed.): Mathematical Biology. IAM, vol. 18. Springer, New York (2003). https://doi.org/10.1007/b98869

29. Neumann, J., Burks, A.W., et al.: Theory of self-reproducing automata, vol. 1102024. University of Illinois Press Urbana (1966). https://doi.org/10.2307/2005041
30. Noble, D.: A modification of the hodgkin-huxley equations applicable to purkinje fibre action and pacemaker potentials. J. Physiol. **160**(2), 317 (1962)
31. Noble, D.: From the hodgkin-huxley axon to the virtual heart. J. Physiol. **580**(1), 15–22 (2007). https://doi.org/10.1113/jphysiol.2006.119370
32. Panfilov, A.V., Keldermann, R.H., Nash, M.P.: Drift and breakup of spiral waves in reaction-diffusion-mechanics systems. Proc. National Acad. Sci. **104**(19), 7922–7926 (2007). https://doi.org/10.1073/pnas.0701895104, https://www.pnas.org/doi/abs/10.1073/pnas.0701895104
33. Peyrat, J.-M., et al.: Statistical comparison of cardiac fibre architectures. In: Sachse, F.B., Seemann, G. (eds.) FIMH 2007. LNCS, vol. 4466, pp. 413–423. Springer, Heidelberg (2007). https://doi.org/10.1007/978-3-540-72907-5_42
34. Piuze, E., Lombaert, H., Sporring, J., Strijkers, G.J., Bakermans, A.J., Siddiqi, K.: Atlases of cardiac fiber differential geometry. In: Ourselin, S., Rueckert, D., Smith, N. (eds.) Functional Imaging and Modeling of the Heart, pp. 442–449. Springer, Berlin, Heidelberg (2013). https://doi.org/10.1007/978-3-642-38899-6_52
35. Regazzoni, F., Dedè, L., Quarteroni, A.: Biophysically detailed mathematical models of multiscale cardiac active mechanics. PLoS Comput. Biol. **16**(10), e1008294 (2020)
36. Ruiz Baier, R., Gizzi, A., Loppini, A., Cherubini, C., Filippi, S.: Modelling thermo-electro-mechanical effects in orthotropic cardiac tissue. Commun. Comput. Phys. **27**(1) (2019). https://doi.org/10.4208/cicp.OA-2018-0253
37. Treml, L.M., Bartocci, E., Gizzi, A.: Modeling and analysis of cardiac hybrid cellular automata via GPU-accelerated monte Carlo simulation. Mathematics **9**(2), 164 (2021)
38. Vasconcellos, E.C., Clua, E.W., Fenton, F.H., Zamith, M.: Accelerating simulations of cardiac electrical dynamics through a multi-GPU platform and an optimized data structure. Concurrency Comput.: Pract. Exper. **32**(5), e5528 (2020). https://doi.org/10.1002/cpe.5528
39. Wolfram, S.: Cellular automata as models of complexity. Nature **311**, 419–424 (1984). https://doi.org/10.1038/311419a0

Fridge Compiler: Optimal Circuits from Molecular Inventories

Lancelot Wathieu(✉), Gus Smith, Luis Ceze, and Chris Thachuk(✉)

Paul G. Allen School of Computer Science and Engineering,
University of Washington, Seattle, USA
{lwathieu,thachuk}@cs.washington.edu

Abstract. Rationally designed molecular circuits describable by well-mixed chemical reaction kinetics can realize arbitrary Boolean function computation yet differ significantly from their electronic counterparts. The design, preparation, and purification of new molecular components poses significant barriers. Consequently, it is desirable to synthesize circuits from an existing "fridge" inventory of distinguishable parts, while satisfying constraints such as component compatibility. Heuristic synthesis techniques intended for large electronic circuits often result in non-optimal molecular circuits, invalid circuits that violate domain-specific constraints, or circuits that cannot be built with the current inventory. Existing "exact" synthesis techniques are able to find minimal feedforward Boolean circuits with complex constraints, but do not map to distinguishable inventory components.

We present the Fridge Compiler, an SMT-based approach to find optimal Boolean circuits within a given molecular inventory. Empirical results demonstrate the Fridge Compiler's versatility in synthesizing arbitrary Boolean functions using three different molecular architectures, while satisfying user-specified constraints. We showcase the successful synthesis of all 256 three-bit and 65,536 four-bit predicate functions using a large custom inventory, with worst-case completion times of only seconds on a modern laptop. In addition, we introduce a unique class of cyclic molecular circuits that cover a larger number of Boolean functions than their conventional counterparts over a common inventory, often with significantly smaller implementations. Importantly, and absent in previous approaches specific to molecular circuits, the Fridge Compiler is logically sound, complete, and optimal for the user-specified cost function and component inventory.

Keywords: Molecular Computing · Exact Logic Synthesis · Cyclic Combinational Circuits

1 Introduction

Molecular circuits are being explored and developed due to their natural interface with chemistry and biochemical systems. They can sense from their environment,

Supported by an NSF grant (CCF 2106695) and a Faculty Early Career Development Award from NSF (CCF 2143227).

J. Pang and J. Niehren (Eds.): CMSB 2023, LNBI 14137, pp. 236–252, 2023.
https://doi.org/10.1007/978-3-031-42697-1_16

perform computation, and actuate a physical response *in situ*. Diffusive molecular circuits capable of arbitrary Boolean function computation—as described by rate-independent, well-mixed chemical reaction networks (CRNs) and our focus here—can be experimentally realized by DNA strand displacement (DSD) architectures.

DSD is a molecular primitive that can realize programmed behavior by the rational design of short DNA oligonucleotide strands. Due to designed sequence similarity 'invader' strands can compete to form Watson-Crick base-pairs with a substrate strand and displace an 'incumbent' strand that was initially hybridized. These reactions are often mediated and thermodynamically driven by the enthalpic gain of additional base pairs formed by an invader in a 'toehold' region of the substrate. Displaced strands can in turn act as invaders in downstream components, creating a network of cascading displacements in the presence of appropriate inputs strands [21]. DSD architectures, also referred to as DSD systems, use DSD to implement modular computing components often designed to robustly implement CRNs [2,4,21]. Figure 1 illustrates a detailed DSD pathway for the seminal two-domain architecture [4], and Sect. 2.2 outlines our approach for representing DSD architectures as Boolean Networks.

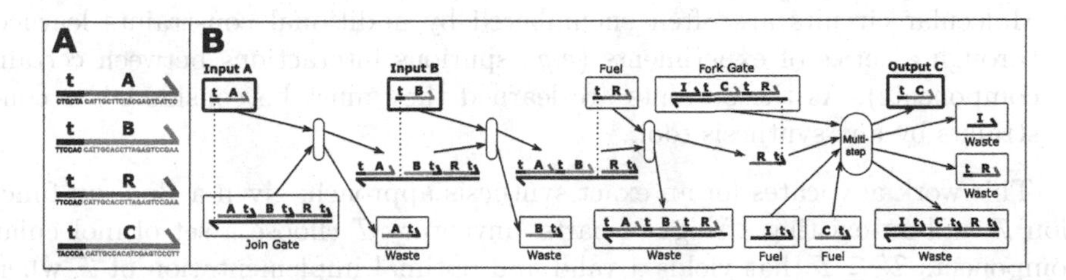

Fig. 1. Example strand displacement implementation of $A + B \rightarrow C$ using the two-domain architecture [4] – interpreted in this paper as $C := \texttt{AND}(A, B)$. DNA is represented with arrows indicating the 3' end. Each step exposes a 'toehold' region labeled `t` in the 'incumbent' strand where the 'invader' initiates binding along the dotted line. **A.** Sequence domains labeled `A`, `B`, `R`, `I`, and `C` serve as distinct signals in our well-mixed solution. Sequences and design insights are derived from [7]. **B.** The DSD pathway shows that output C is generated only if both inputs A and B are present. This process unfolds through a cascade of toehold-mediated reactions that deplete input, gate, and fuel components while generating output and waste components. Given that each circuit preparation can only process one input due to component consumption, we refer to this as a "one-shot" computation.

Given a target function and architecture a molecular programmer must first synthesize a compatible molecular parts list that will correctly compute the intended logic when well-mixed and in the presence of valid input. Manual synthesis is typical; however, automated synthesis becomes necessary to determine minimal size molecular circuits for all but the simplest of functions. Existing approaches have leveraged sophisticated electronic circuit synthesis tools by first synthesizing optimal Boolean circuits over a functionally complete basis (*e.g.* using `AND`, `OR` and `NOT` gates) and then performing technology mapping into the target molecular architecture, as demonstrated in Fig. 2. This strategy falls short for at least four reasons:

1. Traditional circuit synthesis tools are based on heuristic logic synthesis and ignore paradigms like cyclic combinational circuits that can be smaller than their acyclic counterparts.
2. Minimal circuits prior to technology mapping do not necessarily result in minimal molecular circuits; synthesis tools intended for electronic circuit optimization do not consider the unique properties of rate-independent DSD circuits such as it being infeasible to compute negation unless using a dual-rail input encoding, or the relative cost and/or limitation of fan-in and fan-out for particular gate architectures.
3. The molecular programmer must still determine if there is mapping from their existing inventory of molecular parts onto the synthesized circuit. If not, the laborious process of designing and preparing new molecular components would be necessary to realize the synthesized circuit. However, it is entirely possible that an *alternate* circuit could have been realized from an inventory of existing components.
4. Molecular circuits are often encumbered by additional constraints learned through a series of experiments (*e.g.*, spurious interactions between certain components). As these events are learned they must be considered as constraints by any synthesis tool.

This work advocates for an exact synthesis approach: given a Boolean function $\mathcal{F}$ and an existing "fridge of parts" inventory $\mathcal{I}$ choose a set of molecular components $\mathcal{I}' \subseteq \mathcal{I}$ that yields a valid and optimal implementation of $\mathcal{F}$ when mixed together, or give a proof that no such $\mathcal{I}'$ exists. This approach not only leads to a practical tool capable of handling arbitrary constraints and cost functions, it also avoids whenever possible the most costly solutions: those which require the design and/or preparation of molecular components not currently found in the inventory.

Our major contributions in this work are organized as follows. Section 2 defines the molecular circuit synthesis from an inventory problem. Section 3 details the Fridge Compiler tool and an overview of its implementation. Section 4 introduces a new class of cyclic circuits for "one-shot" molecular computation typical in DSD architectures, implementing Boolean functions with fewer or the same number of components than traditional (cyclic or acyclic) circuits. Section 5 provides a number of case studies that demonstrate the compiler is flexible and highly performant. Section 6 explores future directions necessary to meet the current needs of practitioners of molecular programming.

1.1 Relation to Previous Work

Molecular Circuit Synthesis. Tools and compilers exist that support the experimental implementation of arbitrary or specific DNA strand displacement

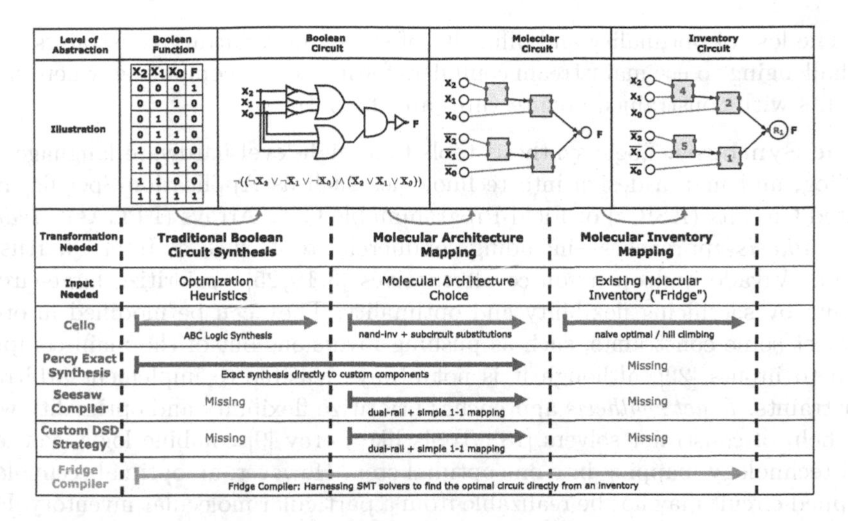

Fig. 2. Overview of circuit synthesis of Boolean functions, contrasting different approaches including Cello for genetic circuits, Percy exact synthesis, Seesaw Compiler, a custom strategy, and the Fridge Compiler. Given an inventory of distinguishable molecular parts and a function specification, other synthesis strategies shown take multiple synthesis steps to find a solution. During these steps optimality, completeness, and/or flexibility may be lost.

architectures, but they typically assume a target chemical reaction network or circuit [2,12,15]. As researchers develop larger and more advanced DSD circuit inventories, an unfulfilled need has arisen for the ability to use this existing material efficiently in a scalable manner while taking into account common DSD constraints. Our focus is automated synthesis of molecular circuits given a Boolean function description and an inventory of well-tested molecular parts.

The `Cello` tool [11,14] workflow is most comparable to our aim in this work, as it seeks to implement a Boolean function using a set of existing genetic circuit components. Genetic circuits use genetic regulatory networks to perform computation and therefore rely on reaction rate differences and repression to implement negation (*e.g.,* to realize NOR logic). In contrast, DSD architectures use rationally designed DNA strands to perform computation and cannot robustly implement negation, often relying on a dual-rail representation for functional completeness.

Both types of circuits share the common challenge of designing and testing limited-resource components. Cello utilizes a heuristic logic synthesis step to create an optimized intermediate circuit, making the technology mapping phase trivial as all gates in the intermediate network can be implemented. The heuristic fridge mapping phase in Cello is also more complex, using analog gate response characterization instead of simple digital abstraction. Although Cello's heuristic logic synthesis offers advantages in genetic circuit design, it cannot implement optimal dual-rail circuits for DSD and lacks the completeness, and constraint-flexibility guarantees of the Fridge Compiler. Additionally, Cello does not support combinational cycles.

The loss of optimality and difficulty of satisfying arbitrary constraints make it challenging to use mainstream compilers for molecular computing, where small circuits with constrained architectures are the norm.

Logic Synthesis. Logic synthesis tools take high-level hardware language like Verilog, and map a design into technologies such as Application-Specific Integrated Circuits (ASICs) or Field-Programmable Gate Arrays (FPGAs). *Heuristic synthesis* approaches—including commercial tools such as Intel Quartus or Xilinix Vivado and open source alternatives [3,19,25]—prioritize large-circuit scaling by sacrificing flexibility and optimality. They can be modified in order to meet some constraints, such as pushing inversions out of the main compute path to inputs [22], although it is not always feasible to implement arbitrary constraints. *Exact synthesis* approaches prioritize flexibility and optimality with the help of constraint solvers [18]. Tools like Percy [9] combine logic synthesis and technology mapping into one optimal step. However, an optimal technology mapped circuit may not be realizable from a particular molecular inventory. Furthermore, both heuristic and exact logic synthesis tools often overlook paradigms such as cyclic combinational circuits, which are required for achieving minimal circuit sizes [16].

Program Synthesis. Program synthesis generates an implementation from its specifications and constraints, often employing SMT solvers such as Z3 [13] to define and search the space of valid solutions. *Satisfiability modulo theory* (SMT) problems, a superset of SAT problems, accommodate variables and constraints from domains (or *theories*) beyond Boolean algrebra, such as bitvectors and uninterpreted functions. SMT solvers have been used to synthesize and verify code in various domains, including DNA computation and synthetic biology for analyzing and verifying biological systems [26]. The synthesized circuits from these tools are also often acyclic, resulting in suboptimal combinational circuits (see Sect. 4).

2 Preliminaries

2.1 Molecular Circuit Synthesis from an Inventory

Molecular circuit synthesis from an inventory aims to find an optimal molecular circuit given a target function, user-defined constraints, and a fridge inventory of distinguishable, freely-diffusing components. Components must be *distinguishable* in diffusive computing, such as those implementations describable as a well-mixed CRN, since relationships between components must be programmed into their designed interactions (and lack of interactions) with other species components due to the lack of spatial organization. In contrast, two `AND` gates, for example, can be *indistinguishable* in an electronic circuit embedded on a surface since their connectivity to other components is entirely controlled by their spatial organization.

Name	CRN Formula	CRN Symbol	Logic Formula	Logic Symbol
Gate	A + B → C + D	G	{C, D} = AND(A,B) = A·B	
Wire	A → C		C = A	
Or / Merge	A → C B → C		C = OR(A,B) = A+B	

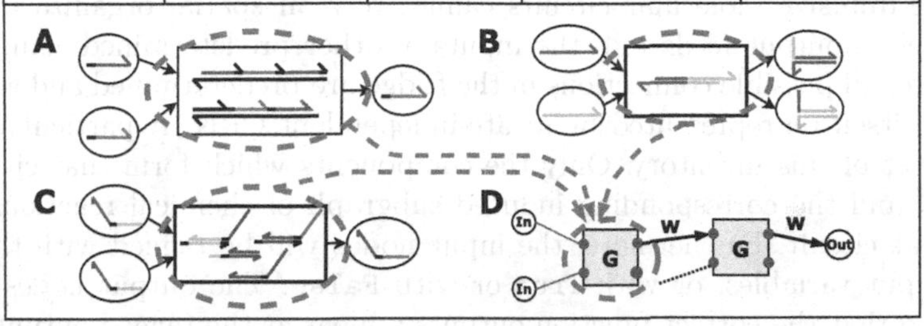

Fig. 3. (Top) Default circuit and inventory representation in this paper: CRNs with components as either 2-input 2-output CRNs (gates) or 1-input 1-output CRNs (wires). OR logic is commonly implemented by multiple wires converging to a common output. (Bottom) Fridge Compiler users can choose among a variety of pre-programmed DSD architectures including (A) two-domain [4], (B) leakless breadboard [23], (C) seesaw [15]. Alternatively, the user can create their own. The Fridge Compiler is capable of supporting a variety of architectures by abstracting the specific implementation into a generic circuit structure (D) while still retaining the important differences in the form of constraints.

2.2 Modelling Molecular Inventories

Rate-independent well-mixed CRNs, which compute correctly without reaction rate assumptions, are an appealing class of reactions due to their relative ease of engineering and ability to perform complex functions [5]. To represent arbitrary Boolean functions using rate-independent CRNs, we utilize a dual-rail input representation with distinct species for positive and negative literals, and consider outputs on upon reachability of output species – thereby treating the CRN representation like a Boolean network [5].

While our primary focus is on DSD-based CRNs our approach could be adapted to other molecular computing frameworks, including protein and genetic circuits, given appropriate consideration for managing complex constraints, inversions/negations, and cycles. The Fridge Compiler, detailed in the next section, is compatible with a variety of DSD architectures and abstracts specific implementation details to the rate-independent CRN level, while maintaining important differences as constraints to the SMT solver.

Figure 3 (top) shows the convention used for different nodes within this Boolean network interpretation of CRNs. A signal represents a single molecule; it is represented by dark blue circle gate ports. A gate represents a reaction/component with (multiple) inputs and (multiple) outputs. Gates in this interpretation perform `AND` logic, like a transition in a Petri net. In many architectures `OR` logic can be performed by mapping multiple distinct signals into a common signal. A wire represents a translator from one signal to another. Figure 3 (bottom) shows the domain-level design of gate nodes in three different DSD architectures. Signals can be tagged as possible circuit outputs or inputs that correspond to fluorescent reporters on hand, or available DNA strands, respectively.

Since diffusive molecular circuits cannot rely on spatial organization, each component's output feeds into the inputs of other predetermined components. Therefore, all possible connections in the fridge are predetermined and an inventory can itself be represented by a rate-independent CRN. A particular circuit is a subset of this inventory. Only the components which form that circuit are enabled, and the corresponding induced subgraph of chemical reactions represents that circuit. Furthermore, the input nodes will be tagged with the function's input variables, or with `True` or with `False`. The output nodes will be tagged so that the correct function outputs appear at the tagged output nodes.

2.3 Desired Features of a (DSD) Fridge Compiler

Along with the distinguishability of components, an arduous development cycle is one of the major reasons for the need and development of an efficient "fridge compiler" with the following properties. *Soundness and Completeness.* An input query consists of a fridge and a Boolean function. Soundness guarantees that if the compiler returns a circuit, it will always correctly implement the input query. Completeness guarantees that if the compiler does not return a circuit, then there does not exist any circuit that fulfills the input query. Most alternative methods are sound, but not complete. *Flexibility of Architecture Choice.* A fridge compiler should be able to support any DSD architecture, and in the future any molecular architecture. *Handling practical constraints.* A fridge compiler should be able to handle arbitrary constraints such as the known incompatibility between stock components. *Optimality.* Optimality is important in molecular computing because molecular components are expensive in their preparation and the probability of spurious interactions increase with system size. Arbitrary cost functions should be possible, since each architecture has different comparative cost for each component. *Supporting Cyclic Circuits.* As shown in Sect. 4, optimal circuits will often contain cycles.

2.4 A Naive Solution

Figure 2 shows the levels of abstraction that traditionally take place to transform a Boolean function description and an inventory into a final circuit. A naive solution to molecular circuit synthesis from an inventory would be to enumerate all possible circuits within that inventory and evaluate whether any implement the desired function. Each enumeration choice requires picking which input node to use for which literal, picking a subset of components to enable, and picking which output node(s) to interpret as output. Checking if the intended function was synthesized requires evaluating all possible input combinations in the worst case. This naive enumeration algorithm has the properties we seek, but is infeasible. The Fridge Compiler keeps these properties, but is efficient in practice.

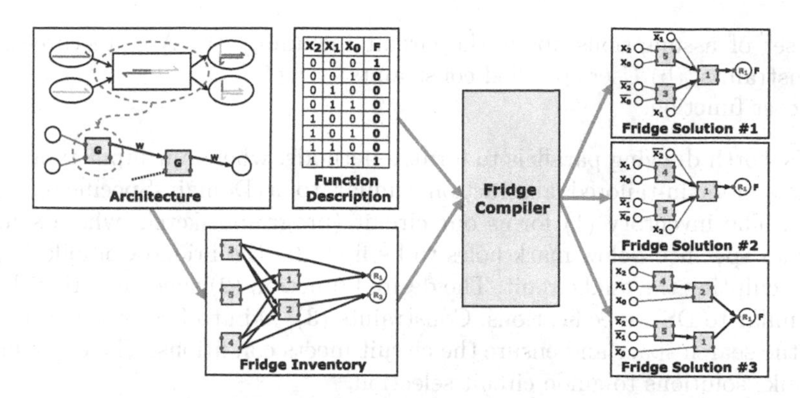

Fig. 4. Given a inventory and function specification, the Fridge Compiler can synthesize all valid circuit solutions (#1, #2, #3, ...), (size) optimal solutions (#1, #2), and optimal solutions that satisfy hard constraints (#1; "don't mix gates 4 and 5, nor gates 1 and 2"). Optimality, completeness, and flexibility are maintained.

3 The "Fridge" Compiler

3.1 General Overview

The Fridge Compiler is an exact synthesis tool similar to those in Sect. 1.1, providing an exact and optimal solution to a circuit description, over a given fridge inventory, and subject to constraints. It uses an SMT solver to describe the large set of enumerations using symbolic datatypes which replace the naive concrete enumerations. This approach fulfills all the desired features enumerated in Sect. 2.3: it is sound, complete, flexible, optimal, supports cycles, and is tractable in practice.

While many exact synthesis tools rely on custom-encoded SAT encoding [9], we build on program synthesis techniques that utilize more general SMT encodings because of their expressive power and ease-of-use [10]. We see this user-first approach as especially valuable in fields like molecular computing where the barrier to entry is already high. Our implementation borrows techniques from program synthesis and formal methods such as program sketches, symbolic execution, and SMT-based variable encoding [10,20,24]. We utilize a "sketching" approach where users with domain expertise write partial programs and leave "holes" for a solver to complete. The following section describes the Fridge Compiler's inputs and implementation.

3.2 Inputs

1. A inventory specified as: (a) A chemical reaction network without rates. (b) The subset of reactants that can be used as circuit inputs. (c) The subset of products that can be used as circuit outputs.
2. A desired output function, $F_n(X) = [F_0(X), F_1(X), ...]$, composed from one or more Boolean predicates.

3. A set of assumptions about the circuit including: (a) Architecture-specific constraints. (b) User-specified constraints.
4. A cost function.

It's worth drawing parallels to formal methods, where our input types above create a domain-tailored abstraction similar to a Domain-Specific Language (DSL). The inventory (1) forms our circuit (program) sketch, where symbolic variables specified below mark holes to be filled by the Fridge Compiler – effectively sculpting the final circuit. The desired function (2) specifies circuit behavior, similar to DSL specifications. Constraints (3), akin to DSL's semantic rules, limit the search space and ensure the circuit meets conditions. The cost function (4) ranks solutions to guide circuit selection.

3.3 Implementation

SMT Variable Encoding. Our approach to variable encoding in SMT is inspired by previous works [10]. *Logical input* variables, denoted as $X = [X_0, X_1, ...]$, serve as the Boolean inputs for the specification function (2). *Circuit selection* variables are Boolean values that dictate whether each component or reaction in the inventory is included in the circuit. Given that all components are distinguishable with pre-determined connections, selection variables define a complete circuit without I/O assignments. We also employ I/O location variables, which use integer values to map the input and output to the circuit [10]. *Input location* variables map our allowed circuit inputs (1b) above to a dual-rail version of our logical input variables. *Output location* variables map our function outputs (2) to our allowed circuit outputs (1b).

While a universal quantifier is performed over the *logical input* variables to ensure the equivalence of the circuit and function across all inputs, the remaining variables act as hole variables in the circuit sketch of the inventory. The solver fills these holes by assigning suitable values, defining a complete circuit including its I/O assignments.

Symbolic Interpreter for the Inventory. Interpreting a molecular circuit begins by assigning Boolean values to circuit inputs and triggering reactions until a steady state is reached, as exemplified in Fig. 5.

To encode this into the solver, we apply symbolic execution [24] that transforms our concrete interpreter into a symbolic expression encapsulating all potential circuits. The interpreter starts from circuit inputs, integrates the SMT variables, and continuously simplifies the expression using Z3's *simplify* command.

In acyclic circuits, the interpreter accumulates logic from input to output, resulting in an expression proportional to the number of reactions in the CRN (1a). For cyclic circuits, which take more steps to reach the steady state depending on the number of cycles, we employ SMT verification to check the expression's equivalence after each gate, compressing the final expression – this step occurs only once per inventory and is typically not costly. Note that even for cyclic circuits a steady state is always reached because once an output is produced it cannot be removed, by the monotonicity property of rate-independent CRNs [5].

This procedure yields SMT expressions for circuit outputs, depending on logical inputs X, circuit selection, and I/O location, designated as $CircuitF_n(X) = [CircuitF_0(X), CircuitF_1(X), ...]$.

Additional Constraints. Additional constraints further guide the solver, focusing the search and limiting the circuit space. These constraints are translated into SMT format and include architecture-specific constraints and user-defined requirements. Examples of architecture-specific constraints include uniqueness of location variables or forcing all outputs to be used by at most one enabled component. Examples of user-specified constraints include component incompatibility, restrictions on fan-in and fan-out, limitations on circuit depth, concentration splitting, or stock depletion. This flexibility in constraint definition allows for more customized and application-specific circuit designs.

Program Synthesis. The program synthesis task is to find an assignment of hole variables – *circuit selection, input location, output location* such that, assuming the *constraints* hold true, the function $F_n(X)$ matches $CircuitFn(X)$ for all possible X – **assuming** *constraints*, **forall** X: **assert** $F_n(X) == CircuitFn(X)$.

Existence vs Optimality. The system functions in two modes: existence mode, which finds the first circuit satisfying the query, and optimize mode, which seeks the most optimal circuit. Both modes are sound and complete. The cost function can be any function that can be evaluated for a specific circuit, allowing for extensive flexibility. For example, if a gate costs twice as much as a wire the cost function could be 2*W+1*G. However, performance is typically best with simple lexicographical cost functions rather than pareto tradeoff cost functions.

Implementation Details. For handling the universal (**forall**) quantifier in our implementation, we provide two options. By default, we use a brute-force mode where the expression $F_n(X) == CircuitFn(X)$ is resolved for each X value ($2^{N_{bits}}$ times), and each result is added as an assertion. This technique is effective for case studies in this paper with fewer input variables. However, when the number of logical inputs is large, we resort to CEGIS [20], a well-established synthesis technique for handling larger search spaces.

We use bitvector variables over integers for the *circuit selection*, *input location*, and *output location* variables, due to their superior performance especially given the relatively small range of these variables.

For optimization, we by default use the built-in SMT optimizer to minimize a set of cost functions. For larger problems where this method may time out, we provide an alternative implementation which uses a cycle of synthesize calls with a decreasing cost constraint.

The Fridge Compiler is currently implemented in Python 3, utilizing Z3 [13] as the SMT solver. An earlier version was implemented in Rosette [24].

3.4 Flexibility of Architecture Choice

The Fridge Compiler can in theory accommodate a variety of architectures, with three common preprogrammed examples serving as guiding templates (see

Fig. 3). These built-in architectures showcase the Fridge Compiler's ability to handle diverse DSD architectures and provide a blueprint for users to specify custom ones or support alternate molecular architectures.

4 One-Shot Cyclic Circuit Synthesis

One-shot, or single-use, computation refers to computation where circuit components are consumed during evaluation of each specific input. While there is ongoing work towards implementing reusable DSD circuits, large-scale DSD systems have implemented one-shot computation in practice [8]. As depicted in Fig. 1, input strands cause a sequence of cascading reactions leading to the depletion of circuit components along with an output readout. Consequently, the circuit must be newly prepared for each desired input.

Cyclic circuits, while usually associated with memory circuits like latches and flip-flops, can also be found in combinational circuits that only depend on current inputs. As early as 1953, Claude Shannon found that allowing cycles in circuits is necessary for obtaining optimal combinational circuits [17]. Significant advancements in the understanding of cyclic combinational circuits have been made by formalizing the terminology, providing comprehensive analysis strategies, proving optimality for some classes of circuits, and developing synthesis techniques to incorporate cycles into smaller combinational circuits [1,6,16]. Despite the potential of cyclic combinational circuits to optimize circuit size, they are not widely supported by mainstream compilers.

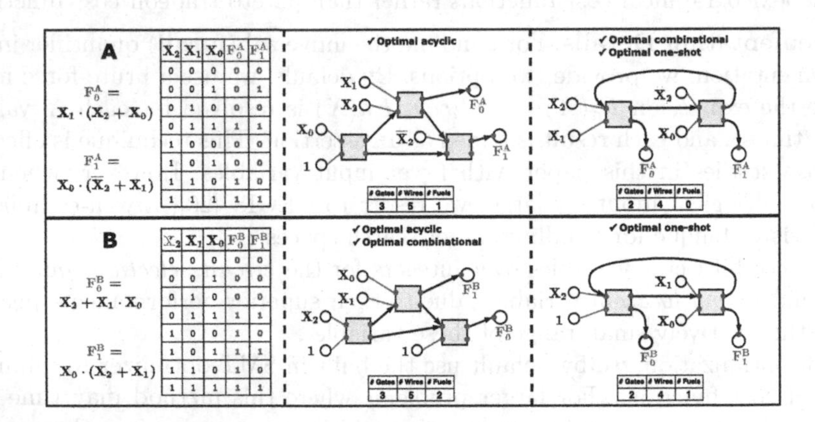

Fig. 5. A. Cyclic combinational circuits. *(Left)* Function definition for F^A; *(Middle)* its optimal acyclic implementation; *(Right)* its optimal cyclic combinational. **B.** *(Left)* Function definition for F^B; *(Middle)* its optimal acyclic implementation; *(Right)* its optimal one-shot cyclic implementation.

To analyze cyclic circuits we can assign an initial value C_i to a wire along each cycle i, and check if the circuit retains any values C_i once the circuit is evaluated on an input. In conventional cyclic combinational circuits, every wire and output is assigned a specific value after circuit evaluation for all input values, ensuring no C_i values are left [16]. Figure 5A shows function F^A implemented through the smallest possible acyclic circuit (left) and through an even smaller combinational cyclic circuit (right). Going through the input combinations we can verify that the outputs are always forced to `True` or `False` – there is no input which leaves the cycle value unresolved, so this circuit is combinational.

We introduce one-shot cyclic circuits, which are cyclic circuits that leverage properties of one-shot circuits to create smaller combinational circuits not possible in traditional settings. These circuits are smaller than both conventional acyclic and cyclic combinational circuits.

CRN-based circuits, characterized by their one-shot assumption (that no previous input exists) have the unique property of initializing all cycle values C_i to `False`, as the component gates start un-reacted without an input. Therefore, any C_i values remaining on the output or along cycles in one-shot CRN-based circuits will actually result in a concrete output of `False`. Figure 5(B) shows function F^B implemented through the smallest possible acyclic circuit (left) and through an even smaller one-shot cyclic circuit (right). Going through the input combinations we find that both outputs F_0^B and F_1^B are dependent on the cycle value for the input $XYZ = 001$. Without the one-shot assumption, this circuit would not be combinational since a previous cycle value would influence the current output. However assuming a one-shot setting, the circuit in this case outputs the initial C_i value of `False`. The resulting truth table and functions are shown. Unlike for F^A, F^B does not have a cyclic combinational circuit smaller than the acyclic implementation, but does in fact have a smaller one-shot cyclic circuit.

This example shows that the one-shot assumption may allow for dual-rail circuits that are smaller than any traditional combinational circuit (acyclic or cyclic). As shown in Sect. 5.3, this property could be verified for any arbitrary inventory using the Fridge Compiler due to its completeness guarantee Any valid combinational circuit (acyclic or cyclic) is also a valid one-shot circuit, since it has outputs that are uninfluenced by their (possible) cycle values, meaning that initializing the cycle values to `False` in a one-shot setting will not impact the circuit's behavior. In contrast, one-shot cyclic circuits are distinct from cyclic combinational circuits as they rely on a previous circuit state, where all gates are initially `False`. Therefore the one-shot assumption creates a larger set of possible solutions, while encompassing both acyclic or cyclic combinational circuits. The Fridge compiler assumes the presence of cycles in its inventory and generates cycles in the resulting circuits when necessary to ensure optimality.

5 Case Studies and Empirical Evaluation

A natural question is: what functions can inventories cover? We chose three sets of Boolean function classes to test: all 256 3-bit predicates, all 65,536 4-bit predicates, and all 32,640 3-bit 2-predicate functions. The 3-bit predicate set was

Test Inventory	Gates	Wires	Function Set	# Functions	# SAT		MIN (sec)		50th % (sec)		99th % (sec)		MAX (sec)	
						Version	SAT	UNSAT	SAT	UNSAT	SAT	UNSAT	SAT	UNSAT
5G 12W [Thachuk]	5	12	3-bit 1-pred	256	246	*Existence*	0.01	0.06	0.02	0.06	0.09	0.10	0.10	0.10
						Optimal	0.01	0.02	0.02	0.03	0.11	0.05	0.20	0.05
5G 12W [Thachuk]	5	12	4-bit 1-pred	65536	22394	*Existence*	0.02	0.05	0.05	0.23	0.29	0.64	0.75	1.69
						Optimal	0.02	0.03	0.07	0.07	0.56	0.19	1.49	0.64
19G 27W (Large Custom)	19	27	4-bit 1-pred	65536	65536	*Existence*	0.04	N/A	0.05	N/A	0.13	N/A	0.52	N/A
						Optimal	0.04	N/A	0.39	N/A	21.60	N/A	181	N/A
3-Gate *Acyclic* Complete	3	15	3-bit 2-pred	32640	3936	*Existence*	0.02	0.02	0.02	0.02	0.02	0.03	0.08	0.09
						Optimal	0.01	0.01	0.01	0.01	0.03	0.02	0.05	0.1
3-Gate *Cyclic* Complete	3	26	3-bit 2-pred	32640	4824	*Existence*	0.02	0.02	0.02	0.02	0.03	0.03	0.04	0.11
						Optimal	0.01	0.01	0.02	0.01	0.03	0.02	0.16	0.07
4-Gate *Acyclic* Complete	4	28	3-bit 2-pred	32640	11384	*Existence*	0.02	0.03	0.03	0.10	0.09	0.25	0.22	0.51
						Optimal	0.02	0.02	0.12	0.07	0.50	0.16	4.13	0.51
4-Gate *Cyclic* Complete	4	50	3-bit 2-pred	32640	13604	*Existence*	0.03	0.05	0.04	0.17	0.18	0.88	0.46	6.66
						Optimal	0.02	0.03	0.14	0.09	2.02	0.53	46.57	3.45

Fig. 6. Runtimes and coverage results for case studies of various inventories and functions. For instance, the last row tells us that for the complete inventory of 4 gates and all possible wires (including feedback wires), the Fridge Compiler was run (in existence and optimality mode) with all functions of 3 input bits and 2 output predicates. Of those 32,640 target functions, 13,604 were implementable with the inventory. Of the 13,604 optimal circuits found, 99% of them were found in 2 s or less.

chosen in order to demonstrate a qualitative comparison to the results from the original Cello work [14] that attempted to synthesize (and experimentally realize) all 3-bit predicate functions. The 4-bit predicate set was chosen as a natural extension and in order to support known designs [15], and the 3-bit 2-predicate set was chosen to highlight the resource-sharing benefits of cyclic circuits. The primary goal of these case studies is to demonstrate a possible use-case of the Fridge Compiler. The current approach addresses problem instance sizes that are currently feasible to build experimentally. Although there is no present need for the Fridge Compiler to handle significantly larger problem instances, many techniques such as those discussed in Sect. 3.3 have shown to improve program synthesis runtime substantially without changing the underlying setup. Future work would include systematically characterizing larger inventories with a larger number of logical inputs.

5.1 Case Study: Synthesizing All k-Bit Predicates

After running the Fridge Compiler on all 256 3-bit predicates, it turns out the breadboard inventory of Fig. 4 covers 246/256 3-bit predicates. One of the functions this inventory cannot cover is odd parity. The Fridge Compiler guarantees that given the architecture, the inventory, the function, and the constraints, no circuit can represent this function. We designed a sum-of-product circuit that covers 3-bit odd parity as a new inventory; that inventory covers all 256 3-bit predicates. This makes intuitive sense: odd parity is one of the most complex functions in terms of circuit complexity.

5.2 Case Study: Inventory for All 4-Bit Predicates

The 5-gate 12-wire inventory from the previous case study covers 22,394/65,536 4-bit predicates. A larger inventory was custom designed to cover all 65,536 4-bit predicates: It contains 19 gates and 27 wires, and contains the odd parity circuit with a few extra gates. In order to test constraints, we allow this inventory to map input variable literals such as X_i or $\overline{X_i}$ to be mapped to more than one gate input. This constraint is valid in some architectures, and can often leads to more function coverage (see Figs. 6, 7).

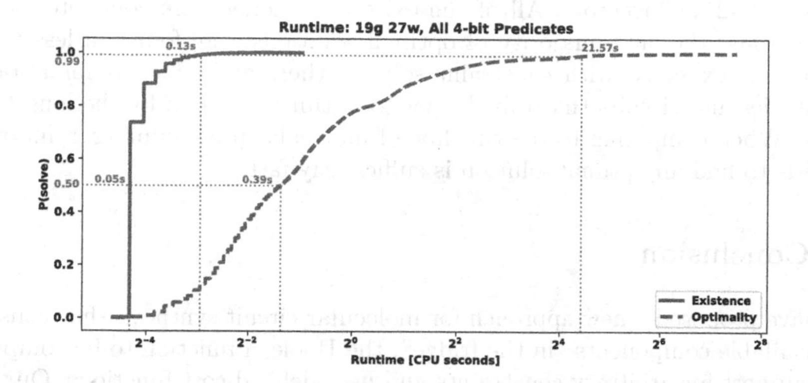

Fig. 7. A cumulative distribution function (CDF) curve illustrating runtime distribution for the test suite from Fig. 6's third row. The X and Y axes represent runtime and proportion of tests completed in that time, respectively. This large custom inventory is able to synthesize all 65,536 4-bit functions. Existence finds any circuit that implements the function, and is typically solved faster than finding the optimal circuit.

5.3 Case Study: Cyclic vs Acyclic

Cyclic circuits excel at optimizing resource-sharing between different parts of the circuit. To demonstrate this, we ask the Fridge Compiler to synthesize all 2-predicate functions with 3-bit inputs using a fridge of 3 Gates and any arbitrary wire connections. Figure 5 (Left) show 2 examples of functions in this set, and 2 implementations for each function. The Fridge Compiler finds 3,936/32,640 of the functions are implementable using 3 gates with acyclic circuits. By enabling permitting combinational cycles, an additional 72 functions can be realized, and 72 existing functions have smaller circuit implementations. For instance, Fig. 5A shows a function that has a smaller circuit when cominational cycles are enabled. One-shot cycles can support even smaller circuits than combinational cyclic circuits. By enabling the Fridge Compiler to find one-shot cycles, we first observe that all 144 functions that were additionally found or improved with combinational cycles are still found, since combinational circuits are a subset of one-shot circuits. An additional 816 functions have circuit implementations, and 48 existing functions have smaller circuit implementations. Figure 5B shows a function where one-shot cycles enabled a smaller implementation while combinational cycles did not. Synthesis over a complete 4-gate inventory shows even better improvement. While 11,384/32640 functions are implementable using acyclic circuits, enabling one-shot cycles increases that amount to 13,604/32,640 functions synthesized.

5.4 Runtime Performance

Even while guaranteeing soundness, completeness, and optimality, Fridge Compiler runtimes are on the order of seconds on a personal laptop. Figure 6 shows the function coverage and runtimes for all case studies listed. In addition, the cumulative distribution function (CDF) in Fig. 7 visually compares the runtimes of the 19G-27W inventory. All of the existence queries were completed within seconds, and the large majority of optimal solutions were found in less than a second. As expected with constraint solvers, there are a few combinations of inventories and circuits that take longer to optimize, as seen by the long tail in Fig. 7. When comparing to the timeline of molecular programming, a matter of seconds to find an optimal solution is sufficiently fast.

6 Conclusion

We have proposed a new approach for molecular circuit synthesis that considers the available components "in the fridge", the Boolean function to be computed, and support for arbitrary constraints and user-defined cost functions. Our approach fills a need not met by the current menagerie of tools that support the development of molecular circuits, nor is the need met by existing circuit synthesis tools that were designed for electronic circuits; the latter misses important distinctions between electrical and molecular circuit properties. Importantly, our approach is sound, complete, and optimal. Empirical evaluations demonstrate that this approach is also efficient in practice with worst-case solutions found on the order of seconds. We have also introduced and supported synthesis of a new class of one-shot cyclic combinational circuits that can cover more functions and can yield smaller circuits for a given inventory.

In future work we intend to demonstrate the one-shot cyclic circuit in Fig. 5B using a DSD circuit. We also plan to explore how to identify the minimum additional components to an existing inventory needed to support a set of functions, identify the smallest necessary inventory for a set of functions, and add a feature to balance stock depletion of fridge components. Furthermore, we plan to extend our work to include other molecular computing architectures such as genetic and protein circuits. We will also try adapting this methodology towards synthesizing analog, probabilistic, and/or continuous piecewise linear functions.

References

1. Backes, J.D., Riedel, M.D.: The synthesis of cyclic dependencies with Boolean satisfiability. ACM Trans. Design Autom. Electr. Syst. **17**, 44:1–44:24 (2012)
2. Badelt, S., Shin, S.W., Johnson, R.F., Dong, Q., Thachuk, C., Winfree, E.: A general-purpose CRN-to-DSD compiler with formal verification, optimization, and simulation capabilities. In: Brijder, R., Qian, L. (eds.) DNA 2017. LNCS, vol. 10467, pp. 232–248. Springer, Cham (2017). https://doi.org/10.1007/978-3-319-66799-7_15

3. Brayton, R.K., Mishchenko, A.: ABC: an academic industrial-strength verification tool. In: International Conference on Computer Aided Verification (2010)
4. Cardelli, L.: Two-domain DNA strand displacement. Math. Struct. Comput. Sci. **23**, 247–271 (2010)
5. Chen, H.L., Doty, D., Soloveichik, D.: Rate-independent computation in continuous chemical reaction networks. In: Proceedings of the 5th Conference on Innovations in Theoretical Computer Science, ITCS 2014, pp. 313–326. Association for Computing Machinery (2014)
6. Chen, J.H., Chen, Y.C., Weng, W.C., Huang, C.Y., Wang, C.Y.: Synthesis and verification of cyclic combinational circuits. 2015 28th IEEE International System-on-Chip Conference (SOCC), pp. 257–262 (2015)
7. Chen, Y.J., et al.: Programmable chemical controllers made from DNA. Nat. Nanotechnol. **8**(10), 755–762 (2013). https://doi.org/10.1038/nnano.2013.189
8. Eshra, A., Shah, S., Song, T., Reif, J.H.: Renewable DNA hairpin-based logic circuits. IEEE Trans. Nanotechnol. **18**, 252–259 (2019)
9. Haaswijk, W., Soeken, M., Mishchenko, A., Micheli, G.D.: SAT-based exact synthesis: encodings, topology families, and parallelism. IEEE Trans. Comput. Aided Des. Integr. Circuits Syst. **39**, 871–884 (2020)
10. Jha, S., Gulwani, S., Seshia, S.A., Tiwari, A.: Oracle-guided component-based program synthesis. In: 2010 ACM/IEEE 32nd International Conference on Software Engineering, vol. 1, pp. 215–224 (2010)
11. Jones, T.S., Oliveira, S.M.D., Myers, C.J., Voigt, C.A., Densmore, D.M.: Genetic circuit design automation with Cello 2.0. Nat. Protocols **17**, 1097–1113 (2022)
12. Lakin, M.R., Youssef, S., Polo, F., Emmott, S., Phillips, A.: Visual DSD: a design and analysis tool for DNA strand displacement systems. Bioinformatics **27**, 3211–3213 (2011)
13. de Moura, L.M., Bjørner, N.S.: Z3: An efficient SMT solver. In: International Conference on Tools and Algorithms for Construction and Analysis of Systems (2008)
14. Nielsen, A.A.K., et al.: Genetic circuit design automation. Science **352** (2016)
15. Qian, L., Winfree, E.: Scaling up digital circuit computation with DNA strand displacement cascades. Science **332**, 1196–1201 (2011)
16. Riedel, M.D.: Cyclic combinational circuits. California Institute of Technology (2004)
17. Shannon, C.E.: Realization of All 16 Switching Functions of Two Variables Requires 18 Contacts: Bell Laboratories Memorandum, pp. 711–714. Wiley-IEEE Press (1953)
18. Soeken, M., et al.: Practical exact synthesis. In: 2018 Design, Automation & Test in Europe Conference & Exhibition (DATE), pp. 309–314 (2018)
19. Soeken, M., Riener, H., Haaswijk, W., Micheli, G.D.: The EPFL logic synthesis libraries. ArXiv abs/1805.05121 (2018)
20. Solar-Lezama, A., Tancau, L., Bodik, R., Seshia, S., Saraswat, V.: Combinatorial sketching for finite programs. In: Proceedings of the 12th international conference on Architectural support for programming languages and operating systems, pp. 404–415 (2006)
21. Soloveichik, D., Seelig, G., Winfree, E.: DNA as a universal substrate for chemical kinetics. Proc. Natl. Acad. Sci. **107**, 5393–5398 (2009)
22. Testa, E., et al.: Inverter propagation and fan-out constraints for beyond-CMOS majority-based technologies. 2017 IEEE Computer Society Annual Symposium on VLSI (ISVLSI), pp. 164–169 (2017)

23. Thachuk, C., Winfree, E.: A fast, robust, and reconfigurable molecular circuit breadboard. In: 15th Annual Conference on Foundations of Nanoscience (2018). https://thachuk.com/talk/2018-fnano-invited/2018-FNANO-invited.pdf, invited Talk
24. Torlak, E., Bodík, R.: Growing solver-aided languages with Rosette. In: SIGPLAN Symposium On New Ideas, New Paradigms, and Reflections on Programming and Software (2013)
25. Wolf, C.: Yosys open synthesis suite (2016)
26. Yordanov, B., Hamadi, Y., Kugler, H., Wintersteiger, C.M.: Z34Bio: an SMT-based framework for analyzing biological computation. In: SMT Workshop 2013 11th International Workshop on Satisfiability Modulo Theories (2013)

Joint Distribution of Protein Concentration and Cell Volume Coupled by Feedback in Dilution

Iryna Zabaikina[1(✉)], Pavol Bokes[1], and Abhyudai Singh[2]

[1] Comenius University, Bratislava 84248, Slovakia
{iryna.zabaikina,pavol.bokes}@fmph.uniba.sk
[2] University of Delaware, Newark, DE 19716, USA
absingh@udel.edu

Abstract. We consider a protein that negatively regulates the rate with which a cell grows. Since less growth means less protein dilution, this mechanism forms a positive feedback loop on the protein concentration. We couple the feedback model with a simple description of the cell cycle, in which a division event is triggered when the cell volume reaches a critical threshold. Following the division we either track only one of the daughter cells (single cell framework) or both cells (population framework). For both frameworks, we find an exact time-independent distribution of protein concentration and cell volume. We explore the consequences of dilution feedback on ergodicity, population growth rate, and the bias of the population distribution towards faster growing cells with less protein.

Keywords: Chapman–Kolmogorov equation · Population balance equation · Supercritical multitype branching process · Measure-valued Markov process · Fourier method · Laplace tranform

1 Introduction

Due to low copy number of molecules, gene expression is a stochastic process [1,21]. The typical question asked of a mathematical model is to determine the distribution of a protein level at steady state [3,45]. Traditionally, the number of cells is treated as a discrete variable which is subject to production and depletion events [4,48]. Protein production is modelled by a compound Poisson process [12,34]. Production events are referred to as bursts and the number of protein molecules produced per event as the burst size [6,7]; these are assumed to be independent and identically geometrically distributed [35,36]. Depletion of protein can be driven by two different mechanisms: degradation or partitioning at cell division [30].

Degradation removes one molecule at a time and leads to tractable models with explicit steady state distributions [13,24,32]. Partitioning based depletion requires that a model for the cell cycle be combined with the model for gene

J. Pang and J. Niehren (Eds.): CMSB 2023, LNBI 14137, pp. 253–268, 2023.
https://doi.org/10.1007/978-3-031-42697-1_17

expression [17,42]. At the end of the cell cycle, each molecule "flips a coin" to decide which of the two daughter cells it goes to [5]. From a single cell line perspective, the partitioning of molecules manifests as a downward jump from the current protein cell count to roughly one half of it. Other types of partitioning than symmetric binomial have also been considered [30]. Models with partitioning errors are more complex than the degradation based models [29,49].

One way to circumvent the complexity due to partitioning is to track the protein concentration instead of the protein count [8,31]. Production is still modelled by a compound Poisson process [28,33], but the geometric distribution of burst sizes is replaced by its coarse-grained continuous counterpart, the exponential distribution [11]. Following other authors, we thereby assume that burst sizes are in units of concentration [22]; this implies a coupling of burst size to cell volume as shown experimentally [40]. The binomial partitioning reduces, by the law of large numbers, to the exact halving of molecules [27]. At the same time, the volume of the mother cell is assumed to split into two exact halves [46,53,54]. Consequently, the concentration (the number of proteins per unit volume) is unaffected by the partitioning at the end of the cell cycle. Instead, it decays continuously over the duration of the cell cycle as the volume of the cell grows [15,52]. In the simplest case, we assume that the volume grows exponentially during the cell cycle, so that the concentration decays exponentially. The basic model without feedback leads to the gamma distribution of protein concentration [10,25,39].

We consider a scenario where the level of a given protein negatively regulates the exponential rate of cell volume expansion [47,55]. Such feedback coupling is often manifested as a result of the burden placed by the protein on the cells' expression/metabolic machinery [16,37,41,51]. In the feedback scenario, low protein cells proliferate faster than high protein cells; it is therefore important to distinguish between the single cell (genealogy) framework and the population framework [19,23,50]. For a single cell line case, the model is a bivariate Markov process, the variables being the protein concentration and cell volume. For the population case, the model is a measure-valued Markov process [14], in which the state of the population is represented by a measure (a generalised density) on the two-dimensional space of possible protein concentrations and cell volumes. Section 2 introduces the two frameworks in detail and states the main results. The results are derived in Sects. 3 and 4 by studying the associated evolution equations (Chapman–Kolmogorov and the population balance equations). Section 5 highlights the implications of the analysis.

2 Model Formulation

For each cell, we track its concentration $x(t)$ and volume $v(t)$ at time t. Protein synthesis occurs in random (memoryless) discrete events with stochastic rate α. Therefore, the interarrival times of synthesis events are i.i.d. exponentially distributed random variables with mean $1/\alpha$. Each burst event creates a jump or burst in the protein concentration $x \to x+B$, where the burst size B is drawn from the exponential distribution with mean β. The cell volume is not affected by a protein synthesis burst.

Between successive bursts, the cell volume increases with a concentration dependent rate and the concentration is diluted according to differential equations

$$\frac{\dot{v}}{v} = -\frac{\dot{x}}{x} = \gamma(x) = \frac{1}{1+kx}. \tag{1}$$

The volume $v(t)$ is strictly increasing during the cell cycle. The concentration $x(t)$ is strictly decreasing between consecutive production bursts. When the cell volume reaches a given threshold $2v^*$, a cell division event occurs, and the volume is immediately halved

$$v = 2v^* \rightarrow v = v^*. \tag{2}$$

Cell division does not affect the protein concentration. Typical protein and volume trajectories of a single cell line are shown in Fig. 1. The reset rule (2) is based on cell division as a Sizer [20].

2.1 Single Cell Model: A Bivariate Markov Process

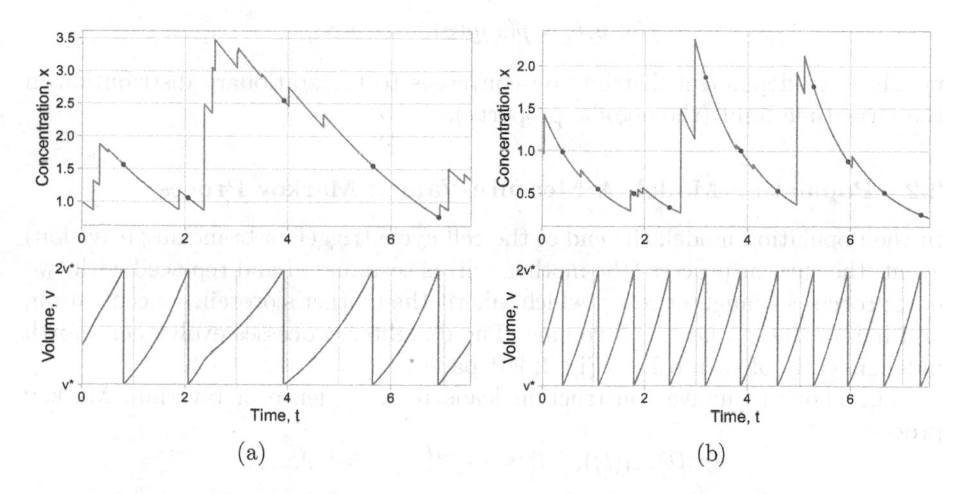

Fig. 1. Sample trajectories of the concentration $x(t)$ and the volume $v(t)$ in a single cell affected by (a) the strong feedback with $k = 2$, (b) the absence of the feedback, i.e., $k = 0$; other parameters of simulations are $\alpha = 0.9$, $\beta = 5$. Each division event is marked by a red dot on the concentration trajectory, which corresponds to an instant drop from $2v^*$ to v^* on the volume trajectory. (Color figure online)

In the single-cell model, we are looking at a single cell line: we do not follow the other daughter cell that is created in a cell division. The model is then a piecewise deterministic bivariate Markov process $(x(t), v(t)) \in R^2_+$. Let $f(x, v, t)$ denote the probability density function (pdf) of the random vector $(x(t), v(t))$

at time t. We have the initial condition $f(x, v, t = 0) = \delta(x - x_0)\delta(v - v_0)$, where $x_0 > 0$ and $v_0 > 0$ are known initial concentration and volume. We thereby assume that the initial volume is less that the critical volume at which a cell division event is triggered: $v_0 < 2v^*$.

For $t > 0$, the pdf $f(x, v, t)$ satisfies a bivariate Chapman–Kolmogorov equation, which is formulated in Sect. 3. We solve the Chapman–Kolmogorov equation at steady state. We find that the stationary distribution has the product form $f(x, v) = p(x)g(v)$, where the marginal distributions are given by

$$p(x) = (1 + kx)\frac{\beta\eta^2}{\Gamma(\alpha)}e^{-\eta x}(\eta x)^{\alpha-1}, \quad x > 0, \tag{3}$$

$$g(v) = \frac{1}{\ln(2)v}, \quad v^* < v < 2v^*. \tag{4}$$

The protein distribution (3) exists only if $\eta = 1/\beta - \alpha k > 0$.

In the absence of feedback ($k = 0$), the cell cycle length is equal to the doubling time $T = \ln 2$ of the exponential function, and the cell volume $v(t)$ is a T-periodic function (Fig. 1b). Because of the periodicity, the time-dependent pdf $f(x, v, t)$ does not converge to the stationary distribution $p(x)g(v)$ if $k = 0$. The inclusion of feedback breaks the periodicity (Fig. 1a). Therefore, for $k > 0$,

$$f(x, v, t) \sim p(x)g(v), \quad t \to \infty,$$

i.e. the time-dependent distribution converges to the stationary distribution in the large-time limit (the ergodic property).

2.2 Population Model: A Measure-Valued Markov Process

In the population model, the end of the cell cycle triggers a branching (division) event: the current process (the mother cell) is terminated and replaced with two new processes (daughter cells), which inherit the mother's protein concentration, and half of the mother's cell volume. The daughter processes evolve henceforth independently of each other (Fig. 2, left panel).

The above recursive construction leads to a sequence of bivariate Markov processes

$$(x_i(t), v_i(t)), \quad t_b^i \le t < t_e^i, \quad i = 1, 2, \ldots, \tag{5}$$

where t_b^i and t_e^i denote the beginning and the end of the cell cycle of the ith cell, and $x_i(t)$ gives the protein concentration at time t of the ith cell and $v_i(t)$ gives its volume. The original cell from which the entire population is derived is indexed by $i = 1$. The ordering of the rest of the sequence depends on the algorithmic implementation of the population process and is immaterial for our purposes.

The composition of the population at time t can be represented by the empirical population density (or measure)

$$m(x, v, t) = \sum_{i:t_b^i < t < t_e^i} \delta(x - x_i(t))\delta(v - v_i(t)). \tag{6}$$

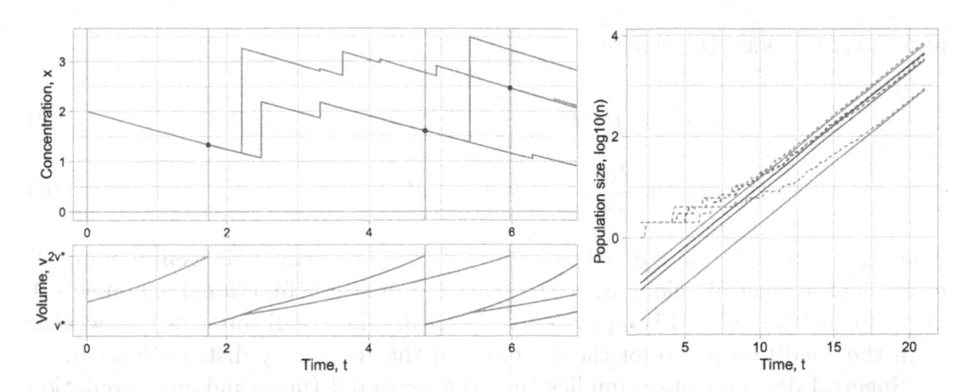

Fig. 2. *Left*: the time evolution of a sample cell population, where an individual color is assigned to each cell. We plot the protein concentration in the cell, its volume, and the division time (vertical lines). *Right:* The time evolution of the log-scaled sample population size in four different simulation runs with the same initial conditions and parameters values $\alpha = 0.9$, $\beta = 0.5$, $k = 2$, and $v^* = 1.5$ (dashed lines); the solid lines correspond to the exponential growth with the rate constant (8), shifted by the random factor $\log_{10} W(x_0, v_0)$, whose specific values were estimated for each simulation using linear regression. Each simulation has random fluctuations at the beginning and afterwards converges to the expected exponential growth.

The sum in (6) extends over the cells that exist at time t. Each existing cell contributes to the empirical population density (6) by a unit mass placed on top of $(x_i(t), v_i(t)) \in (0, \infty) \times [v^*, 2v^*]$. The integral of the empirical population density over the whole state space,

$$n(t) = \int_{v^*}^{2v^*} \int_0^{\infty} m(x, v, t)\mathrm{d}x\mathrm{d}v = \#\{i : t_b^i < t < t_e^i\}, \tag{7}$$

gives the total number of cells existing at time t and is finite. The empirical population density (6) is an example of a measure-valued Markov process [14, 18].

Let us consider the expected value of the empirical population density

$$h(x, v, t) = \mathbb{E}\, m(x, v, t),$$

the expectation being taken over all possible realisations of the population process. At the initial time $t = 0$, we have a single cell with non-random initial concentration x_0 and initial volume v_0, so that

$$h(x, v, 0) = m(x, v, 0) = \delta(x - x_0)\delta(v - v_0).$$

For $t > 0$, the expected population density $h(x, v, t)$ satisfies a population balance equation. This is formulated in Sect. 4.

The large-time behaviour of the population balance equation is characterised by its principal eigenvalue λ, the associated eigenfunction $\phi(x, v)$, and the adjoint eigenfunction $w(x, v)$. In Sect. 4, we show that

$$\lambda = 1 - \frac{\alpha k \beta}{k\beta + 1} \tag{8}$$

and $\phi(x, v) = p(x)g(v)$, where

$$p(x) = (1 + kx)\frac{\beta\sigma^2}{\Gamma(\xi)}e^{-\sigma x}(\sigma x)^{\xi-1}, \quad x > 0, \tag{9}$$

$$g(v) = \frac{2v^*}{v^2}, \quad v^* < v < 2v^*, \tag{10}$$

where $\xi = \alpha/(1+k\beta)$ and $\sigma = 1/\beta - \alpha k/(1+k\beta)$. The principal eigenfunction (9) exists (in the sense of belonging to the space of integrable functions) only if $\sigma > 0$. We also see that $\sigma > 0$ is equivalent to $\lambda > 0$. The condition $\sigma > 0$ is weaker than the condition $\eta > 0$ for the existence of the stationary distribution (3).

Spectral decomposition implies that the expected (non-random) population density satisfies

$$h(x, v, t) \sim w(x_0, v_0)e^{\lambda t}p(x)g(v), \quad t \to \infty. \tag{11}$$

The adjoint eigenfunction $w(x_0, v_0) > 0$, whose functional form we do not determine, thus characterises the influence of the initial condition on the large-time asymptotics (11). Analogously to what was said in the single cell model, (11) holds only in the aperiodic case ($k > 0$).

By the theory of supercritical branching processes (see e.g. Theorem 2 in Section V.7 of [2] or Theorem 3.1 in [26]), the (random) empirical population density satisfies

$$m(x, v, t) \sim W(x_0, v_0)e^{\lambda t}p(x)g(v), \quad t \to \infty, \tag{12}$$

where $W(x_0, v_0) > 0$ is a random variable dependent on initial data such that $\mathbb{E}\,W(x_0, v_0) = w(x_0, v_0)$. Equation (12) can equivalently be written in terms of the population size (7) and the normalised protein distribution as

$$n(t) \sim W(x_0, v_0)e^{\lambda t}, \quad \frac{m(x, v, t)}{n(t)} \sim p(x)g(v), \quad t \to \infty. \tag{13}$$

By the first relation in (13), the population $n(t)$ increases eventually exponentially with the rate constant (8); the initial condition and the low-population noise affects the large-time behaviour of the population only through the random pre-exponential factor $W(x_0, v_0)$ (Fig. 2, right panel). By the second relation in (13), the distribution of protein concentration and cell volume among a large population is (nearly) nonrandom and given by (9)–(10) (Fig. 3).

3 Chapman–Kolmogorov Equation

The goal of this section is to derive the single-cell stationary distributions $p(x)$ of protein concentration x and $g(v)$ of cell volume v. We start by formulating the time-dependent problem. The probability for the cell to be of volume $v^* \leq v \leq 2v^*$ and to have the protein concentration $x > 0$ at time $t > 0$ is

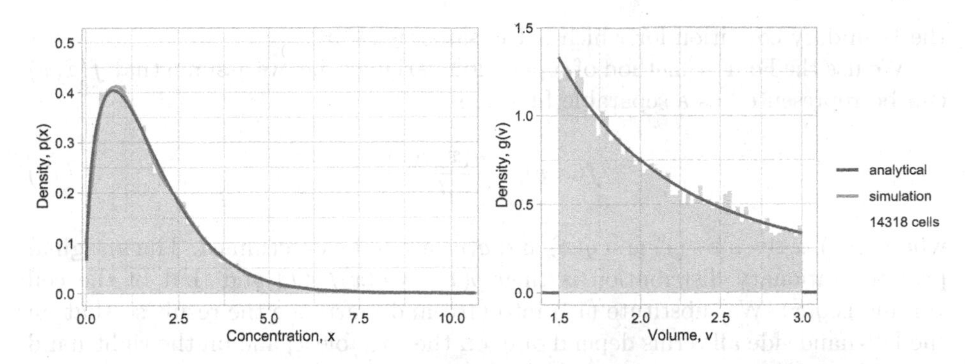

Fig. 3. Comparison of the analytical distributions (9) and (10) to the results of a large-time kinetic Monte Carlo simulation with parameters $\alpha = 2$, $\beta = 0.45$, $k = 0.7$, $v^* = 1.5$, a simulation endpoint $t = 20$, and initial conditions $x_0 = v_0 = 2$.

given by the joint probability density function is $f(x, v, t)$; its time evolution is described by Chapman-Kolmogorov equation:

$$\begin{aligned}\frac{\partial f(x,v,t)}{\partial t} &= \frac{\partial}{\partial x}\left(x\gamma(x)f(x,v,t)\right) - \frac{\partial}{\partial v}\left(v\gamma(x)f(x,v,t)\right)\\ &\quad + \frac{\alpha}{\beta}\int_0^x e^{-\frac{x-y}{\beta}} f(y,v,t)\mathrm{d}y - \alpha f(x,v,t).\end{aligned} \tag{14}$$

The initial and boundary conditions are:

$$f(x, v, 0) = \delta(x - x_0)\delta(v - v_0), \tag{15a}$$

$$f(x, 2v^*, t) = \frac{1}{2} f(x, v^*, t), \tag{15b}$$

where x_0 is initial protein concentration in the cell, v_0 is its initial volume. The Chapman–Kolmogorov equation (14) is a partial integro-differential equation. The differential operator on the right hand side of (14) drives the drift of the probability mass due to the deterministic flow (1). The integral operator in (14) provides the transfer of probability mass due to instantaneous protein bursts. The boundary condition (15b) captures the halving of cell volume in cell division. Since large cells immediately before a division grow twice as fast as small cells immediately after a division, a factor of one half is needed on the right hand side of (15b) in order that the probability fluxes at the left and right boundaries of the volume interval balance out. Indeed, integrating (14) over the state space $(x, v) \in (0, \infty) \times [v^*, 2v^*]$ confirms that the total probability $\int_0^\infty \int_{v^*}^{2v^*} f(x, v, t)\mathrm{d}x\mathrm{d}v$ remains constant over time for solutions $f(x, v, t)$ to (14).

Since our aim is to find a stationary distribution $f(x, v)$, we set $\partial f/\partial t = 0$ and subsequently apply the Leibniz integral rule to the last two terms, which yields an integral equation:

$$\frac{\partial}{\partial x}\left(x\gamma(x)f(x,v)\right) - \frac{\partial}{\partial v}\left(v\gamma(x)f(x,v)\right) - \alpha\frac{\partial}{\partial x}\left(\int_0^x e^{-\frac{x-y}{\beta}} f(y,v)\mathrm{d}y\right) = 0, \tag{16}$$

the boundary condition for which is the same as (15b).

We use the Fourier method of separation variables, i.e. we assume that $f(x,v)$ can be represented as a separable function

$$f(x,v) = \frac{q(x)g(v)}{\gamma(x)}, \tag{17}$$

where $\gamma(x)$ is given by (1) and $q(x)$ and $g(v)$ are to be determined. The marginal protein stationary distribution is then $p(x) = q(x)/\gamma(x)$ and that of the cell volume is $g(x)$. We substitute (17) into (16) and rearrange the result so that on the left-hand side all terms depend only on the variable x, and on the right-hand side only on v:

$$\frac{(xq(x))'}{q(x)} - \frac{\alpha}{q(x)}\frac{\mathrm{d}}{\mathrm{d}x}\int_0^x e^{-\frac{x-y}{\beta}}\frac{q(y)}{\gamma(y)}\mathrm{d}y = \frac{(vg(v))'}{g(v)} = \xi. \tag{18}$$

Since both sides are independent from one another, each side must be equal to a constant, which we denoted as ξ. We rewrite (18) and add the boundary condition on $g(v)$:

$$(vg(v))' - \xi g(v) = 0, \quad g(2v^*) = \frac{1}{2}g(v^*), \tag{19a}$$

$$(xq(x))' - \xi q(x) - \alpha\frac{\mathrm{d}}{\mathrm{d}x}\left(\int_0^x e^{-\frac{x-y}{\beta}}\frac{q(y)}{\gamma(y)}\mathrm{d}y\right) = 0. \tag{19b}$$

First, we solve equation (19a), which is an ordinary differential equation of the first order. The general solution is

$$g(v) = Cv^{\xi-1}; \tag{20}$$

upon applying the boundary condition, we find that $\xi = 0$. Since $g(v)$ is a probability distribution, the integral of $g(v)$ over the interval $(v^*, 2v^*)$ must be equal to one. The stationary distribution of the cell volume is thus (4).

We proceed with the second equation (19b): we substitute $\xi = 0$ and integrate, obtaining

$$xq(x) = \alpha\int_0^x e^{-\frac{x-y}{\beta}}(1+ky)q(y)\mathrm{d}y. \tag{21}$$

The homogeneous Volterra integral equation (21) be solved by standard methods [55]. There is no integration constant in (21) because the terms on both sides are zero for $x = 0$. Solving and returning to the protein distribution via $p(x) = q(x)/\gamma(x)$ leads to (3). Thus, we found a solution to (16) in the product form (17) as advertised in Sect. 2.1.

4 Population Balance Equation

The goal of this section is to derive the whole-population time-invariant distributions $p(x)$ of protein concentration x and $g(v)$ of cell volume v as well as the large-time population growth rate constant λ. We start by formulating the time-dependent problem. The expected population density $h(x,v,t)$ of cells with concentration $x>0$ and volume $v^* \le v \le 2v^*$ at a particular point of time $t \ge 0$ satisfies a population balance equation. In the interior of the state space ($x>0$, $v^*<v<2v^*$), the population balance equation coincides with the Chapman–Kolmogorov equation (14):

$$\begin{aligned}\frac{\partial h(x,v,t)}{\partial t} &= \frac{\partial}{\partial x}\left(x\gamma(x)h(x,v,t)\right) - \frac{\partial}{\partial v}\left(v\gamma(x)h(x,v,t)\right)\\ &\quad + \frac{\alpha}{\beta}\int_0^x e^{-\frac{x-y}{\beta}} h(y,v,t)\mathrm{d}y - \alpha h(x,v,t).\end{aligned} \tag{22}$$

The initial and boundary conditions are:

$$h(x,v,0) = h_0(x,v), \tag{23a}$$

$$h(x,2v^*,t) = \frac{1}{4}h(x,v^*,t), \tag{23b}$$

where $h_0(x,v)$ in (23a) is an initial population density. The extra factor of two in the boundary condition (23b) compared to the previous boundary condition (15b) reflects the cell doubling at the end of a cell cycle in the population scenario.

The population grows in time, and we assume that $h(x,v,t)$ has the following separable form:

$$h(x,v,t) = e^{\lambda t}p(x)g(v), \tag{24}$$

where

$$p(x) = \frac{q(x)}{\gamma(x)}. \tag{25}$$

The function $\gamma(x)$ is given by (1) while λ, $q(x)$, and $g(v)$ are to be determined. Substituting (24) and (25) into (22) and (23b) yields

$$(vg(v))' - \xi g(v) = 0, \qquad g(2v_0) = \frac{1}{4}g(v_0), \tag{26a}$$

$$(xq(x))' - \left(\frac{\lambda}{\gamma(x)} + \xi\right)q(x) - \alpha\frac{\mathrm{d}}{\mathrm{d}x}\int_0^x \frac{q(y)}{\gamma(y)}e^{-\frac{x-y}{\beta}}\mathrm{d}y = 0, \tag{26b}$$

where ξ is a constant independent of x and v. Equation (26a) is equivalent to (19a) except the boundary condition. The general solution remains to be (20), but the boundary condition now yields $\xi = -1$; from this, the population volume distribution (10) follows immediately.

Substituting $\xi = -1$ into (26b), we obtain the integro-differential equation

$$q(x) - \lambda(1+kx)q(x) + (xq(x))' - \alpha\frac{\mathrm{d}}{\mathrm{d}x}\int_0^x (1+ky)q(y)e^{-\frac{x-y}{\beta}}\mathrm{d}y = 0. \tag{27}$$

Applying the Laplace transform to (27) yields

$$Q(s) - \lambda P(s) - sQ'(s) - \alpha s B(s) P(s) = 0, \tag{28}$$

where $P(s)$ and $Q(s)$ are the Laplace images of functions $p(x)$ and $q(x)$, respectively, and $B(s) = (s + \frac{1}{\beta})^{-1}$ is the Laplace transform of the complementary cumulative distribution function of the (exponential) burst size distribution. Applying the Laplace transform directly to (25), one obtains

$$P(s) = Q(s) - kQ'(s),$$

which is used to rewrite (28) into an ODE for $Q(s)$:

$$\frac{\mathrm{d}Q(s)}{Q(s)} = \frac{\lambda + \alpha s B(s) - 1}{\lambda k + \alpha k s B(s) - s} \mathrm{d}s. \tag{29}$$

In the case of exponentially distributed burst sizes, the right-hand side of (29) simplifies to a rational fraction:

$$\frac{\mathrm{d}Q(s)}{Q(s)} = \frac{\frac{1-\lambda}{\beta} + s(1-\rho)}{-\frac{\lambda k}{\beta} - s(k\rho - \frac{1}{\beta}) + s^2} \mathrm{d}s, \quad \rho = \alpha + \lambda. \tag{30}$$

The solution approach requires the partial fraction decomposition of the right-hand side of (30). The quadratic in the denominator has two real roots

$$s_{1,2} = \frac{1}{2}\left(k\rho - \frac{1}{\beta} \pm \sqrt{D}\right), \quad D = (k\rho + \frac{1}{\beta})^2 - \frac{4\alpha k}{\beta}, \tag{31}$$

where s_1 is strictly positive and s_2 is strictly negative for any positive values of parameters α, β, λ and k. The decomposition of (30) takes the form:

$$\frac{\mathrm{d}Q(s)}{Q(s)} = \frac{A_1}{s - s_1} + \frac{A_2}{s - s_2}, \tag{32}$$

where A_1 and A_2 are defined by

$$A_{1,2} = \frac{1-\rho}{2} \pm \frac{1 - \rho + 2\alpha + \beta k \rho (1-\rho)}{2\beta\sqrt{D}}. \tag{33}$$

The solution of the ODE (32) is

$$Q(s) = C(s - s_1)^{A_1}(s - s_2)^{A_2}. \tag{34}$$

The Laplace transform (34) has to be analytic in the complex half-plane, i.e., $\mathrm{Re}(s) > 0$. Therefore, $A_1 \in \{0, 1, 2, \ldots\}$. In particular, the principal eigenvalue is obtained by setting

$$A_1 = 0,$$

which implies

$$\lambda = 1 - \frac{\alpha k \beta}{k\beta + 1}. \tag{35}$$

We expect that setting $A_1 = 1, 2, \ldots$ leads to additional eigenvalues.

We substitute (35) into (31) and (33) and simplify; the resulting values are strictly negative, and we introduce additional symbols for their opposite values, which are strictly positive:

$$\begin{aligned} s_2 &= \frac{\alpha k}{k\beta + 1} - \frac{1}{\beta}, & \sigma &= -s_2 > 0, \\ A_2 &= -\frac{\alpha}{k\beta + 1}, & \xi &= -A_2 > 0. \end{aligned} \tag{36}$$

Inserting $s = 0$ into (28) and using the normalisation condition $P(0) = 1$ yield

$$Q(0) = \lambda, \tag{37}$$

which is used to find the value of C in (34). Applying the inverse Laplace transform to (34), we obtain

$$q(x) = \frac{\beta\sigma^2}{\Gamma(\xi)} e^{-\sigma x} (\sigma x)^{\xi - 1},$$

from which, by relation (25), the population protein concentration distribution (9) immediately follows.

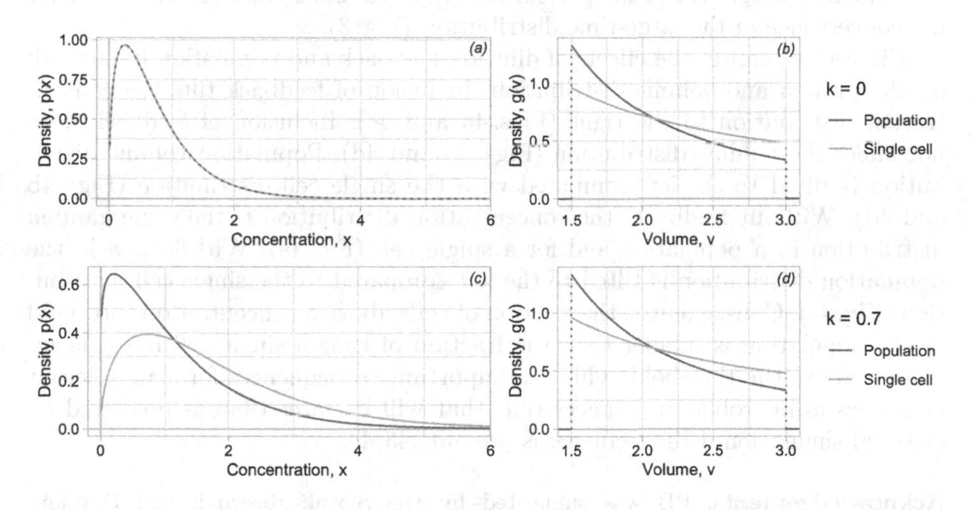

Fig. 4. The effect of absence (the first row) and presence (the second row) of feedback in dilution on the large-time distributions of protein concentration x and cell volume v. Parameters are as follow: $\alpha = 1.5$, $\beta = 0.5$, $k = 0.6$, $v^* = 1.5$.

5 Conclusions

In this paper, we studied a stochastic model for a protein that inhibits the growth of cell volume. The main result is the large-time distribution of protein concentration and cell volume in a single cell over time (3)–(4) and in an expanding cell population (9)–(10). Interestingly, the two are independent in the large-time limit (but interdependent transiently in the presence of the dilution feedback). We expect that the large-time independence carries over to more complex models of cell division than the reset rule (2). However, additional coupling between protein and cell size (e.g. volume-dependent production, partitioning noise) may introduce a dependence between the two variables [38].

The single-cell stationary distribution (3)–(4) exists if the product $\alpha\beta$ of burst frequency and burst size is less than the maximal dilution rate $1/k = \lim_{x\to\infty} x/(1+kx)$. Clearly, in the alternative case ($\alpha\beta > 1/k$), the build up of protein prevents stationarity [55]. In the population scenario, the large-time distribution (9)–(10) exists if the large-time population growth rate constant λ (8) is positive. In the alternative scenario ($(\alpha - 1)\beta > 1/k$), the build up of protein overburdens the cells and stalls the population growth. We note that this cannot happen in the low burst frequency (high noise) scenario $\alpha < 1$.

In the absence of dilution feedback, the volume process is periodic and hence does not converge to its stationary distribution (Fig. 1b). Periodicity also appears for more complex cell division mechanisms than (2) as long as (i) the volume grows exponentially and (ii) the cell divides into two equal halves [9]. Feedback in dilution makes the growth protein-dependent and hence non-exponential. As a consequence, we get rid of the periodicity (Figs. 1a and 2) and obtain ergodicity i.e. convergence of the large-time distribution (Fig. 3).

Figure 4 visualises the effects of dilution feedback and population framework on the protein and volume distribution. Inclusion of feedback tilts the concentration distribution to the right (Figs. 4a and 4c). Inclusion of feedback does not affect the volume distribution (Figs. 4b and 4d). Population volume distribution is tilted to the left compared with the single cell distribution (Figs. 4b and 4d). Without feedback, the concentration distribution is the same gamma distribution in a population and for a single cell (Fig. 4a). With feedback, the population distribution is tilted to the left compared to the single cell distribution (Fig. 4c). Consequently, the fraction of cells above a concentration threshold in the population is smaller than the fraction of time a single cell has concentration above the threshold. This has important consequences for drug-tolerant persisters in microbial and cancer cells that will be rarer than as predicted by classical simulation if the feedback is present [43,44].

Acknowledgements. PB was supported by the Slovak Research and Development Agency under contract no. APVV-18-0308, and VEGA grants 1/0339/21 and 1/0755/22.

References

1. Albayrak, C., et al.: Digital quantification of proteins and mRNA in single mammalian cells. Mol. Cell **61**(6), 914–924 (2016). https://doi.org/10.1016/j.molcel.2016.02.030
2. Athreya, K.B., Ney, P.E., Ney, P.: Branching processes. Courier Corporation (2004)

3. Backenköhler, M., Bortolussi, L., Großmann, G., Wolf, V.: Abstraction-guided truncations for stationary distributions of Markov population models. In: Quantitative Evaluation of Systems: 18th International Conference, QEST 2021, Paris, France, August 23–27, 2021, Proceedings 18, pp. 351–371. Springer (2021). https://doi.org/10.1007/978-3-030-85172-9_19
4. Backenköhler, M., Bortolussi, L., Wolf, V.: Variance reduction in stochastic reaction networks using control variates. In: Principles of Systems Design: Essays Dedicated to Thomas A. Henzinger on the Occasion of His 60th Birthday, pp. 456–474. Springer (2022)
5. Beentjes, C.H., Perez-Carrasco, R., Grima, R.: Exact solution of stochastic gene expression models with bursting, cell cycle and replication dynamics. Phys. Rev. E **101**(3), 032403 (2020). https://doi.org/10.1103/PhysRevE.101.032403
6. Bokes, P.: Heavy-tailed distributions in a stochastic gene autoregulation model. J. Stat. Mech: Theory Exp. **2021**(11), 113403 (2021). https://doi.org/10.1088/1742-5468/ac2edb
7. Bokes, P.: Stationary and time-dependent molecular distributions in slow-fast feedback circuits. SIAM J. Appl. Dyn. Syst. **21**(2), 903–931 (2022). https://doi.org/10.1137/21M1404338
8. Bokes, P., Singh, A.: Protein copy number distributions for a self-regulating gene in the presence of decoy binding sites. PLoS ONE **10**(3), e0120555 (2015). https://doi.org/10.1371/journal.pone.0120555
9. Bokes, P., Singh, A.: Cell volume distributions in exponentially growing populations. In: Computational Methods in Systems Biology: 17th International Conference, CMSB 2019, Trieste, Italy, September 18–20, 2019, Proceedings 17. pp. 140–154. Springer (2019). https://doi.org/10.1007/978-3-030-31304-3_8
10. Bokes, P., Singh, A.: Controlling noisy expression through auto regulation of burst frequency and protein stability. In: Hybrid Systems Biology: 6th International Workshop, HSB 2019, Prague, Czech Republic, April 6–7, 2019, Selected Papers 6, pp. 80–97. Springer (2019). https://doi.org/10.1007/978-3-030-28042-0_6
11. Cai, L., Friedman, N., Xie, X.S.: Stochastic protein expression in individual cells at the single molecule level. Nature **440**(7082), 358–362 (2006). https://doi.org/10.1038/nature04599
12. Çelik, C., Bokes, P., Singh, A.: Stationary distributions and metastable behaviour for self-regulating proteins with general lifetime distributions. In: Computational Methods in Systems Biology: 18th International Conference, CMSB 2020, Konstanz, Germany, September 23–25, 2020, Proceedings, pp. 27–43. Springer (2020). DOI: https://doi.org/10.1007/978-3-030-60327-4_2
13. Çelik, C., Bokes, P., Singh, A.: Protein noise and distribution in a two-stage gene-expression model extended by an mRNA inactivation loop. In: Computational Methods in Systems Biology: 19th International Conference, CMSB 2021, Bordeaux, France, September 22–24, 2021, Proceedings 19, pp. 215–229. Springer (2021). https://doi.org/10.1007/978-3-030-85633-5_13
14. Dawson, D.A., Maisonneuve, B., Spencer, J., Dawson, D.: Measure-valued Markov processes. Springer (1993)
15. De Jong, H., et al.: Mathematical modelling of microbes: metabolism, gene expression and growth. J. R. Soc. Interface **14**(136), 20170502 (2017). https://doi.org/10.1098/rsif.2017.0502
16. Dekel, E., Alon, U.: Optimality and evolutionary tuning of the expression level of a protein. Nature **436**(7050), 588–592 (2005). https://doi.org/10.1038/nature03842

17. Desoeuvres, A., Szmolyan, P., Radulescu, O.: Qualitative dynamics of chemical reaction networks: an investigation using partial tropical equilibrations. In: Computational Methods in Systems Biology: 20th International Conference, CMSB 2022, Bucharest, Romania, September 14–16, 2022, Proceedings, pp. 61–85. Springer (2022). https://doi.org/10.1007/978-3-031-15034-0_4
18. Doumic, M., Hoffmann, M.: Individual and population approaches for calibrating division rates in population dynamics: Application to the bacterial cell cycle. arXiv preprint arXiv:2108.13155 (2021).
19. Duso, L., Zechner, C.: Stochastic reaction networks in dynamic compartment populations. Proc. Natl. Acad. Sci. **117**(37), 22674–22683 (2020). https://doi.org/10.1073/pnas.2003734117
20. Facchetti, G., Chang, F., Howard, M.: Controlling cell size through sizer mechanisms. Curr. Opinion Syst. Biol. **5**, 86–92 (2017). https://doi.org/10.1016/j.coisb.2017.08.010
21. Fraser, L.C., Dikdan, R.J., Dey, S., Singh, A., Tyagi, S.: Reduction in gene expression noise by targeted increase in accessibility at gene loci. Proc. Natl. Acad. Sci. **118**(42), e2018640118 (2021). https://doi.org/10.1073/pnas.2018640118
22. Friedman, N., Cai, L., Xie, X.S.: Linking stochastic dynamics to population distribution: an analytical framework of gene expression. Phys. Rev. Lett. **97**(16), 168302 (2006). https://doi.org/10.1103/PhysRevLett.97.168302
23. Genthon, A.: Analytical cell size distribution: lineage-population bias and parameter inference. J. R. Soc. Interface **19**(196), 20220405 (2022). https://doi.org/10.1098/rsif.2022.0405
24. Holehouse, J., Cao, Z., Grima, R.: Stochastic modeling of autoregulatory genetic feedback loops: a review and comparative study. Biophys. J . **118**(7), 1517–1525 (2020). https://doi.org/10.1016/j.bpj.2020.02.016
25. Huang, G.R., Saakian, D.B., Rozanova, O., Yu, J.L., Hu, C.K.: Exact solution of master equation with Gaussian and compound Poisson noises. J. Stat. Mech: Theor. Exp. **2014**(11), P11033 (2014). https://doi.org/10.1088/1742-5468/2014/11/P11033
26. Janson, S.: Functional limit theorems for multitype branching processes and generalized Pólya urns. Stochastic Process. Appl. **110**(2), 177–245 (2004). https://doi.org/10.1016/j.spa.2003.12.002
27. Jędrak, J., Kwiatkowski, M., Ochab-Marcinek, A.: Exactly solvable model of gene expression in a proliferating bacterial cell population with stochastic protein bursts and protein partitioning. Phys. Rev. E **99**(4), 042416 (2019). https://doi.org/10.1103/PhysRevE.99.042416
28. Jędrak, J., Ochab-Marcinek, A.: Time-dependent solutions for a stochastic model of gene expression with molecule production in the form of a compound Poisson process. Phys. Rev. E **94**(3), 032401 (2016). https://doi.org/10.1103/PhysRevE.94.032401
29. Jia, C., Grima, R.: Coupling gene expression dynamics to cell size dynamics and cell cycle events: exact and approximate solutions of the extended telegraph model. Iscience **26**(1), 105746 (2023). https://doi.org/10.1016/j.isci.2022.105746
30. Jia, C., Singh, A., Grima, R.: Cell size distribution of lineage data: analytic results and parameter inference. Iscience **24**(3), 102220 (2021). https://doi.org/10.1016/j.isci.2021.102220
31. Jia, C., Singh, A., Grima, R.: Concentration fluctuations in growing and dividing cells: insights into the emergence of concentration homeostasis. PLoS Comput. Biol. **18**(10), e1010574 (2022). https://doi.org/10.1371/journal.pcbi.1010574

32. Jia, C., Xie, P., Chen, M., Zhang, M.Q.: Stochastic fluctuations can reveal the feedback signs of gene regulatory networks at the single-molecule level. Sci. Rep. **7**(1), 1–9 (2017). https://doi.org/10.1038/s41598-017-15464-9
33. Jia, C., Zhang, M.Q., Qian, H.: Emergent Lévy behavior in single-cell stochastic gene expression. Phys. Rev. E **96**(4), 040402 (2017). https://doi.org/10.1103/PhysRevE.96.040402
34. Kumar, N., Platini, T., Kulkarni, R.V.: Exact distributions for stochastic gene expression models with bursting and feedback. Phys. Rev. Lett. **113**(26), 268105 (2014). https://doi.org/10.1103/PhysRevLett.113.268105
35. Kumar, N., Singh, A., Kulkarni, R.V.: Transcriptional bursting in gene expression: analytical results for general stochastic models. PLoS Comput. Biol. **11**(10), e1004292 (2015). https://doi.org/10.1371/journal.pcbi.1004292
36. Mackey, M.C., Tyran-Kaminska, M., Yvinec, R.: Dynamic behavior of stochastic gene expression models in the presence of bursting. SIAM J. Appl. Math. **73**(5), 1830–1852 (2013). https://doi.org/10.1137/12090229X
37. Molenaar, D., Van Berlo, R., De Ridder, D., Teusink, B.: Shifts in growth strategies reflect tradeoffs in cellular economics. Mol. Syst. Biol. **5**(1), 323 (2009). https://doi.org/10.1038/msb.2009.82
38. Nieto-Acuña, C., Arias-Castro, J.C., Vargas-García, C., Sánchez, C., Pedraza, J.M.: Correlation between protein concentration and bacterial cell size can reveal mechanisms of gene expression. Phys. Biol. **17**(4), 045002 (2020). https://doi.org/10.1088/1478-3975/ab891c
39. Ochab-Marcinek, A., Tabaka, M.: Bimodal gene expression in noncooperative regulatory systems. Proc. Natl. Acad. Sci. **107**(51), 22096–22101 (2010). https://doi.org/10.1073/pnas.1008965107
40. Padovan-Merhar, O., et al.: Single mammalian cells compensate for differences in cellular volume and DNA copy number through independent global transcriptional mechanisms. Mol. Cell **58**(2), 339–352 (2015). https://doi.org/10.1016/j.molcel.2015.03.005
41. Patange, O., et al.: Escherichia coli can survive stress by noisy growth modulation. Nat. Commun. **9**(1), 5333 (2018). https://doi.org/10.1038/s41467-018-07702-z
42. Romanel, A., Jensen, L.J., Cardelli, L., Csikász-Nagy, A.: Transcriptional regulation is a major controller of cell cycle transition dynamics. PLoS ONE **7**(1), e29716 (2012). https://doi.org/10.1371/journal.pone.0029716
43. Rotem, E., et al.: Regulation of phenotypic variability by a threshold-based mechanism underlies bacterial persistence. Proc. Natl. Acad. Sci. **107**(28), 12541–12546 (2010). https://doi.org/10.1073/pnas.1004333107
44. Shaffer, S.M., et al.: Rare cell variability and drug-induced reprogramming as a mode of cancer drug resistance. Nature **546**(7658), 431–435 (2017). https://doi.org/10.1038/nature22794
45. Shahrezaei, V., Swain, P.S.: Analytical distributions for stochastic gene expression. Proc. Natl. Acad. Sci. **105**(45), 17256–17261 (2008). https://doi.org/10.1073/pnas.0803850105
46. Taheri-Araghi, S., et al.: Cell-size control and homeostasis in bacteria. Curr. Biol. **25**(3), 385–391 (2015). https://doi.org/10.1016/j.cub.2014.12.009
47. Tan, C., Marguet, P., You, L.: Emergent bistability by a growth-modulating positive feedback circuit. Nat. Chem. Biol. **5**(11), 842–848 (2009). https://doi.org/10.1038/nchembio.218
48. Thattai, M., Van Oudenaarden, A.: Intrinsic noise in gene regulatory networks. Proc. Natl. Acad. Sci. **98**(15), 8614–8619 (2001). https://doi.org/10.1073/pnas.151588598

49. Thomas, P., Shahrezaei, V.: Coordination of gene expression noise with cell size: analytical results for agent-based models of growing cell populations. J. R. Soc. Interface **18**(178), 20210274 (2021). https://doi.org/10.1098/rsif.2021.0274
50. Turpin, B., Bijman, E.Y., Kaltenbach, H.M., Stelling, J.: Population design for synthetic gene circuits. In: Computational Methods in Systems Biology: 19th International Conference, CMSB 2021, Bordeaux, France, September 22–24, 2021, Proceedings 19, pp. 181–197. Springer (2021). https://doi.org/10.1007/978-3-030-85633-5_11
51. Vadia, S., Levin, P.A.: Growth rate and cell size: a re-examination of the growth law. Curr. Opin. Microbiol. **24**, 96–103 (2015). https://doi.org/10.1016/j.mib.2015.01.011
52. Van Heerden, J.H., Kempe, H., Doerr, A., Maarleveld, T., Nordholt, N., Bruggeman, F.J.: Statistics and simulation of growth of single bacterial cells: illustrations with B. subtilis and E. coli. Sci. Reports **7**(1), 16094 (2017). https://doi.org/10.1038/s41598-017-15895-4
53. Vargas-Garcia, C.A., Ghusinga, K.R., Singh, A.: Cell size control and gene expression homeostasis in single-cells. Curr. Opinion Syst. Biol. **8**, 109–116 (2018). https://doi.org/10.1016/j.coisb.2018.01.002
54. Xia, M., Greenman, C.D., Chou, T.: PDE models of adder mechanisms in cellular proliferation. SIAM J. Appl. Math. **80**(3), 1307–1335 (2020). https://doi.org/10.1137/19M1246754
55. Zabaikina, I., Zhang, Z., Nieto, C., Bokes, P., Singh, A.: Quantifying noise modulation from coupling of stochastic expression to cellular growth: An analytical approach. bioRxiv, pp. 2022–10 (2022). https://doi.org/10.1101/2022.10.03.510723

Author Index

J. Pang and J. Niehren (Eds.): CMSB 2023, LNBI 14137, pp. 269–270, 2023.
https://doi.org/10.1007/978-3-031-42697-1

Printed in the United States
by Baker & Taylor Publisher Services

Printed in the United States
by Baker & Taylor Publisher Services